高等院校信息技术规划教材

Android高级开发技术案例教程

毋建军　编著

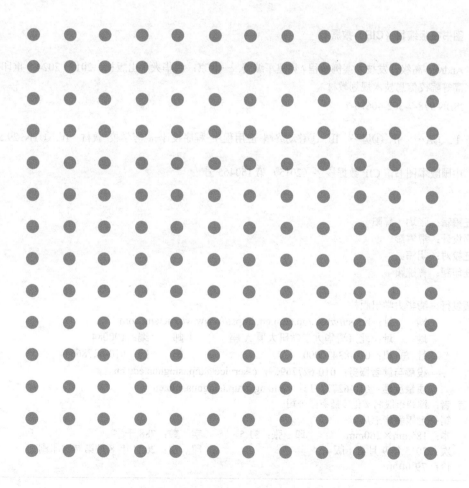

清华大学出版社
北京

内 容 简 介

本书从 Android 基础开始，由浅入深，采用"项目导向"的内容组织模式，理论和实践结合，通过完整的移动 Android 系统项目和 Android 物联网系统项目全方位地介绍了 Android 应用开发高级技术中的理论和方法。全书共 10 章，可分为三大部分，第一部分介绍 4G 智能手机发展、Android 开发基础、Android 开发环境搭建、Android NDK 开发环境搭建及开发、Android 应用程序、Fragement 与 Activity、Android 界面设计基础、Android 界面基础控件；第二部分介绍 Android 界面系统高级控件、Android 界面菜单及对话框、Android 组件消息通信及服务、Android 数据存储及应用、Google 位置应用服务开发；第三部分介绍 Android 物联网应用开发基础及综合应用。

本书作为 Android 应用开发原理与技术应用的教材，内容全面且通俗易懂，对 Android 技术应用及其与物联网结合应用所涉及的关键核心技术进行了全面的详解，提供了详细的实例进行学习导引，通过真实的系统应用项目有机地组织 Android 技术开发、物联网应用开发所涉及的知识内容，着重于对应用开发能力的渐进式培养。

本书可作为有 Java 基础的高等院校计算机、物联网、移动软件开发专业本、专科相关专业的教材，也可作为 Android 开发人员的参考书。

本书封面贴有清华大学出版社防伪标签，无标签者不得销售。
版权所有，侵权必究。举报：010-62782989，beiqinquan@tup.tsinghua.edu.cn。

图书在版编目（CIP）数据

Android 高级开发技术案例教程 / 毋建军编著. —北京：清华大学出版社，2015（2024.8 重印）
高等院校信息技术规划教材
ISBN 978-7-302-40616-7

Ⅰ. ①A… Ⅱ. ①毋… Ⅲ. ①移动终端-应用程序-程序设计-高等学校-教材 Ⅳ. ①TN929.53

中国版本图书馆 CIP 数据核字（2015）第 150465 号

责任编辑：张玥　薛阳
封面设计：常雪影
责任校对：胡伟民
责任印制：曹婉颖

出版发行：清华大学出版社
网　　址：https://www.tup.com.cn, https://www.wqxuetang.com
地　　址：北京清华大学学研大厦 A 座　　邮　编：100084
社 总 机：010-83470000　　邮　购：010-62786544
投稿与读者服务：010-62776969，c-service@tup.tsinghua.edu.cn
质量反馈：010-62772015，zhiliang@tup.tsinghua.edu.cn

印 装 者：涿州市般润文化传播有限公司
经　　销：全国新华书店
开　　本：185mm×260mm　　印　张：31.5　　字　数：788 千字
版　　次：2015 年 9 月第 1 版　　　　　印　次：2024 年 8 月第 7 次印刷
定　　价：79.00 元

产品编号：062265-02

前言

近年来，随着移动互联网和物联网技术的快速发展及应用，移动4G及物联网技术不断成熟并发展完善，传统的软件开发基础技术已经远远不能满足当前社会的需求。当前，不论是以Android为代表的Google移动软件生态链，还是以iOS为代表的苹果移动开发技术，都催生和孵化了许多 App Market 应用，尤其是以Android系统为基础的移动应用，在开源代码、开源框架的有力推动下，得到了快速的发展，影响并改变着整个移动互联网技术生态链条，也深刻地影响着大学院校专业的建设和学生的教育。21世纪的今天，是移动互联网、移动物联网技术引领的时代，也是莘莘学子追寻新技术、培养新能力、适应未来市场需求的过程。

现在，移动互联技术在企业项目开发中应用越来越广泛，围绕Andoid衍生的JNI技术、NDK技术、设计模式、移动UI设计、Map应用、3D图形应用、音视频等已经成为技术开发研究者深入研究的领域。同时，基于Android系统的移动物联网相关的一些核心技术也已经成为院校计算机、物联网相关专业学生未来就业、移动软件开发人员快速提升的必备技术，被许多开发人员作为一项专项技能进行学习和掌握。因而，深入学习基于Android的移动应用开发的核心技术、物联网开发技术对很多人而言非常重要。

目前基于Android的移动互联软件开发，通常都为厚重的技术类书籍，非常烦琐且没有将理论分析和实践技术进行结合，更没有对整个Android开发涉及的核心技术进行全面、整体、由浅入深的介绍。没有适合院校专业教学使用的书籍。尤其是从近年来软件工程技术领域发展和移动软件、物联网专业建设来看，Android核心技术比较通俗易懂，适合没有技术背景的人员阅读，但比较全面的Android开发、Android与物联网结合应用开发方面的书籍较少，能够应用于专业教学、符合专业人才培养、实践能力培养的则更少。如何有效地解决这些问题，编写符合移动软件、物联网应用开发课程教学特点和需求的教材，已成为不容避开的现实问题，急待解决。

　　源于此，本书中介绍的 Android 高级技术包含初级、中级、高级、物联网应用 4 个方面的核心技术及应用，初级技术包含 Android 开发环境搭建、创建应用程序、NDK 应用、Fragement 与 Activity、Android 组件；中级技术包含 Android UI 设计基础、Android 界面基础控件；高级技术包含 Android 界面系统高级控件、Android 界面菜单及对话框、Android 组件消息通信及服务、Android 数据存储及应用、Google 位置应用服务开发；物联网应用技术包含物联网终端设备、传感器、终端数据采集及存储、服务器通信、数据图形控制及展示，其目的是为了使初学者和读者对整个 Android 技术从架构到初、中、高级技术有个了解和认识性的循序渐进的学习过程。学习者在阅读中会发现，技术的讲解是一方面，更为重要的是符合人们认知规律的螺旋式渐进技术体系安排，有利于读者培养理论和技术应用有效结合的学习模式，通过项目技术引导可以使读者明白为什么而学技术（学习的目标性），技术核心要点及原理之间的关系及衔接（学哪些内容及学习内容之间先后次序关系），同时，也可以通过技术应用了解自己学习的深入程度及效果。

　　此外，由于移动互联网和物联网方面技术及应用的快速发展，Android 核心技术在物联网领域也得到了快速的发展，本书后续部分专门对基于 Android 系统的物联网应用进行了详细介绍，对移动互联网操作系统和物联网操作系统进行了对比，并针对物联网传感数据采集、传输、移动客户端图形展示应用进行了深入的技术解析和详解。

　　基于教学和学习需要，本书配有教学课件和书中所有的案例代码，读者可从清华大学出版社的网站下载。

　　由于编者的水平所限，书中难免有遗漏和不足之处，敬请广大读者指正和反馈。

编　者
2015 年 1 月

目录

第 1 章 Android 开发基础 1
1.1 4G 智能手机发展 1
1.2 Android 简介 6
1.3 搭建 Android 开发环境 11
1.3.1 Android 开发环境系统要求 12
1.3.2 Windows 系统平台下搭建开发环境 12
1.3.3 Linux 系统平台下搭建开发环境 19
1.4 搭建 Android NDK 开发环境 21
1.5 Android Studio 和 SDK 概述 24
1.5.1 Android Studio 简介 24
1.5.2 Android SDK 简介 24
1.5.3 Android 常用开发工具 25
1.6 创建 Android 程序和 Android NDK 程序 27
1.6.1 创建和使用虚拟设备 27
1.6.2 在 Eclipse 下创建 Android 程序 30
1.6.3 命令行创建 Android 程序 33
1.6.4 调试 Android 程序 40
1.6.5 创建 Android NDK 程序 45
1.6.6 Android 应用程序签名、打包、发布 48
习题 51

第 2 章 Android 项目及程序 52
2.1 Android 项目构成 52
2.1.1 目录结构 52
2.1.2 AndroidManifest.xml 文件简介 54
2.1.3 gen 目录 56
2.1.4 res 目录 57

  2.1.5 layout 目录 ·· 58
  2.1.6 values 目录 ·· 59
  2.1.7 project.properties 文件 ·································· 59
 2.2 Android 应用程序组成 ··· 60
  2.2.1 Android 应用程序概述 ······································ 60
  2.2.2 Activity 组件 ··· 60
  2.2.3 Service 组件 ·· 61
  2.2.4 Intent 和 Intent Filter 组件 ·································· 61
  2.2.5 BroadcastReceiver 组件 ···································· 63
  2.2.6 ContentProvider 组件 ······································ 63
 2.3 Fragment 与 Activity ·· 64
  2.3.1 Fragment 简介 ·· 64
  2.3.2 Fragment 的生命周期 ······································ 65
  2.3.3 Fragment 继承 ·· 67
  2.3.4 Fragment 创建方式 ·· 67
  2.3.5 Fragment 应用 ·· 69
 2.4 Android 生命周期 ··· 73
  2.4.1 程序生命周期 ·· 73
  2.4.2 组件生命周期 ·· 74
 2.5 项目案例 ·· 87
习题 ··· 91

第 3 章 Android 界面设计基础 ··· 92

 3.1 Android 界面设计简介 ··· 92
  3.1.1 移动和触摸设备设计原则 ·································· 93
  3.1.2 触摸屏与物理按键设计 ···································· 93
 3.2 Android 界面框架及部件 ··· 93
  3.2.1 Android 用户界面结构 ····································· 94
  3.2.2 Android 与 MVC 设计 ····································· 95
  3.2.3 视图树模型 ·· 96
 3.3 Android 界面控件类简介 ··· 96
  3.3.1 View 类 ·· 96
  3.3.2 ViewGroup 类 ·· 97
  3.3.3 界面控件 ·· 98
 3.4 Android 界面布局 ··· 98
  3.4.1 Android 布局策略 ··· 98
  3.4.2 线性布局 LinearLayout 及应用 ······························ 99
  3.4.3 相对布局 RelativeLayout 及应用 ···························· 103

3.4.4　表格布局 TableLayout 及应用 ……………………………………… 106
　　　3.4.5　帧布局 FrameLayout 及应用 ……………………………………… 109
　　　3.4.6　绝对布局 AbsoluteLayout 及应用 …………………………………… 113
　　　3.4.7　网格布局 GridLayout 及应用 ……………………………………… 115
　3.5　项目案例 ………………………………………………………………………… 120
　习题 …………………………………………………………………………………… 125

第 4 章　Android 界面基础控件 ……………………………………………………… 126

　4.1　文本控件简介 …………………………………………………………………… 126
　　　4.1.1　文本框 TextView 及应用 …………………………………………… 127
　　　4.1.2　编辑框 EditText 及应用 …………………………………………… 130
　4.2　按钮控件简介 …………………………………………………………………… 132
　　　4.2.1　按钮 Button 及应用 ………………………………………………… 132
　　　4.2.2　图片按钮 ImageButton 及应用 ……………………………………… 135
　4.3　单选与复选按钮简介 …………………………………………………………… 139
　　　4.3.1　单选按钮 RadioButton ……………………………………………… 139
　　　4.3.2　复选按钮 CheckBox ………………………………………………… 141
　　　4.3.3　RadioButton 和 CheckBox 综合应用 ……………………………… 143
　4.4　时间与日期控件简介 …………………………………………………………… 145
　　　4.4.1　时间选择器 TimePicker …………………………………………… 145
　　　4.4.2　日期选择器 DatePicker …………………………………………… 146
　　　4.4.3　时间与日期控件综合应用 ………………………………………… 147
　4.5　图片控件简介 …………………………………………………………………… 151
　　　4.5.1　图片控件 ImageView 及应用 ………………………………………… 151
　　　4.5.2　切换图片控件 ImageSwitcher、Gallery 应用 ……………………… 156
　4.6　时钟控件简介 …………………………………………………………………… 162
　　　4.6.1　模拟时钟 AnalogClock 与数字时钟 DigitalClock ………………… 162
　　　4.6.2　AnalogClock 和 DigitalClock 应用 ………………………………… 163
　4.7　项目案例 ………………………………………………………………………… 167
　习题 …………………………………………………………………………………… 171

第 5 章　Android 界面系统高级控件 ………………………………………………… 172

　5.1　列表控件简介 …………………………………………………………………… 172
　　　5.1.1　列表控件 ListView 及应用 ………………………………………… 172
　　　5.1.2　下拉列表控件 Spinner 及应用 ……………………………………… 176
　5.2　进度条与滑块控件简介 ………………………………………………………… 181
　　　5.2.1　进度条 ProgressBar 及应用 ………………………………………… 181
　　　5.2.2　滑块 SeekBar 及应用 ………………………………………………… 184

- 5.3 评分控件及应用 187
- 5.4 自动完成文本控件及应用 188
- 5.5 Tabhost 控件及应用 191
- 5.6 视图控件应用 194
  - 5.6.1 滚动视图控件 ScrollView 及应用 194
  - 5.6.2 网格视图控件 GridView 及应用 196
- 5.7 Android 事件处理 200
  - 5.7.1 Android 事件和监听器 200
  - 5.7.2 Android 事件处理机制 201
  - 5.7.3 Android 事件处理机制应用 205
  - 5.7.4 按键事件应用 210
  - 5.7.5 触摸事件应用 213
- 5.8 Android 消息传递机制 217
  - 5.8.1 异步任务 217
  - 5.8.2 Handler 类应用 224
- 5.9 Android 音视频播录应用 227
  - 5.9.1 音频播放应用 227
  - 5.9.2 视频播放应用 231
  - 5.9.3 音视频录制应用 235
- 5.10 Android 图形应用 239
  - 5.10.1 Canvas 组件图形应用 239
  - 5.10.2 OpenGL ES 包组件图形应用 240
- 5.11 项目案例 242
- 习题 251

第 6 章 Android 界面菜单、对话框 253

- 6.1 菜单控件 Menu 253
  - 6.1.1 Menu 概述 253
  - 6.1.2 选项菜单及应用 254
  - 6.1.3 子菜单及应用 259
  - 6.1.4 快捷菜单及应用 262
- 6.2 对话框控件 Dialog 267
  - 6.2.1 对话框 Dialog 简介 267
  - 6.2.2 警告（提示）对话框 AlertDialog 及应用 268
  - 6.2.3 日期选择对话框 DatePickerDialog 及应用 271
  - 6.2.4 时间选择对话框 TimePickerDialog 及应用 274
  - 6.2.5 进度对话框 ProgressDialog 及应用 278
- 6.3 信息提示控件 281

　　　　6.3.1　Toast 控件及应用 ·········· 281
　　　　6.3.2　Notification 控件及应用 ·········· 284
　6.4　项目案例 ·········· 289
　习题 ·········· 294

第 7 章　Android 组件消息通信与服务 ·········· 296

　7.1　Intent 消息通信 ·········· 296
　　　7.1.1　Intent 组件及通信 ·········· 296
　　　7.1.2　使用 Intent 启动 Activity ·········· 299
　　　7.1.3　获取 Activity 返回值 ·········· 306
　　　7.1.4　Intent Filter 原理与匹配机制 ·········· 309
　7.2　Intent 广播消息 ·········· 313
　　　7.2.1　广播消息 ·········· 313
　　　7.2.2　BroadcastReceiver 监听广播消息及应用 ·········· 313
　7.3　E-mail 邮件应用 ·········· 319
　7.4　手机短信发送应用 ·········· 328
　7.5　网络访问及通信 ·········· 336
　　　7.5.1　使用 URL 读取网络资源及应用 ·········· 337
　　　7.5.2　使用 HTTP 访问网络资源及应用 ·········· 341
　7.6　电话拨打服务及应用 ·········· 351
　7.7　Service 组件服务 ·········· 355
　7.8　项目案例 ·········· 356
　习题 ·········· 366

第 8 章　Android 数据存储及应用 ·········· 367

　8.1　SharedPreferences 存储及访问 ·········· 368
　　　8.1.1　SharedPreferences 简介 ·········· 368
　　　8.1.2　访问本程序数据 ·········· 371
　　　8.1.3　读取其他应用程序数据 ·········· 374
　8.2　SQLite 数据库存储及操作 ·········· 376
　　　8.2.1　SQLite 数据库简介 ·········· 376
　　　8.2.2　创建 SQLite 数据库方式 ·········· 377
　　　8.2.3　SQLite 数据库操作 ·········· 380
　　　8.2.4　SQLite 数据库管理及应用 ·········· 383
　8.3　文件存储及读写 ·········· 395
　　　8.3.1　文件存储及应用 ·········· 395
　　　8.3.2　SD 卡存储及应用 ·········· 405
　8.4　数据共享访问 ·········· 412

8.4.1 ContentProvider 简介 ·················· 412
8.4.2 Uri、UriMatcher 和 ContentUris 简介 ·················· 413
8.4.3 创建 ContentProvider ·················· 416
8.4.4 ContentResolver 操作数据 ·················· 417
8.4.5 ContentProvider 应用 ·················· 418
8.5 网络存储应用 ·················· 421
8.6 数据存储项目案例 ·················· 424
习题 ·················· 437

第 9 章 Google 位置应用服务开发 ·················· 438

9.1 地理位置定位服务 ·················· 438
9.1.1 Android Location API 简介 ·················· 439
9.1.2 获取位置定位 ·················· 442
9.2 Google Map 应用 ·················· 445
9.2.1 Google Map API 简介 ·················· 445
9.2.2 申请 Map API KEY 和创建 AVD ·················· 446
9.3 项目案例 ·················· 449
习题 ·················· 452

第 10 章 Android 物联网应用开发基础 ·················· 453

10.1 物联网概述 ·················· 453
10.1.1 物联网简介 ·················· 453
10.1.2 物联网体系框架及应用协议 ·················· 454
10.1.3 物联网关键技术 ·················· 455
10.1.4 物联网操作系统与移动互联网 ·················· 456
10.1.5 物联网未来发展 ·················· 457
10.2 物联网设备 ·················· 458
10.2.1 物联网终端 ·················· 458
10.2.2 物联网网关 ·················· 459
10.3 Android 硬件传感器 ·················· 460
10.4 物联网终端数据采集应用开发 ·················· 463
10.5 物联网传感数据图形应用 ·················· 475
10.6 项目案例 ·················· 482
习题 ·················· 489

第 1 章

Android 开发基础

学习目标

本章主要介绍 4G 智能手机的发展、智能手机操作系统类型、Android 移动操作系统版本的发展历史、过程；同时讲述 Android 开发环境的搭建过程、Android Studio、Android SDK，以及 Android NDK 开发环境构建，创建 Android 程序、NDK 程序的方法和工具。通过本章的学习，帮助读者掌握以下知识要点。

（1）4G 智能手机及智能手机操作系统及类别。
（2）Android 系统发展历史、体系结构、特征及未来发展方向。
（3）Android 系统在不同平台下开发环境搭建。
（4）Android NDK 开发环境搭建及应用。
（5）Android Studio、SDK 结构、构成及工具。
（6）使用不同方式和方法创建、调试 Android 应用程序。

1.1 4G 智能手机发展

4G 是英文 4rd Generation 的缩写，指 4G 国际标准（我国具有自主知识产权的通信标准 TD-LTE）。相对第一代模拟制式手机（1G）和第二代 GSM、TDMA 等数字手机（2G），第三代移动通信技术，随着第 4 代通信牌照 TD-LTE 在 2013 年 11 月的发放，标志着我国通信进入 4G 时代。4G 最大的技术特点是提供了更高的无线下载速度，在静态下理论无线下载速度可以达到 1Gb/s，动态情况下可以达到 100Mb/s，可以对无线数据提供更好的支持，有效满足用户的需求。其主要特点如下。

（1）容量、速率更高。最低数据传输速率为 2Mb/s，最高可达 100Mb/s。
（2）兼容性更好。4G 系统开放了接口，能实现与各种网络的互联，同时能与二代、三代手机兼容。它能在不同系统间进行无缝切换，并提供多媒体高速传送业务。
（3）数据处理更灵活。智能技术在 4G 系统中的应用，能自适应地分配资源。智能信号处理技术将实现任何信道条件下的信号收发。
（4）用户共存。4G 系统会根据信道条件、网络状况自动进行处理，实现高速用户、低速用户、用户设备的互通与并存。

（5）自适应网络。针对系统结构，4G 系统将实现自适应管理，它可根据用户业务进行动态调整，从而最大程度地满足用户需求。

随着移动运营商对 4G 的商业投用，智能手机逐步取代 3G 成为市场主流，在中国市场中，安卓系统手机市场份额已达 80%以上，IOS、微软 Windows Phone、黑莓等手机系统也分别占据了一定份额。下面就智能手机操作系统及其发展进行简要介绍。

目前，市场上的智能手机操作系统有很多，各大操作系统之间的争夺将更加突出，并逐渐以联盟阵营的方式来推动智能手机的普及。曾经以及现在比较有影响力的智能手机操作系统主要有以下几种。

1. Android 系统

Android 的原意指"机器人"，是 Google 于 2007 年推出的基于 Linux 平台的开源手机操作系统。Android 系统平台由操作系统、中间件、用户界面和应用软件组成，是首个为移动终端打造的真正开放和完整的移动软件系统平台。

目前，市场上采用 Android 系统的主要手机厂商包括宏达电子（HTC）、三星、联想、华为、摩托罗拉、LG、Sony Ericsson、小米、酷派等，国内厂商有小米、华为、中兴、联想、酷派等。第一款使用谷歌 Android 操作系统的手机 G1，如图 1-1 所示。Android 系统不但应用于智能手机，也在平板电脑市场急速扩张。目前，它已成为全球最受欢迎的智能手机平台。

图 1-1　第一款 Android 操作系统的手机（G1）

随着 Android 4.4 的诞生及发展，Android 系统的发展及应用越来越广泛。Android 平台资源下载安装与 Symbian 类似，不同的是，Android 系统平台是一个开源系统，所安装的程序不需要进行证书检查。

2. Symbian 系统

Symbian（塞班）是一个实时性、多任务的操作系统，其起源于 1998 年，在英国伦敦，多家手机厂商为了对抗微软即将推出的智能手机系统，诺基亚、摩托罗拉、爱立信和宝意昂公司在英国伦敦共同投资成立了 Symbian 公司。1999 年，支持 Symbian 系统的爱立信 R380 上市，当时由于 R380 系统正处于实践阶段，并未得到很好的推广。世界上第一款采用 Symbian 操作系统的手机，如图 1-2 所示。

2001 年，诺基亚推出了第一款 Symbian PDA 手机，如图 1-3 所示，型号为 9110，采用了 AMD 公司的内嵌式 CPU，内置 8MB 存储空间。9110 已经集成了网络、PIM、网页浏览、电子邮件等功能，并且已经开始支持 Java，这使得它已经能够运行小型的第三方软件。

图 1-2 世界上第一款采用 Symbian 系统的手机　　图 1-3 诺基亚推出第一款 Symbian PDA 手机

此后，Nokia 主导的 S60v3、S60v5、Symbian^3 相继发布和推广，在 2008 年之前，Symbian 系统在市场占主导地位，接着 Symbian 公司被诺基亚全资收购，成为诺基亚旗下公司。参与开发的多家厂商最初仍然有系统的使用权，但到后来都纷纷宣布退出 Symbian 的阵营，取而代之的是安卓，目前 Symbian 的支持厂商只有诺基亚。

2011 年，诺基亚正式宣布与微软达成全球战略合作伙伴关系，双方在智能手机领域进行深度合作。微软的 Windows Phone 7 系统成为诺基亚的主要手机操作系统。2012 年诺基亚 Lumia 智能手机——Lumia 820 和 Lumia 920，配置了 Windows Phone 8 操作系统及双核处理器，并在市场推广应用。

3. iOS

iPhone OS 是由苹果公司为 iPhone 开发的操作系统，后来套用在 iPod touch、iPad 以及 Apple TV 产品上使用。iPhone 将移动电话、可触摸宽屏以及具有桌面级电子邮件、网页浏览、搜索和地图功能的因特网通信设备这三种产品完美地融为一体，重新定义了移动电话的功能。iOS 拥有简单易用的界面，良好的操作体验。就像其基于 Mac OS X

操作系统一样，它也是以 Darwin 为基础的。在 2010 年 6 月 WWDC 大会上，苹果宣布 iPhone、iPod touch 和 iPad 使用的 iPhone OS 操作系统更名为 iOS，统一了苹果的移动设备名称。

iOS 的系统架构分为 4 个层次：核心操作系统层（the Core OS Layer）、核心服务层（the Core Services Layer）、媒体层（the Media Layer）、可轻触层（the Cocoa Touch Layer），如图 1-4 所示。

2014 年 6 月 3 日，苹果公司在 WWDC 2014 上发布了 iOS 8，并提供了开发者预览版更新，同年 9 月 17 日，发布了 iOS 8 的正式版，其机型 iPhone 6 和 iPhone 6 Plus 如图 1-5 所示。

图 1-4 iOS 6 运行于 iPhone 4S　　　　图 1-5 iPhone 6 和 iPhone 6 Plus

由于 iOS 是从 Mac OS X 核心演变而来，因此开发工具也是基于 Xcode。Xcode 是 iPhone 软件开发工具包的开发环境，iOS 的 SDK 需要拥有英特尔处理器且运行 Mac OS X Leopard 系统的 Mac 才能使用。其他的操作系统，包括微软的 Windows 操作系统和旧版本的 Mac OS X 都不支持。SDK 本身是可以免费下载的，但为了发布软件，开发人员必须加入 iPhone 开发者计划，其中有一步需要付款以获得苹果的批准。加入之后，开发人员将会得到一个牌照，他们可以用这个牌照将他们编写的软件发布到苹果的 App Store 上。发布软件一共有三种方法：通过 App Store，通过企业配置仅在企业内部员工间应用，也可通过基于 Ad-hoc 而上载至多达 100 部 iPhones。

4. Windows Mobile

Windows Mobile（WM）是微软针对移动设备而开发的操作系统。该操作系统的设计初衷是尽量接近于桌面版本的 Windows。微软按照计算机操作系统的模式来设计 WM，以便能使得 WM 与计算机操作系统一模一样。WM 的应用软件以 Microsoft Win32 API 为基础。在 Windows Mobile 6.5 发布的同时，微软宣布以后的 Windows Mobile 产品将改名为 Windows Phone，以改变其落后的形象。Windows Mobile 捆绑了一系列针对移动设备而开发的应用软件，这些应用软件创建在 Microsoft Win32 API 的基础上。可以运行 Windows Mobile 的设备包括 Pocket PC、SmartPhone 和 Portable Media Center。2010 年 10 月，微软宣布终止对 WM 的所有技术支持，Windows Mobile 系列正式退出手机系统市场。

5．Windows Phone

2010 年 2 月，微软公司正式发布 Windows Phone 7 智能手机操作系统，简称 WP7，并于 2010 年年底发布了基于此平台的移动设备。全新的 Windows Phone 把网络、个人计算机和手机的优势集于一身，让人们可以随时随地享受到想要的体验。它具有桌面定制、图标拖曳、滑动控制等一系列前卫的操作体验，其主屏幕通过提供类似仪表盘的体验来显示新的电子邮件、短信、未接来电、日历约会等，让人们对重要信息保持时刻更新。Windows Phone 力图打破人们与信息和应用之间的隔阂，提供适用于人们包括工作和娱乐在内完整生活的方方面面，实现最优秀的端到端体验。

2012 年 9 月，诺基亚与微软在纽约召开联合发布会，发布 Windows Phone 8 系统手机 Lumia 920 及 820。Lumia 920 实现了手机摄影、高清屏显示、无线充电三大技术突破，这也是诺基亚旗下首款 Windows Phone 8 系统手机。如图 1-6 所示为 Lumia 1520。

2014 年 4 月，Build 2014 开发者大会发布 Windows Phone 8.1，在 Windows Phone 8.1 系统中，下载的 XAP 文件可以存放在 download 文件夹内，为不支持 SD 卡扩展的 Windows Phone 手机实现了本地安装 app。

图 1-6　诺基亚 Lumia 1520

6．黑莓

黑莓（BlackBerry）是加拿大一家手提无线通信设备品牌，于 1999 年创立。其系统特色是支持 PushMail 电子邮件、移动电话、文字短信、互联网传真、网页浏览及其他无线资讯服务。

2007 年 7 月在中国大陆地区引进第一款设备 BlackBerry 8700g 以及国外流行的 BlackBerry Passport，如图 1-7 所示，由 TCL 代工生产，中国移动同时也向企业用户开始推广 BlackBerry 业务。

由于黑莓与中国移动推出手机邮箱有冲突、GPRS 的网络速度限制及中国用户当时没有使用电子邮件的习惯，因此中国移动引进黑莓没有成功地在大范围内得到推广。

7．其他手机操作系统

（1）OMS：中国移动在 Android 系统上定制开发的系统。

（2）OS：加拿大 RIM 公司开发。

（3）MeeGo：Nokia 与 Intel 联合开发。

(4) Bada：SamSung 研发。
(5) BrewMP：高通公司开发。

图 1-7　BlackBerry 8700g 和 BlackBerry Passport

1.2　Android 简介

1. Android 发展历史

Android 一词的本义指"机器人"，是 Google 于 2007 年推出的以 Linux 为基础的开放源代码操作系统，主要使用于便携设备。目前尚未有统一中文名称，中国大陆地区较多人使用"安卓"或"安致"。Android 操作系统最初由 Andy Rubin 开发，最初主要支持手机。2011 年第一季度，Android 在全球的市场份额首次超过塞班系统，跃居全球第一。2012 年 7 月数据显示，Android 占据全球智能手机操作系统市场 59%的份额，中国市场占有率为 76.7%。

Android 平台由操作系统、中间件、用户界面和应用软件组成。目前，最新版本为 Android 5.0。

2007 年 11 月 5 日，谷歌公司推出 Android 操作系统，并且在这一天宣布建立一个全球性的联盟组织，该组织由 34 家手机制造商、软件开发商、电信运营商以及芯片制造商共同组成，并与 84 家硬件制造商、软件开发商及电信营运商组成开放手持设备联盟（Open Handset Alliance）来共同研发改良 Android 系统，这一联盟将支持谷歌发布的手机操作系统以及应用软件。Google 以 Apache 免费开源许可证的授权方式，发布了 Android 的源代码。

2008 年，在 Google I/O 大会上，谷歌提出了 Android HAL 架构图；同年 8 月 18 日，

Android 获得了美国联邦通信委员会(FCC)的批准；2008 年 9 月，谷歌正式发布了 Android 1.0 系统，这也是 Android 系统最早的版本。

2009 年 4 月，谷歌正式推出了 Android 1.5 手机。从 Android 1.5 版本开始，谷歌开始将 Android 的版本以甜品的名字命名，Android 1.5 命名为 Cupcake（纸杯蛋糕）。该系统与 Android 1.0 相比有了很大的改进。

2009 年 9 月，谷歌发布了 Android 1.6 的正式版 Donut（甜甜圈），并且推出了搭载 Android 1.6 正式版的手机 HTC Hero(G3)，凭借着出色的外观设计以及全新的 Android 1.6 操作系统，HTC Hero(G3)成为当时全球最受欢迎的手机。

2010 年 2 月，Linux 内核开发者 Greg Kroah-Hartman 将 Android 的驱动程序从 Linux 内核"状态树"上除去，从此，Android 与 Linux 开发主流分开发展。在同年 5 月，谷歌正式发布了 Android 2.2 操作系统。

2010 年 12 月，谷歌正式发布了 Android 2.3 操作系统 Gingerbread（姜饼）。

2013 年 9 月，谷歌推出了 Android 4.4 KitKat 奇巧版本。2014 年 6 月，Google I/O 2014 开发者大会在旧金山正式召开，发布了 Android 5.0 的前身 L（Lollipop）版 Android 开发者预览。

2014 年 10 月，Google 发布全新 Android 操作系统 Android 5.0。Nexus 设备 Nexus 6、Nexus 9 平板及 Nexus Player 安装了 Android 5.0。2014 年 11 月，Android 的最新版本 5.0 Lollipop 面向用户正式推出。开发者可以下载 Android 5.0 Platform（API Level 21）来开发和测试自己的 Android 5.0 应用，并能向 Google Play 发布 Android 5.0 所专属的应用程序。

Google Play 的前身是谷歌 Android Market，是一个由 Google 为 Android 设备开发的在线应用程序商店。2012 年 9 月 27 日，Google Play App 下载量突破 250 亿次。2013 年 4 月 10 日，Google 提前正式发布 Google Play 4.0 版本，新版本采用全新 UI 设计，并且取消了"精品商品"页面。2014 年 6 月，谷歌 I/O 开发者大会上，首批可以在 Android Wear 系统中运行的应用程序已经正式开始陆续在 GooglePlay 应用商店中上架。

2. Android 系统版本及功能发展

Android 在正式发行之前，最开始拥有两个内部测试版本，并且以著名的机器人名称来对其进行命名，它们分别是：阿童木（Android Beta），发条机器人（Android 1.0）。后来谷歌将其命名规则变更为用甜点作为它们系统版本的代号的命名方法。甜点命名法开始于 Android 1.5 发布。作为每个版本代表的甜点的尺寸越变越大，然后按照 26 个字母排序：纸杯蛋糕（Android 1.5，Cupcake），甜甜圈（Android 1.6，Donut），松饼（Android 2.0/2.1，Eclair），冻酸奶（Android 2.2，Froyo），姜饼（Android 2.3，Gingerbread），蜂巢（Android 3.0，Honeycomb），冰激凌三明治（Android 4.0，ICE Crean Sandwich），Android 果冻豆（Jelly Bean，Android 4.1），最新的版本为 Android 5.0 Lollipop。Android 4.0 之前各版本发布的 Logo 如图 1-8 所示。

1）Android 4.4（KitKat）奇巧

（1）支持两种编译模式；

（2）RAM 优化；

图 1-8　Android 4.0 之前各版本发布的 Logo

（3）新图标、锁屏、启动动画和配色方案；

（4）新的拨号和智能来电显示；

（5）加强主动式语音功能；

（6）集成 Hangouts IM 软件；

（7）全屏模式；

（8）支持 Emoji 键盘；

（9）轻松访问在线存储；

（10）无线打印；屏幕录像功能；内置字幕管理功能；计步器应用；低功耗音频和定位模式；

（11）新的接触式支付系统；

（12）新的蓝牙配置文件和红外兼容性。

2）Android 5.0 Lollipop

（1）Android 5.0 Lollipop 成为首个支持 64 位 ARM、X86 和 MIPS 架构 SoC 的 Android 操作系统。

（2）省电助手（Battery Saver）功能正式启用。

（3）更强大的优先模式，用户可以设置定时，或者直接通过音量键来激活优先模式。在该模式下，只有被用户设置为优先模式的应用和联系人才可以在本机上显示通知。

（4）迁移数据的功能，用户可以在设置 5.0 新系统的设备时，在 Tap & Go 流程当中将新旧两台设备贴合在一起，新设备会自动检索旧设备登录的 Google 账号等信息，并开始恢复。

(5)锁屏通知显示功能,而且用户可以直接在锁屏上与通知内容进行互动,比如回复短信和电子邮件、消音闹钟、查看推文等。

(6)面部识别解锁功能。

3. Android 系统的优势及缺点

1)Android 系统与其他系统相比的优势

与 Symbian 相比,Android 为开源系统,系统发展更具前景;快速增长的海量第三方免费软件;无"证书"限制,安装软件更自由。

与 iPhone 相比,Android 更开放;风格更自由,简洁;开源系统,更多第三方免费软件;软件安装卸载更方便,无须第三方平台软件。

与 Windows Mobile 相比,其更方便,简捷。

2)Android 系统缺点

Android 系统手机电池普遍不耐用,更多依赖于手机网络,不同厂商之间手机系统应用软件兼容性、移动设备等都没有统一的标准,其系统开发设计及应用依赖于手机厂商。

4. Android 体系结构

Android 作为一个移动设备系统平台,其采用软件堆层的架构,共分为 4 层,自下而上分别是 Linux 内核(操作系统,OS)、中间件层、应用程序框架、应用程序,如图 1-9 所示。

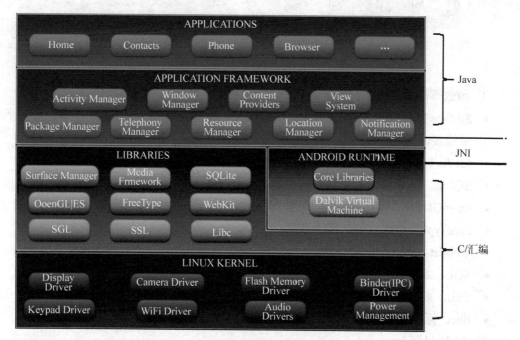

图 1-9　Android 系统结构

（1）底层以 Linux 核心为基础，由 C 语言开发，只提供基本功能，是硬件和其他软件堆层之间的一个抽象隔离层，提供安全机制、内存管理、进程管理、网络协议堆栈和驱动程序等，如图 1-10 所示。

图 1-10　Linux 内核

（2）中间层包括函数库 Library 和虚拟机 Virtual Machine，由 C++开发，由函数库和 Android 运行时构成。如图 1-11 所示。

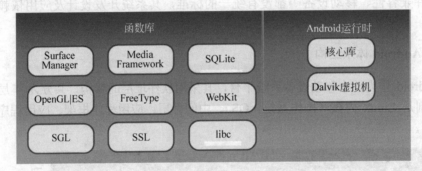

图 1-11　Android 系统中间层

① 函数库，主要提供一组基于 C/C++的函数库，包括以下内容。
- Surface Manager，支持显示子系统的访问，提供应用程序与 2D、3D 图像层的平滑连接。
- Media Framework，实现音视频的播放和录制功能。
- SQLite，轻量级的关系数据库引擎。
- OpenGL ES，基于 3D 图像加速。
- FreeType，位图与矢量字体渲染。
- WebKit，Web 浏览器引擎。
- SGL，2D 图像引擎。
- SSL，数据加密与安全传输的函数库。
- libc，标准 C 运行库，Linux 系统中底层应用程序开发接口。

② Android 运行时
- 核心库，提供 Android 系统的特有函数功能和 Java 语言函数功能。

- Dalvik 虚拟机，实现基于 Linux 内核的线程管理和底层内存管理。

（3）应用框架层包含操作系统的各种管理程序，提供 Android 平台基本的管理功能和组件重用机制，主要包含以下内容，如图 1-12 所示。

图 1-12　Android 系统应用程序框架

① Activity Manager，管理应用程序的生命周期。
② Window Manager，启动应用程序的窗体。
③ Content Providers，共享私有数据，实现跨进程的数据访问。
④ Package Manager，管理安装在 Android 系统内的应用程序。
⑤ Teleghony Manager，管理与拨打和接听电话的相关功能。
⑥ Resource Manager，允许应用程序使用非代码资源。
⑦ Location Manager，管理与地图相关的服务功能。
⑧ Notification Manager，允许应用程序在状态栏中显示提示信息。

（4）应用层是最上层，由各种应用软件包括通话程序、短信程序、邮件客户端、浏览器、通讯录和日历等构成，应用软件则由各公司自行开发，以 Java 编写，如图 1-13 所示。

图 1-13　Android 系统应用层

1.3　搭建 Android 开发环境

由于操作系统的不同，关于 Android 系统开发环境的搭建过程也有很大的不同，本部分针对 Windows 系统环境和 Linux 环境下不同的操作系统简要介绍其开发环境的搭建过程。

1.3.1 Android 开发环境系统要求

Android 开发环境的设置，针对不同的 Windows 操作系统、Eclipse 版本、Android 开发工具包以及虚拟机 JDK，有许多可以采用的组合方式，具体选择如表 1-1 所示。

表 1-1 Android 开发环境系统要求

操作系统/IDE	版 本	备 注
Windows 系列	XP、Vista、Windows 7、Windows 8、Server 版本	
Eclipse IDE（for Java Developer）	Eclipse 3.x,3.3.x（Europa）、3.4.x（Ganymede）、3.5.x（Galileo）、3.6.x（Helios）、3.7.x(indigo)	
ADT（Android Development Tools）	12.0.0—23.0.4	23.0.4 必须使用 Eclipse 3.7.2 以上版本
Java SE		JDK1.5、JDK1.6、JDK1.7、JDK1.8

1.3.2 Windows 系统平台下搭建开发环境

在 Windows 系列操作系统中搭建配置 Android 开发环境，需要支持的软件有 JDK、Eclipse、Android SDK、ADT（Android Development Tools），具体开发环境搭建配置过程如下。

1. JDK 的下载、安装

在浏览器地址栏中输入 URL "http://www.oracle.com/technetwork/java/javase/downloads/index.html"，选择 "JDK 下载"，然后进行默认安装或者自选安装路径，安装完成后在控制台输入命令 "java"，测试 JDK 是否安装成功。

2. Eclipse 软件的下载

在 IE 地址栏中输入 URL 地址 "http://www.eclipse.org/downloads"，选择 Eclipse IDE for Java Developers 进行下载，然后进行解压即可。注意：Eclipse 软件不需要安装，在安装完 JDK 之后，只需解压 Eclipse 即可使用。

3. Android SDK 的下载、配置

（1）在浏览器地址栏中输入 "http://developer.android.com/sdk/index.html"，如图 1-14 所示，选择 Android SDK 成熟版本 android-sdk_r24.0.2 进行下载，并在本地目录 h:\android\ 下进行解压，以便后续开发使用。

（2）双击 h:\android\android-sdk-windows 文件中的 SDK Manager.exe，SDK 程序自动从互联网上检测当前 SDK 可选安装包的版本，如图 1-15 所示。

SDK Tools Only

If you prefer to use a different IDE or run the tools from the command line or with build scripts, you can instead download the stand-alone Android SDK Tools. These packages provide the basic SDK tools for app development, without an IDE. Also see the SDK tools release notes.

Platform	Package	Size	SHA-1 Checksum
Windows	installer_r24.0.2-windows.exe (Recommended)	91428280 bytes	edac14e1541e97d68821fa3a709b4ea8c659e676
	android-sdk_r24.0.2-windows.zip	139473113 bytes	51269c8336f936fc9b9538f9b9ca236b78fb4e4b
Mac OS X	android-sdk_r24.0.2-macosx.zip	87262823 bytes	3ab5e0ab0db5e7c45de9da7ff525dee6cfa97455
Linux	android-sdk_r24.0.2-linux.tgz	140097024 bytes	b6fd75e8b06b0028c2427e6da7d8a09d8f956a86

All Android Studio Packages

Select a specific Android Studio package for your platform. Also see the Android Studio release notes.

Platform	Package	Size	SHA-1 Checksum
Windows	android-studio-bundle-135.1641136.exe (Recommended)	868344232 bytes	9c1c8ea6aa17fb74e0593c62fd48ee62a8950be7
	android-studio-ide-135.1641136.exe (No SDK tools included)	260272840 bytes	464d1c5497ab3d1bdef441365791ab36c89cd5ae

图 1-14 Android SDK 下载

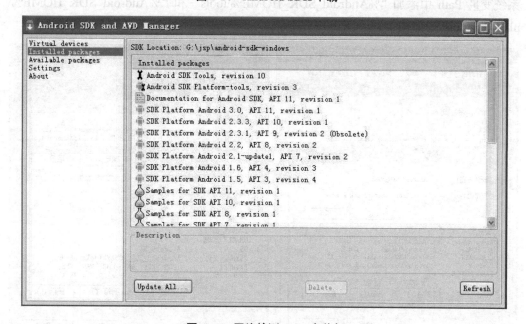

图 1-15 网络检测 SDK 安装包

然后既可以选择部分安装，也可以选择全部安装不同版本的 SDK 软件包，接着单击 Install 按钮进行下载，安装选择的 SDK 版本，程序下载安装到 h:\android\android-sdk-windows 目录文件中，如图 1-16 所示。

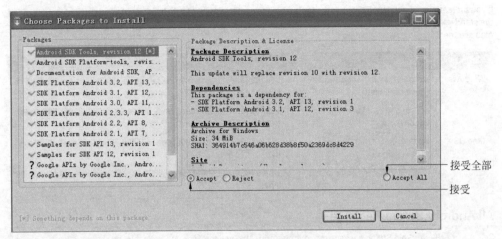

图 1-16　选择 SDK 安装包

4．设置 Android SDK HOME

如果需要在控制台界面（cmd 运行）运行 SDK 命令，则需要进行环境变量配置，右击"我的电脑"选择"属性"选项命令，选择"高级"→"环境变量"→"系统变量"，新建系统变量 Android_SDK_HOME，添加变量值为 h:\android\android-sdk-windows，在系统变量 Path 中添加"%Android_SDK_HOME%\tools"和"%Android_SDK_HOME%\platform-tools"，如图 1-17～图 1-19 所示。

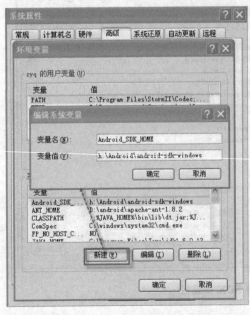

图 1-17　设置 Android SDK HOME 环境变量 1

图 1-18　设置 Android SDK HOME 环境变量 2

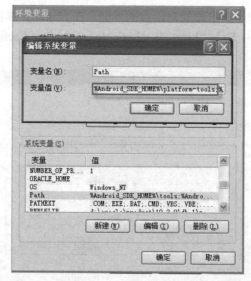

图 1-19 设置 Android SDK HOME 环境变量 3

5. ADT 安装、配置

ADT（Android Development Tools）是开发 Android 的工具插件，在 Eclipse 环境下开发 Android 程序，必须使用 ADT 插件。

ADT 安装步骤如下。

（1）打开 Eclipse 软件，执行 Help→Install New Software 命令，如图 1-20 所示。

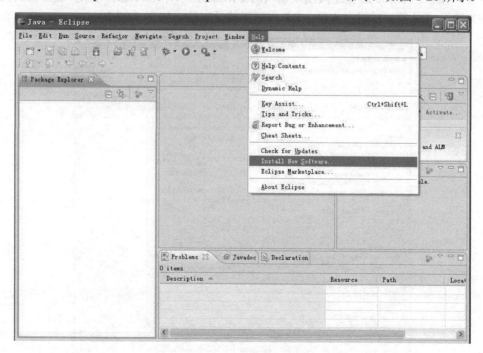

图 1-20 选择安装新软件

（2）单击 Add 按钮，弹出添加新站点的界面，在 Name 文本框中填写自己确定的名称，在 Location 文本框中输入"http://dl-ssl.google.com/android/eclipse"，然后单击 OK 按钮，如图 1-21 所示。

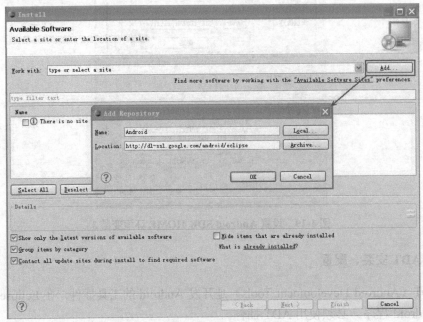

图 1-21　添加下载软件的新站点

（3）然后 Eclipse 远程连接刚才输入的 URL 站点，并在线显示可以安装的工具插件（DDMS、ADT 等）或进行本地 ADT 本地下载后安装，如图 1-22 所示。

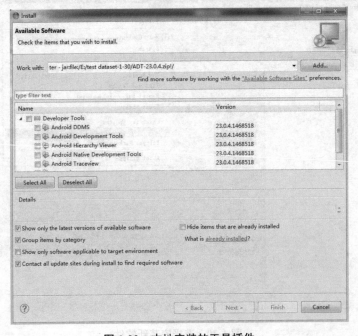

图 1-22　本地安装的工具插件

(4) 选中 DDMS、ADT 工具插件，单击 Next 按钮，出现插件安装界面，如图 1-23 所示。

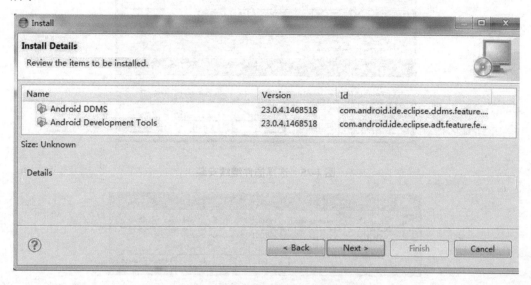

图 1-23　选择工具插件后的插件安装界面

(5) 单击 Next，弹出 Review Licenses 对话框，选择 I accept the terms of the license agreements 单选按钮，单击 Finish 进行安装，如图 1-24 所示。

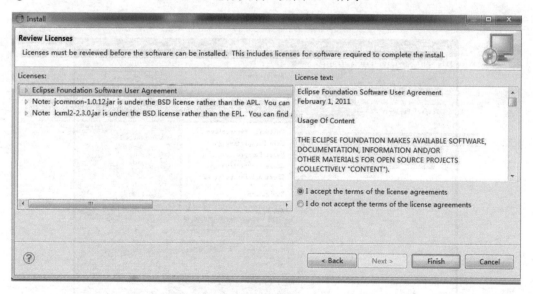

图 1-24　接受安装

(6) 在安装过程中，会出现安装软件包中包含未签名的内容警告对话框，如图 1-25 所示。

单击 OK 按钮继续安装，安装完成后，在弹出的询问是否重启 Eclipse 对话框中单击 Yes 按钮以便安装软件生效，如图 1-26 所示。

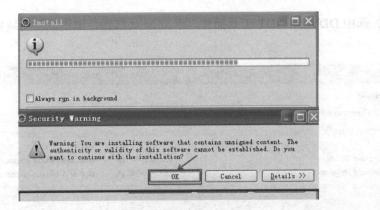

图 1-25　选择插件继续安装

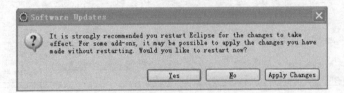

图 1-26　重启生效

6. Eclipse 中设置 Android SDK HOME

（1）执行 Window→Preferences 命令，如图 1-27 所示。

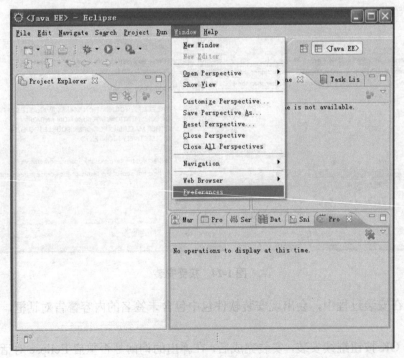

图 1-27　选择 Preferences

（2）在弹出的对话框中，在左侧选择 Android，然后在右侧的 SDK Location 文本框中选择下载的 SDK 解压目录，在对话框中右侧下方，会出现 SDK Target 的列表，然后单击 OK 按钮，完成 SDK 的配置，如图 1-28 所示。

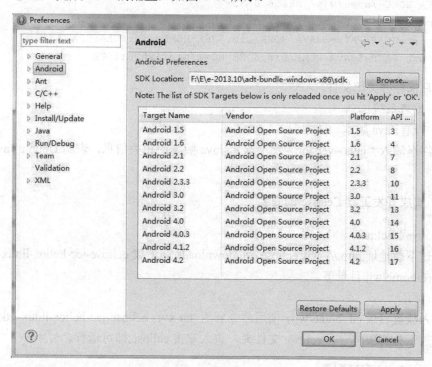

图 1-28 设置 SDK 路径

1.3.3 Linux 系统平台下搭建开发环境

1．Ubuntu Java 安装配置的详细步骤

1）下载 JDK（Linux 版本）

通过浏览器打开：http://www.oracle.com/technetwork/java/javase/downloads/index.html，选择 jdk-6u33-linux-i586.bin 并下载，将 jdk-6u33-linux-i586.bin 放置于目录/home/wjj/下。

2）解压文件

（1）打开终端，进入放置 jdk 的目录 cd/home/wjj/。

（2）使用命令"chmod u+x jdk-6u33-linux-i586.bin"，更改文件权限为可执行。

（3）使用命令"sudo./jdk-6u33-linux-i586.bin"解压文件,则在 wjj 目录下面可以看到解压的文件夹 jdk1.6.0_33。

3）配置环境变量

以 root 身份使用命令"sudo gedit /etc/profile"，打开并编辑 profile 文件。

在 profile 文件最后添加：

```
#set java environment
JAVA_HOME=/home/wjj/jdk1.6.0_33
export JRE_HOME=/home/wjj/jdk1.6.0_33/jre
export CLASSPATH=$JAVA_HOME/lib:$JRE_HOME/lib:$CLASSPATH
export PATH=$JAVA_HOME/bin:$JRE_HOME/bin:$PATH:$SDK
```

然后保存并关闭文件。

4）重启系统

5）查看 Java 版本

在终端输入"java –version"将会显示 Java 版本的相关信息，表明 Ubuntu Java 安装成功。

2．集成开发工具 Eclipse 的安装

1）下载 Eclipse

通过下载地址 http://www.eclipse.org/downloads/，下载 eclipse-jee-helios-linux-gtk.tar.gz，放在/home/wjj 目录下。

2）解压

进入放置目录 cd /home/wjj，使用命令解压 tar xvfz eclipse-jee-helios-linux-gtk.tar.gz，在此目录下会解压出一个 eclipse 文件夹，进入双击 eclipse 即可运行。

3．Android 安装配置

1）下载

通过下载地址 http://developer.android.com/sdk/index.html，下载 android-sdk_r20-linux_86.tgz。

2）解压文件放在目录/home/wjj 下

使用命令进入目录 cd /home/wjj，使用命令"tar zxvf android-sdk_r20-linux_86.tgz"解压文件。

3）配置环境变量

以 root 身份使用命"令 sudo gedit /etc/profile"，打开并编辑 profile 文件。

在 profile 文件最后添加：

```
export SDK=${PATH}:<your_sdk_dir>/tools
```

例如：

```
export SDK=$/home/wjj/android-sdk-linux_86/tools
```

4）下载和配置 ADT

安装和配置过程与 Windows 环境下配置过程相同。

1.4 搭建 Android NDK 开发环境

NDK 是一个在 Android 系统中支持开发者重用 C 和 C++语言开发的代码库的工具集合，在游戏引擎、信号处理等对 CPU 负载要求比较高的方面应用较多。搭建 Android NDK 开发环境需要安装 Cygwin、NDK，并整合 Android NDK 的开发环境 Eclipse,使 Eclipse 与 NDK 进行整合。NDK 目前最近的版本包为 android-ndk-r10d。

1. 安装 Cygwin

（1）打开 Cygwin 官方网站 http://www.cygwin.com/，根据自己的操作系统平台选择 32 位或 64 位的版本。

Cygwin 可以方便地在命令行编译 jni 代码，NDK 在 r8e 版本之后已经不再需要 Cygwin 的支持。

Cygwin 的安装有两种方式可供选择：一种是本地安装，另一种是在线安装方式。

可以选择在线安装或本地安装，或者下载不安装，本书选择本地安装方式，然后选择需要安装的包文件，如图 1-29 所示。

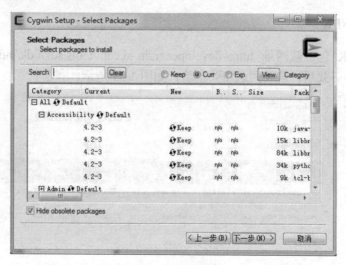

图 1-29 选择安装包

为了避免选包之间的依赖关系错误，在默认设置下，只需要选择 Devel。单击列表中的 Devel，将后面的 Default 改为 Install，其余均默认，这样会安装全部的开发工具。也可以只安装 NDK 需要用到的包：autoconf2.1、automake1.10、binutils、gcc-core、gcc-、g++、gcc4-core、gcc4-g++、gdb、pcre、pcre-devel、gawk、make，不进行全部安装。

（2）启动 Cygwin，输入命令"cygcheck –c cygwin"检查测试安装是否成功，输入命令"make –v,gcc –v"出现 make 和 gcc 的版本信息则表明 make 和 gcc 已经安装成功，如图 1-30 所示。

图 1-30　测试安装版本

2．下载 NDK

打开 NDK 包下载网址 http://developer.android.com/tools/sdk/ndk/index.html，下载 NDK 开发包，下载后，需要同意相关条款，如图 1-31 所示。

图 1-31　NDK 下载网站

第1章 Android开发基础

NDK解压后的目录结构如图1-32所示，platforms中是Android API不同平台，samples中是样例。

build	2014/8/15 23:00	文件夹	
docs	2014/8/15 23:03	文件夹	
platforms	2014/8/15 23:02	文件夹	
prebuilt	2014/8/15 23:04	文件夹	
samples	2014/8/15 23:04	文件夹	
sources	2014/8/15 23:05	文件夹	
tests	2014/8/15 23:04	文件夹	
toolchains	2014/8/15 23:04	文件夹	
documentation.html	2012/8/21 13:23	HTML 文档	1 KB
find-win-host.cmd	2014/2/11 9:20	Windows 命令脚本	1 KB
GNUmakefile	2012/8/21 13:23	文件	2 KB
ndk-build	2013/11/13 15:40	文件	10 KB
ndk-build.cmd	2014/2/11 9:20	Windows 命令脚本	1 KB
ndk-depends.exe	2014/5/14 23:17	应用程序	134 KB
ndk-gdb	2014/4/2 11:40	文件	25 KB
ndk-gdb.py	2014/1/28 11:29	Python File	33 KB
ndk-gdb-py	2013/11/13 15:40	文件	1 KB
ndk-gdb-py.cmd	2014/2/11 9:20	Windows 命令脚本	1 KB
ndk-stack.exe	2014/5/14 23:17	应用程序	642 KB
ndk-which	2012/9/10 12:06	文件	2 KB
README.TXT	2012/8/21 13:23	文本文档	2 KB
RELEASE.TXT	2014/7/3 3:41	文本文档	1 KB
remove-windows-symlink.sh	2014/1/10 15:52	SH 文件	2 KB

图1-32 NDK解压目录

3. NDK 与 Eclipse 集成

解压下载后的 android bundle r10，在 Eclipse 中需要配置 NDK 的路径，执行 Window → Preferences 命令，打开如图 1-33 所示窗口进行设置。

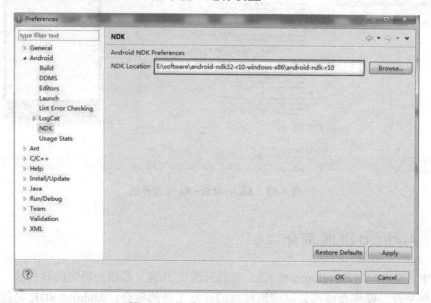

图1-33 Eclipse 中 NDK 路径指定

1.5　Android Studio 和 SDK 概述

1.5.1　Android Studio 简介

Android Studio 是基于 Intelli JIDEA 的 Android 应用开发的官方 IDE，Intelli JIDEA 是 Intelligent Java IDE 的简称，是一个企业级的框架开发工具，支持 Java EE、Spring、GWT、Grails、Play 等框架，包含 Maven、Gradle、STS 及 Git、SVN、Mercurial 等工具，支持的移动开发平台有 Android、PhoneGap、Cordova 和 Ionic，它目前最新的版本是 14。而 Android Studio 提供了丰富的布局拖曳、动态布局展示、多 API 版本布局效果同时展示、Google 云平台支持、语法提示检查等功能。其创建的工程目录结构和工程如图 1-34 和图 1-35 所示。

图 1-34　Android Studio 工程目录结构

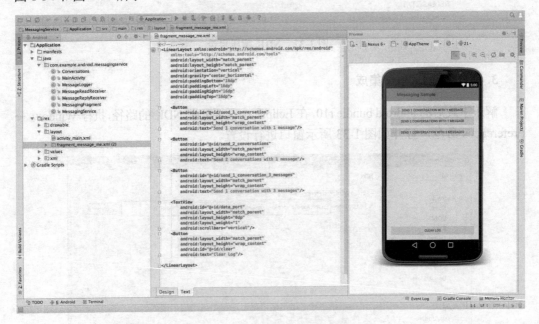

图 1-35　Android Studio 工程界面

1.5.2　Android SDK 简介

SDK（Software Development Kit，软件开发工具包）是指为特定的软件包、软件框架、硬件平台、操作系统等建立应用软件的开发工具的集合。Android SDK 是指专门用于 Android 手机操作系统创建应用软件的软件开发工具包。Android SDK 不用安装，下载后，直接解压到指定的位置即可。

解压完 Android SDK 后，打开目录，目录结构如图 1-36 所示。

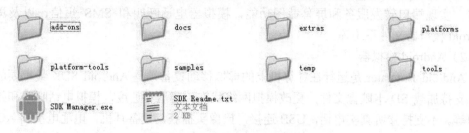

图 1-36 Android SDK 目录结构

（1）add-ons：目录下的文件是用来开发应用的 Google API，支持基于 Google Map 的地图开发，及系统模拟器图片。

（2）docs：目录下的文件是 Android SDK 的帮助文档和说明文档，通过根目录下的 index.html 文件启动。

（3）platforms：存在子目录 android-3、android-4 等，分别用来保存 1.5、1.6、2.0、2.1、2.2、2.3、3.0、4.0、4.1、4.2、4.4 版的 Android SDK 的库函数、外观样式、程序示例和辅助工具等。

（4）tools：该目录下是通用的 Android 开发和调试工具。

（5）extras：包含 android 和 google 两个目录，分别用来存放 support library 和 usb drivers，usb_driver 目录下存放 AMD64 和 x86 平台的 USB 驱动程序。

（6）platform-tools：包含开发平台需要的开发工具和测试工具。

（7）temp：包含了一些常用的文件模板。

（8）samples：包含不同版本的 SDK 演示实例。

（9）SDK Manager：SDK 管理，用于安装和更新 SDK 组件。

1.5.3 Android 常用开发工具

Android SDK 中包含许多开发工具，这些工具帮助程序开发者在 Android 开发平台上开发移动应用，这些开发工具被划分为两类：SDK 工具和平台工具。SDK 工具独立于平台，适用不同的平台；平台工具通常支持最新的 Android 开发平台。

1．SDK 工具

最重要的 SDK 工具有 SDK Manager（Android SDK）、AVD Manager（Android AVD）、Emulator、DDMS（Dalvik Debug Monitor Server），下面就简要介绍一些常用的 SDK 工具。

1）Android SDK

Android SDK 用于管理 AVDs、工程和安装 SDK 的组件，可以创建、删除 AVD。

在 Eclipse 中使用 Window→Android SDK Manager 按钮就会调用管理 SDK。

DDMS（Dalvik Debug Monitor Server）：调试监视服务，用于调试 Android 应用程序，管理运行在设备或模拟器上的进程，监视 Android 系统中的进程、堆栈信息，查看 logcat 日志，实现端口转发服务和屏幕截图功能，模拟器电话呼叫和 SMS 短信，以及浏览 Android 模拟器文件系统等。

2）Android 模拟器

Android Emulator 是运行在计算机上的虚拟移动设备，是 Android SDK 最重要的工具，支持加载 SD 卡映像文件，更改模拟网络状态，延迟和速度，模拟电话呼叫和接收短信等。不支持接听真实电话，USB 连接，摄像头捕获，设备耳机，电池电量和 AC 电源检测，SD 卡插拔检查和使用蓝牙设备。

3）dmtracedump

该工具可以使用跟踪日志生成图形化的方法生成调用图及堆栈图。该工具使用 Graphviz Dot 组件来生成图形，所以要运行 dmtracedump 就必须先安装 Graphviz。

4）Draw9 Patch

Draw9 Patch 是 Android 提供的可伸缩的图形文件格式，基于 PNG 文件。Draw9 Patch 工具可以使用 WYSIWYG 编辑器建立 NinePatch 文件。它也可以预览经过拉伸的图像，高亮显示内容区域。

5）Hierarchy Viewer

层级观察器工具允许用户调试和优化自己的用户界面。它用可视的方法把视图（View）的布局层次展现出来，此外还给当前界面提供了一个具有像素栅格（Grid）的放大镜观察器，以便正确地布局。

6）Traceview

Traceview 跟踪显示工具是 Android 平台配备的一个很好的性能分析工具。以图形化的方式显示应用程序，帮助用户了解要跟踪的程序的性能，并且能具体到方法。用来调试应用程序，分析执行效率。

7）mksdcard

创建 SD 卡工具，该工具创建一个 FAT32 格式的磁盘镜像，主要用于模拟手机 SD 卡。在创建 AVD 过程中，可以选择该文件作为 SD 卡。

8）zipalign.exe

该工具优化了应用程序的打包方式。这样做使 Android 与用户的应用程序交互更加有效和简便，有可能提高应用程序和整个系统的运行速度。

9）hprof-conv

主要用于转换文件格式，转换 hprof 文件为一个标准的文件格式，以便浏览。

10）layoutopt

用于快速分析开发的应用程序布局，优化、提高效率。

11）Monkey

Android 的压力测试工具，是在模拟器上或设备上运行的一个小程序，它能够产生随机的用户事件流，如点击、触摸、挥手，还有一系列的系统级事件。可以使用 Monkey

来给正在开发的程序做随机的但可重复的压力测试。

Monkeyrunner：Android 的自动测试工具。

12）JUnit Test

Android 集成了 JUnit 测试框架，通过它可以在软件开发过程中不断地进行测试，如果要添加其包需要在 AndroidManifest.xml 中<application>标签中增加对 android.test.runner 的声明和引用，如下：

```
<uses-library android:name="android.test.runner" />
```

13）SQlite3

SQLite3 数据库能够让用户方便地访问 SQLite 数据文件，用来创建和管理 SQLite 数据库。

14）ProGuard

这个工具是一个 Java 代码混淆的工具。在 2.3 版本的 SDK 中可以看到在 android-sdk-windows/tools/下面多了一个 proguard 文件夹，表明 Google 已经把 ProGuard 技术放在了 Android SDK 里面，可以通过正常的编译方式实现代码混淆了。

2．Android 平台工具

通常，开发者直接使用的平台工具是 ADB（Android Debug Bridge，Android 调试桥），ADB 工具可以让用户在模拟器或设备上安装应用程序的.apk 文件，并从命令行访问模拟器或设备。用户也可以用它把 Android 模拟器或设备上的应用程序代码和一个标准的调试器连接在一起。

其他的平台工具，如 AIDL（Android Interface Description Language，Android 接口描述语言）、AAPT（Android Asset Packaging Tool，Android 资源打包工具）、dexdump、dx，很少由开发者直接使用，一般都是通过 Android 编译工具或者 ADT 调用来完成任务。

1.6　创建 Android 程序和 Android NDK 程序

1.6.1　创建和使用虚拟设备

AVD（模拟器或模拟设备）是 Android 程序测试运行的虚拟平台，每一个 AVD 都模拟了一个独立的虚拟设备，来运行 Android 的程序。自 Android SDK 1.5 版本之后，Android 程序测试运行必须创建 AVD 虚拟运行平台，才能进行运行及测试。下面分别介绍在 Eclipse 环境中和命令行下 AVD 的创建和使用。

1．Eclipse 环境下使用和创建 AVD 模拟器

（1）在 Eclipse 菜单栏中，执行 Window→Android SDK and AVD Manager 命令，弹出对话框，默认选择 Virtual devices，如图 1-37 所示。

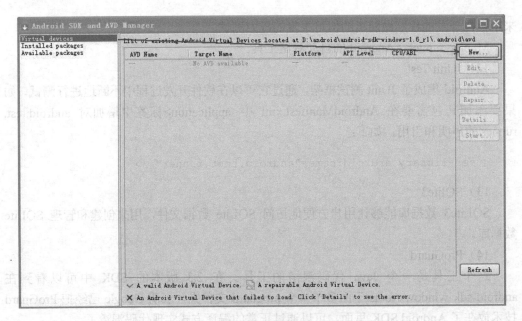

图 1-37 新建 AVD

（2）单击 New 按钮，创建新的 AVD，弹出对话框，设置 AVD 的名称、目标平台 API 版本、SD 卡的大小、模拟设备默认的皮肤等参数，如图 1-38 所示。

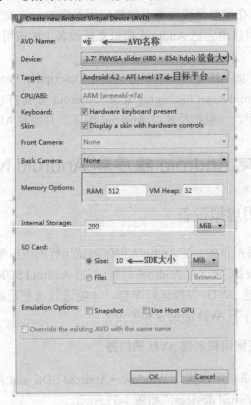

图 1-38 设置创建 AVD 的参数

（3）单击 Create AVD 按钮，创建 AVD，成功创建的 AVD 在对话框右下方显示，如图 1-39 所示。

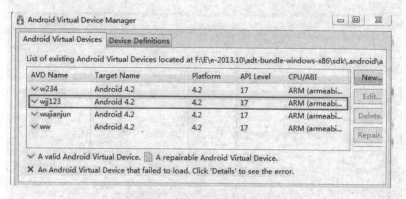

图 1-39　创建后的 AVD

（4）单击 Start 按钮，启动运行 AVD 模拟设备，运行界面如图 1-40 所示。

图 1-40　启动后 AVD 模拟器界面

2. 在命令行下创建和使用 AVD 模拟器

（1）在控制台窗口中，输入 android list targets，查看可用的目标设备，如图 1-41 所示。

（2）选择其中一个目标平台 API 版本，输入命令：

```
android create avd --name wjj --target 10
```

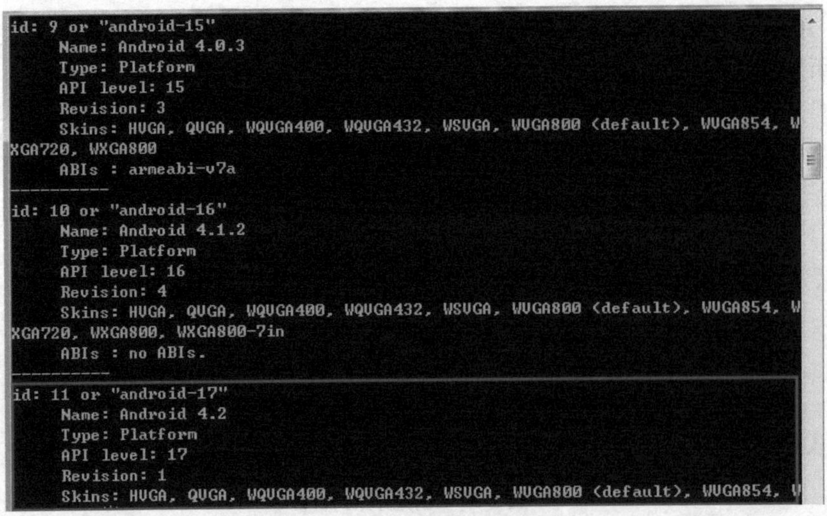

图 1-41 查看可用的目标平台

其中，wjj 是虚拟设备的名称，10 是选择的目标平台 API 版本 id 值，创建虚拟设备 AVD，如图 1-42 所示。

图 1-42 创建后的 AVD

（3）启动虚拟设备，输入"emulator –avd wjj"命令，启动上面创建的虚拟设备 wjj，如图 1-43 所示。

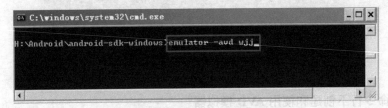

图 1-43 启动后的 AVD 界面

1.6.2 在 Eclipse 下创建 Android 程序

前面的章节中已经讲解了在 Eclipse 中搭建 Android 开发环境和虚拟设备 AVD 的创建，下面介绍在 Eclipse 中如何创建 Android 应用程序的步骤。

(1) 打开 Eclipse，执行 File→new→Android Application Project 命令，如图 1-44 所示，或者在右边的面板中单击右键，在弹出的菜单中执行 New→Android Application Project 命令。

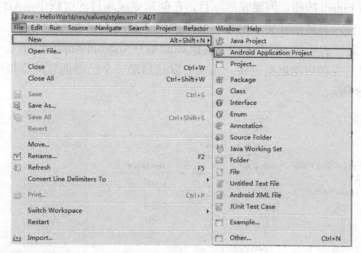

图 1-44　创建 Android 工程

特别提示：如果发现没有找到 Android Application Project 项，可以选择 File→New→Other 命令，在弹出的对话框中的 Android 下，可以找到 Android Application Project 项。

(2) 弹出 Android 项目创建界面，填写项目名称、项目默认的存储路径、目标版本、应用程序名（默认与项目名一致）、包名、创建的 Activity 的名字、最小 SDK 版本（默认与目标版本 API 一致，不要修改），如图 1-45 所示。

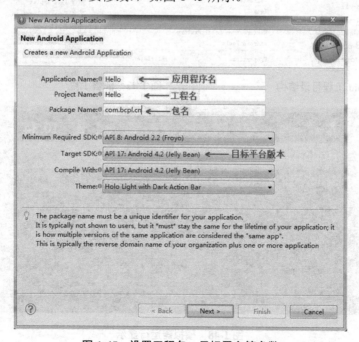

图 1-45　设置工程名、目标平台等参数

特别提示：Min SDK Version（最小 SDK 版本号）中的版本号是根据选择的目标版本 API 自动生成的，不要进行修改。

（3）单击 Finish 按钮，创建完成项目，在创建完成项目的过程中，ADT 会自动生成一些目录和文件，具体后续生成的项目目录结构，如图 1-46 所示。

（4）需要运行的项目，右击选中，执行 Run As→Android Application 命令，如图 1-47 所示。如果没有提前启动虚拟设备，系统会默认启动一个已经创建的虚拟设备，运行项目，如图 1-48 所示。

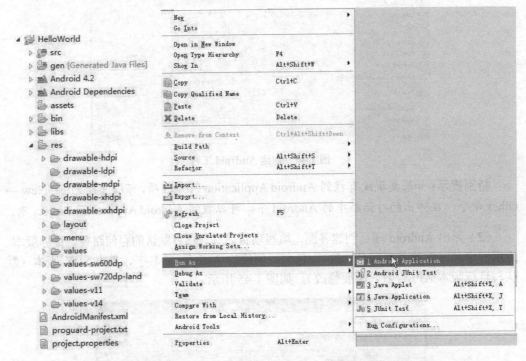

图 1-46　Android 工程目录结构　　　　图 1-47　部署工程

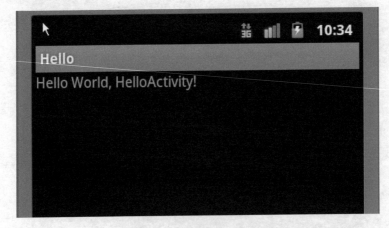

图 1-48　模拟器运行显示

特别提示：在运行项目之前，必须先创建虚拟设备 AVD（按前面章节所述方法创建）。创建虚拟设备过程中选择的目标版本和创建项目中选择的目标版本必须一致或者向下兼容。否则会提示没有兼容的目标版本，需要创建新的 AVD。

1.6.3　命令行创建 Android 程序

命令行工具保存在<sdk>/platform-tools/和<sdk>/tools/目录下，使用命令行工具创建 Android 程序，需要使用 tools 目录下的 android.bat 和 platform-tools 目录下的 adb.exe 工具及 Apache Ant 软件。

1. Android 批处理工具

Android.bat 是一个批处理文件，可以用来建立和更新 Android 工程，同时也管理 AVD，能够创建 Android 工程所需要的目录结构和文件。具体命令和参数如表 1-2 所示。

表 1-2　Android.bat 建立和更新 Android 工程的命令和参数说明

命令	参数	说明	备注
android create project	-k <package>	包名称	必备参数
	-n <name>	工程名称	
	-a <activity>	Activity 名称	
	-t <target>	新工程的编译目标	必备参数
	-p <path>	新工程的保存路径	必备参数
android update project	-t <targe>	设定工程的编译目标	必备参数
	-p <path>	工程的保存路径	必备参数
	-n <name>	工程名称	

2. Apache Ant 工具

Apache Ant 是一个将软件编译、测试、部署等步骤联系在一起的自动化工具，多用于 Java 环境中的软件开发。若在构建 Android 程序时使用 Apache Ant，可以简化程序的编译和 apk 打包过程。Apache Ant 下载网址为 http://ant.apache.org/bindownload.cgi，网站提供 zip、tar.gz 和 tar.bz2 三种格式下载，Windows 系统用户推荐下载 zip 格式的二进制包。目前最新下载的 Apache Ant 压缩包为 apache-ant-1.8.2-bin.zip，版本号为 1.8.2。

3. ADB 工具

ADB(Android Debug Bridge)是多种用途的命令行工具，利用它可以与模拟器或带电的 Android 设备进行连接通信、管理模拟器的状态及调试程序。

利用命令行工具开发 Android 程序，创建 HelloWorld 工程的步骤如下。

（1）使用 android.bat 建立 HelloWorld 工程所需的目录和文件。打开 cmd 控制台，然后进入<sdk>/tools 目录下（如本书 android.bat 存放路径为 H:\Android\android-sdk-windows\tools），输入命令：

```
android create project -n HelloWorld -k www.bcpl.HelloWorld -a HelloWorld
-t 9 -p d:\Android\workplace\HelloWorld
```

或者

```
android create project -name HelloWorld --package www.bcpl.HelloWorld
--activity HelloWorldActivity --target 9 --path d:\Android\workplace\
HelloWorld
```

上述命令中创建的新工程的名称为 HelloWorld，包名称为 www.hisoft.HelloWorld，Activity 名称是 HelloWorldActivity，编译目标的 ID 为 9，新工程的保存路径是 d:\Android\workplace\HelloWorld。如图 1-49 所示，在新建立的工程目录中，发现其中一些是使用 Eclipse 开发环境时创建同样的工程不会出现的文件，例如 build.xml、local.properties。这些新文件的出现，主要是为了在构建 Android 程序时使用 Apache Ant。

图 1-49 创建的工程目录

新创建的 HelloWorld 工程文件和目录列表说明如表 1-3 所示。

表 1-3 HelloWorld 工程文件和目录列表说明

文件	说明
AndroidManifest.xml	应用程序声明文件
build.xml	Ant 的构建文件
default.properties	保存编译目标，由 Android 工具自动建立，不可手工修改
build.properties	保存自定义的编译属性
local.properties	保存 Android SDK 的路径，仅供 Ant 使用
src\www\bcpl\HelloWorld\ HelloWorld.java	Activity 文件
bin\	编译脚本输出目录
gen\	保存 Ant 自动生成文件的目录，例如 R.java
libs\	私有函数库目录，在工程创建初期是空目录
res\	资源目录
src\	源代码目录

运行结果如图 1-50 所示。

图 1-50　工程运行后结果

特别注意：如果已经在环境变量中设置了 SDK 路径，则可以直接在 cmd 中输入命令，不用进入<SDK>/tools 后再输入命令。

（2）设置和测试 Apache Ant 环境变量。

① 在 Windows 系统中添加新的环境变量，Apache 才能正常运行。右击"我的电脑"执行"属性"→"高级"→"环境变量"→"系统变量"命令，新建 ANT_HOME，变量值为解压后的 apache-ant-1.8.2 安装存放目录，本书是放在 d:\android\apache-ant-1.8.2 下（可以根据自己实际安放位置进行修改），如图 1-51 所示。

图 1-51　设置 ANT_HOME 变量

然后在"变量值"文本框中,添加":%ANT_HOME%\bin",如图 1-52 所示。

② 测试判断设置的环境变量正确性。在 CMD 中运行输入 ant 命令,通过命令的输出信息判断环境变量是否设置正确。

如果输出的提示包含"Unable to locate tools.jar. Expected to find it in…",则表明设置环境变量不正确。

如果环境变量设置正确,ant 命令的输出结果如图 1-53 所示。

图 1-52　设置 Path 变量

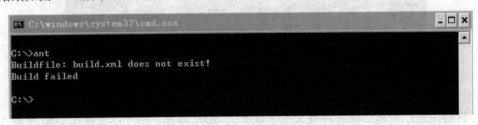

图 1-53　测试 ant 环境变量设置

(3) 应用程序数字签名。

在 Android 平台上开发的所有应用程序都必须进行数字签名后,才能安装到模拟器或手机上,否则将返回错误提示:

Failure [INSTALL_PARSE_FAILED_NO_CERTIFICATERS]

特别注意:在 Eclipse 开发环境中,ADT 在将 Android 程序安装到模拟器前,已经利用内置的 debug key 为 apk 文件自动做了数字签名,这使用户无须自己生产数字签名的私钥,而能够利于 debug key 快速完成程序调试。但是,如果用户希望正式发布自己的应用程序,则不能使用 debug key,必须使用私有密钥对 Android 程序进行数字签名。

Apache Ant 构建 Android 应用程序支持 Debug 模式和 Release 模式两种构建模式。

Debug 模式是供调试使用的构建模式,是用于快速测试开发的应用程序。Debug 模式自动使用 debug key 完成数字签名。

Release 模式是正式发布应用程序时使用的构建模式,生成没有数字签名的 apk 文件。

Debug 模式对 HelloWorld 工程进行编译,生成具有 debug key 的 apk 打包文件。

步骤:使用 CMD,在工程的根目录 D:\android\workplace\HelloWorld 下,输入"ant debug",结果如图 1-54 所示。

命令运行后,Apache Ant 在工程 bin 目录中生成打包文件 HelloWorld-debug.apk。

如果需要使用 Release 模式,则需在 CMD 中输入 ant release,运行后会在 bin 目录中生成打包文件 HelloWorld-unsigned.apk。

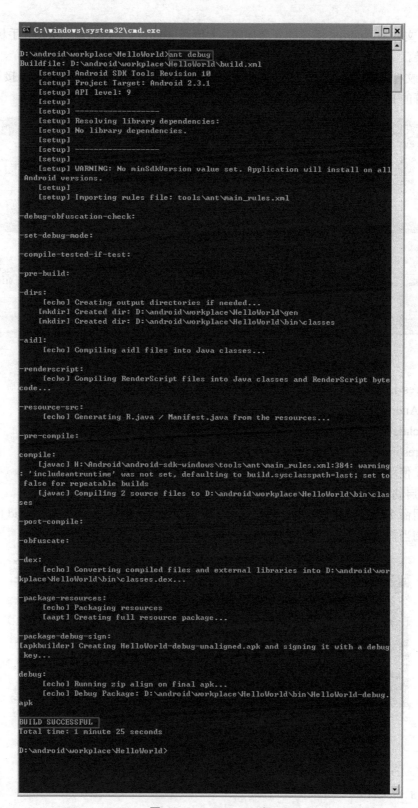

图 1-54　ant debug 工程

apk 文件是 Android 系统的安装程序，上传到 Android 模拟器或 Android 手机后可以进行安装。

apk 文件本身是一个 zip 压缩文件，能够使用 WinRAR、UnZip 等软件直接打开。打开的 HelloWorld-debug.apk 文件，如图 1-55 所示。

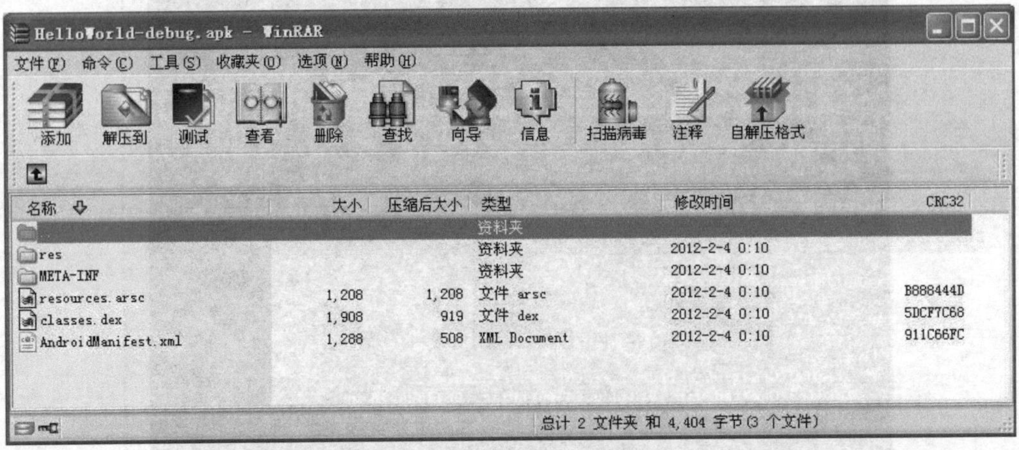

图 1-55　HelloWorld-debug.apk 文件结构

① res\目录用来存放资源文件。

② AndroidManifest.xml 是 Android 声明文件。

③ classes.dex 是 Dalvik 虚拟机的可执行程序。

④ resources.arsc 是编译后的二进制资源文件。

（4）使用 adb.exe 将 HelloWorld 工程上传到 Android 模拟器中。

① 启动 AVD。

使用命令行启动模拟器时，需要先指定所使用的 AVD。可以使用"android list avds"命令查询当前系统所有已经创建的 AVD，如图 1-56 所示。

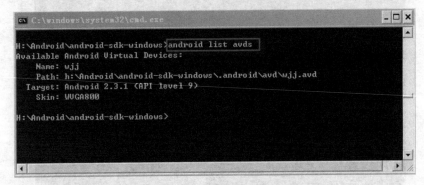

图 1-56　查看已创建的 AVD 设备

在 CMD 中输入"emulator - avd　wjj"，启动 AVD 虚拟设备。

② 上传文件。

Android 模拟器正常启动后，使用 adb.exe 工具把 HelloWorld-debug.apk 文件上传到

模拟器中。

adb.exe 工具除了能够在 Android 模拟器中上传和下载文件，还能够管理模拟器状态，是调试程序时不可缺少的工具。

在 CMD 中，进入工程 HelloWorld/bin 目录，输入命令"adb install HelloWorld-debug.apk"。完成 apk 程序上传到模拟器的过程。

如果上传成功，结果如图 1-57 所示。

图 1-57　安装新创建的 apk 文件

③ 启动应用程序。

apk 文件上传后，需手工启动 HelloWorld 应用程序。

单击模拟器界面左下角上刚安装的 HelloWorld 应用程序图标，即可手工启动，如图 1-58 所示。

图 1-58　查看程序运行结果

如果在模拟器界面中看不见新安装的程序，则可以单击模拟器右侧的 menu 按钮，找到新安装的应用程序图标，如果在模拟器中找不到新安装的程序，可以尝试重新启动

Android 模拟器。因为 Android 的包管理器经常仅在模拟器启动时候检查应用程序的 AndroidManifest.xml 文件，这就导致部分上传的 Android 应用程序不能立即启动。

④ 编译和打包应用程序。修改 HelloWorld 工程代码后，需要使用 Apache Ant 重新编译和打包应用程序，并将新生成的 apk 文件上传到 Android 模拟器中。

如果新程序的包名称没有改变，则在使用 adb.exe 上传 apk 文件到模拟器时，会出现如图 1-59 所示的错误提示。此时，需要在模拟器中先删除原有 apk 文件，再使用 adb.exe 工具上传新的 apk 文件。

图 1-59 安装失败

删除 apk 文件有以下两种方法。

使用 adb uninstall <包名称>的方法。例如，删除 Hello World 工程中的 apk 文件，则可在 CMD 中输入"adb uninstall www.bcpl.HelloWorld"，提示"Success"则表示成功删除。

使用 adb shell rm /data/app/<包名称>.apk 的方法。同样以删除 HelloWorld 工程中的 apk 文件为例，在 CMD 中输入下面的命令，没有任何提示则表示删除成功。

```
adb shell rm /data/app/cn.bcpl.HelloWorld.apk
```

特别注意：如果仅有一个 Android 模拟器在运行，按照上述操作步骤可以用一条命令完成 Android 工程编译、apk 打包和上传过程。如果同时有两个或两个以上的 Android 模拟器存在，这种方法将会失败，因为 adb.exe 不能够确定应该将 apk 文件上传到哪一个 Android 模拟器中。此外，多次使用这种方法时，同样需要先删除模拟器中已有的 apk 文件。

1.6.4 调试 Android 程序

Android SDK 提供了大部分的测试工具，它们分别放在 SDK 的 tools 和 platform-tools 目录下，tools 目录下的测试工具有 DDMS、Hierarchy Viewer、layoutopt、Traceview、dmtracedump；platform-tools 目录下的测试工具有 ADB，此外还有 Dev Tools Android application 应用测试等。

一个典型的 Android 应用程序测试环境，主要由 DDMS（Dalvik Debug Monitor Server）、ADB、设备或者 AVD（Device or Android Virtual Device）、JDWP debugger 这几

部分组成。

在一个 Android 应用程序开发后期，需要对其进行测试，根据 Android 应用程序开发环境和应用程序本身不同的情况，可以选择不同的测试方法和测试环境工具，其主要有：Eclipse 加 ADT 插件开发环境下应用程序测试、其他 Java 集成开发环境（IDEs）的应用程序测试、DDMS 测试、ADB 和 logcat 组合测试、用 Hierarchy Viewer 工具测试优化用户界面、用 layoutopt 工具优化布局、使用 Traceview 和 dmtracdedump 工具以图形的方式展现日志和分析程序性能。

下面对 Android 应用程序的测试所常用的方法和测试工具进行介绍，其他的不再赘述，如感兴趣可以参考 SDK 文档说明。

1．Eclipse 加 ADT 插件开发环境下测试 Android 应用

在 Eclipse 加入 ADT 插件的开发环境下，Android 应用程序的测试可以从两个方面展开：一个是使用 Eclipse 内置的 Java 调试器，进行程序调试、设置断点、查看代码执行中变量变化及使用 logcat 实时查看系统日志；另外一个是使用 DDMS，通过 DDMS 视图中 Devices、Eemulator Control、logcat、Threads、Heap、Allocation Tracker 和 File Explorer 这些面板，查看和调试 Android 应用程序。

1）使用 Eclipse 内置的 Java 调试器

（1）断点设置。

Android 应用程序中断点的设置方法与一般的 Java 程序一样，都是通过在代码区需要设置断点的代码行号前方左侧区域双击或者右击选择 Toggle Breakpoint，设置断点，如图 1-60 所示。

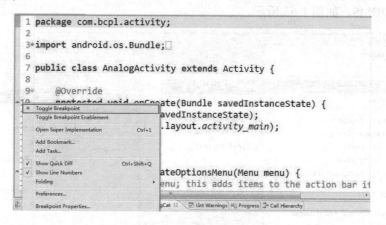

图 1-60　设置调试断点

（2）运行项目调试。

选中上述设置好断点的项目名称，单击右键在弹出的菜单中执行 Debug As→Android Application 命令，然后执行项目调试，在调试界面中，可以通过 Variables、Breakpoints、Debug 等面板参数，查看程序的每一步调试执行情况及变化。

2）DDMS 调试

在 Eclipse 集成开发环境中，除了上述的 Java 调试器之外，还可以使用 DDMS 调试视图工具，对 Android 应用程序进行调试，DDMS 调试视图工具包含：Devices 面板、Emulator Control 面板、LogCat、Thread、Heap、Allocation Tracker、File Explorer 这些调试面板，如图 1-61 所示。

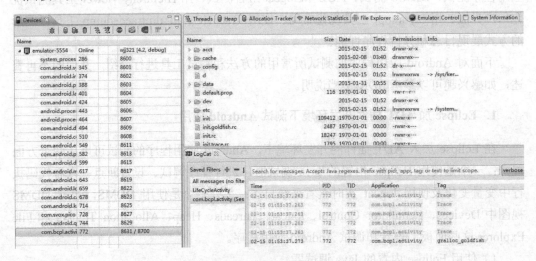

图 1-61　DDMS 界面

一般情况下，可以通过在 Eclipse 中，执行 Window→Open Perspective→DDMS 命令，打开 DDMS 调试视图，或者在 Eclipse 界面的右上角，选择 Open Perspective→DDMS，如图 1-62 所示。如果没有找到 DDMS，则可以通过 Window→Open Perspective→Other 命令找到 DDMS，如图 1-63 所示。

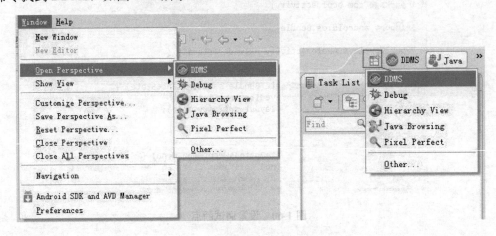

图 1-62　打开 DDMS 调试视图命令

特别注意：启动 DDMS 的另外一种方式，是从命令行启动 DDMS，通过运行保存在 SDK 目录下的 tools 文件夹中的批处理文件 ddms.bat 来启动。

（1）Devices 面板。

DDMS 工具中的 Devices 面板，显示所有的连接到 ADB 的设备、AVDs 信息。当 DDMS 启动后，它会与一个设备的 ADB 相连接，并在 DDMS 与 ADB 之间创建一个 VM 监控服务，用于监控当前设备启动或终止时，使 DDMS 能够及时得到通知信息。DDMS 通过 ADBD（ADB Daemon）获取 VM 处理进程 ID 号，进而连接 VM 调试器。DDMS 给每一个设备的 VM 都分配一个调试端口，通常把 8600 端口分配给第一个调试的 VM，紧接着是 8601，以此类推，默认情况下，DDMS 还对 8700 端口（基端口）进行监听，8700 端口是转发端口，它能接受来自任何调试端口的 VM 流量，并通过 8700 端口转发到调试器。当前程序的调试端口及转发，如图 1-64 所示。

图 1-63　找到 DDMS

图 1-64　Devices 面板

（2）Emulator Control 面板。

在 Emulator Control 面板中，可以输入电话号码，向模拟器打电话或发短信息，并可以显示虚拟地理位置信息，如图 1-65 所示。

（3）Threads。

通过单击 Devices 面板上的 Update Threads 按钮，在右侧的 Threads 面板上显示应用程序名字及当前执行线程的状态、Tid 号等信息，如图 1-66 所示。

（4）Heap、Allocation Tracker 和 File Explorer。

Heap：在 Devices 面板上单击 Update Heap 按钮，在右侧的 Heap 面板上显示堆内存的分配和执行情况；在 Heap 面板上单击 Cause GC 按钮，可以激活堆数据的垃圾回收机制。

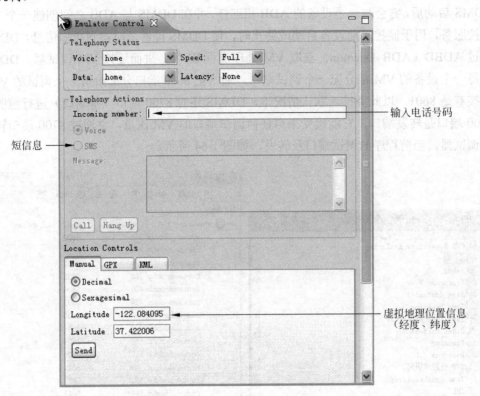

图 1-65　Emulator Control 面板参数设置

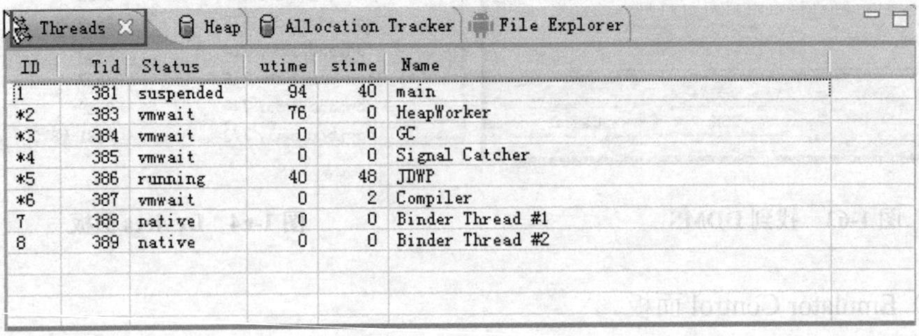

图 1-66　Threads 面板

Allocation Tracker：追踪对象的内存分配，并可以看到哪些类、线程分配到对象，在 Allocation Tracker 面板上，通过单击 Start Tracking 按钮和 Get Allocations 按钮，追踪对象的内存分配。

File Explorer：显示模拟器中的文件，复制、删除，如果启动时加载了 SD 卡，可以查看 SD 卡信息，以及文件的复制操作，如图 1-67 所示。

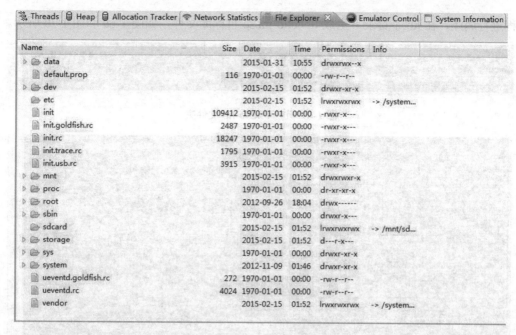

图 1-67　File Explorer 面板

注意事项：文件浏览必须在模拟器启动并部署工程后，才能查看和浏览。

2．其他 Java 集成开发环境（IDEs）的 Android 应用程序调试

通常在这种条件下，Android 应用程序调试主要采用 SDK 提供的调试工具和 Java 调试器，主要有 ADB、DDMS、Java 调试器。Java 调试器采用符合 JDWP 规范的调试器，如 JDB。

采用命令行的方式启动调试环境，主要步骤如下。

（1）先启动 AVD。

（2）进入 SDK 路径下的 tools 目录中，启动 DDMS 和 ADB 工具。

（3）安装 apk 文件到虚拟设备 AVD。

（4）Java 调试器附加到调试端口 8700，或者应用程序在 DDMS 中的特定端口。

1.6.5　创建 Android NDK 程序

本节通过 NDK 自带的样例，创建 NDK 的应用及 Hello-JNI 展示。

使用 cd 命令进入 NDK 解压后的根目录，如图 1-68 所示。

图 1-68　进入 NDK 目录

进入 samples/hello-jni 目录下，使用 NDK 自带的命令脚本 ndk-build 对 Hello-jni 工

程进行编译，在 lib 子目录下生成 libhello-jni.so 目标库文件，如图 1-69 和图 1-70 所示。

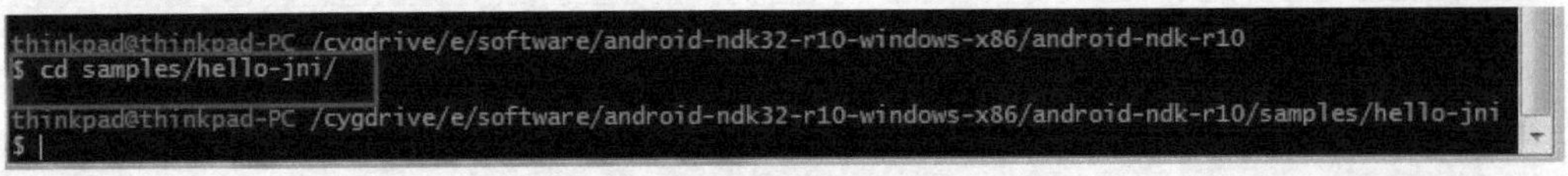

图 1-69 进入 hello-jni 文件夹

图 1-70 生成 .so 目标库文件

打开 Eclipse，执行 New->other 命令，打开 New 对话框，选择 Android 文件夹目录下的 Android Project from Existing Code 项，如图 1-71 所示。

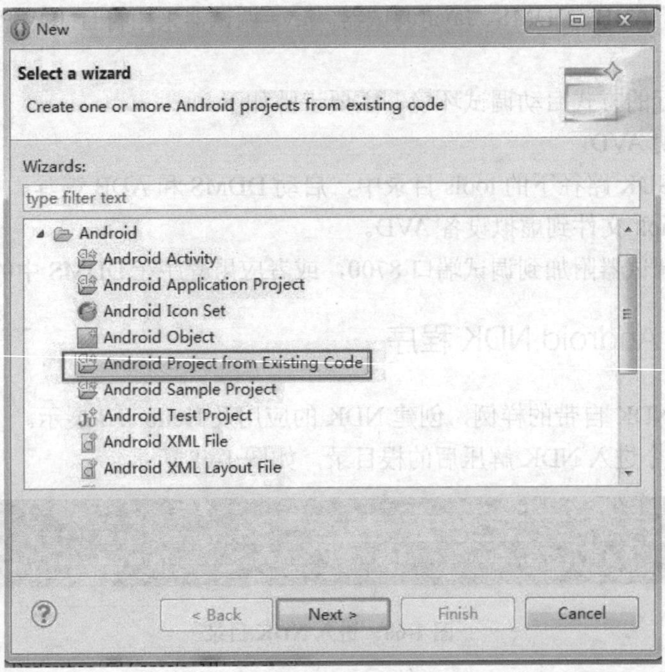

图 1-71 导入已经存在的工程

在打开的对话框中单击 Browser 按钮，选择指定 NDK 解压的目录，并复制到工作空间，单击 Finish 按钮完成目标工程的导入，如图 1-72 所示。

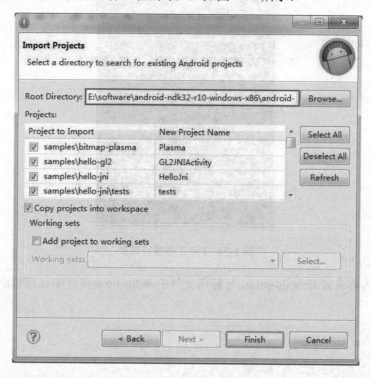

图 1-72　找到已存在的工程目录

选中工程 HelloJni，右击，在弹出的菜单中选择 Run As 位→Android Application，注意在运行前必须先创建 AVD 设备，如图 1-73 所示。

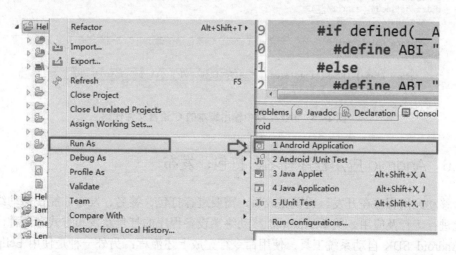

图 1-73　运行 JNI 工程

运行结果在模拟器界面中的显示如图 1-74 所示。

图 1-74　JNI 工程运行结果

显示的内容来自 libhello-jni.so 目标库文件中 hello-jni.c 文件中代码输出，如图 1-75 所示。

```
#else
  #define ABI "armeabi"
  #endif
#elif defined(__i386__)
  #define ABI "x86"
#elif defined(__x86_64__)
  #define ABI "x86_64"
#elif defined(__mips64)    /* mips64el-* toolchain defines __mips__ too */
  #define ABI "mips64"
#elif defined(__mips__)
  #define ABI "mips"
#elif defined(__aarch64__)
  #define ABI "arm64-v8a"
#else
  #define ABI "unknown"
#endif

    return (*env)->NewStringUTF(env, "Hello from JNI !  Compiled with ABI " ABI ".");
}
```

图 1-75　JNI 工程中输出脚本的 C 语言文件代码

1.6.6　Android 应用程序签名、打包、发布

移动应用程序在开发、调试完成后，需要进行打包、签名、发布才能在移动终端设备上进行运行及应用。关于 Android 移动终端设备程序的打包及发布方式有两种，一种是 Android SDK 自动系统工具，使用命令行完成上述流程；另外一种是使用 Eclipse 集成开发工具，完成上述打包、发布流程，1.6.3 节中已经讲述了在命令行下如何创建、开发、打包、发布程序的流程，本节只是采用 Eclipse 集成开发工具完成移动应用程序的打包、签名、发布流程，具体步骤如下：

（1）右击开发完成的移动应用端程序项目 HelloJni，选择 Android Tools→Export Signed Application Package，如图 1-76 和图 1-77 所示。

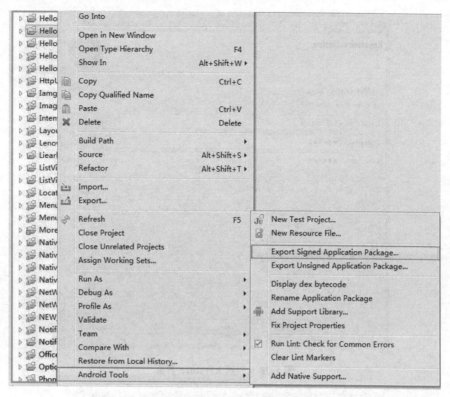

图 1-76　选择导出签名的应用程序包

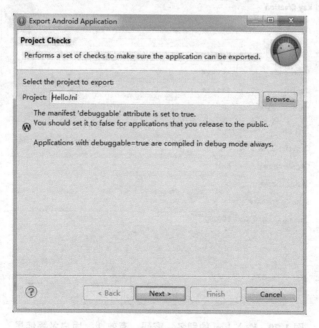

图 1-77　导出的项目工程名称

（2）创建新的 keystore，如果已经存在 keystore，则可以选择使用现有的或者创建新的，输入 keystore 的存储路径及密码和确认密码，以及相关信息，如图 1-78 和图 1-79 所示。

图 1-78　创建 keystore 界面

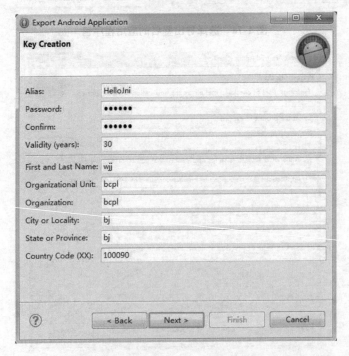

图 1-79　输入 key 的别名、密码、有效期、用户名等信息

(3)输入导出的 apk 文件的存储路径及文件名称和导出的 apk 文件和 key，如图 1-80 所示，单击 Finish 按钮，完成 apk 的导出。

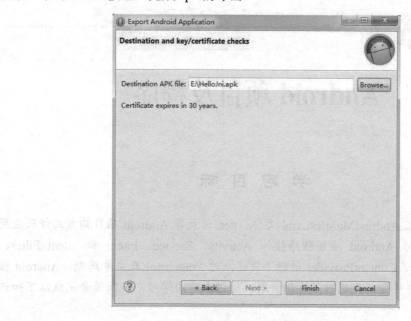

图 1-80　导出的 apk 路径及名称

注意：在 Eclipse 中还可以导出未签名的 APK 包，然后使用 jarsigner 命令进行签名，同时，使用 zipalign.exe 工具优化 APK 安装包。

习　题

一、简答题

1．什么是 4G 智能机？4G 智能机技术与 3G、2G 关键技术有何不同？

2．智能手机操作系统关键技术有哪些？

3．围绕 Android 系统平台，Google 打造了哪些技术生态链？未来的技术热点会是什么？

4．Android Studio、SDK、NDK、ADK（后续会介绍）各自的优劣是什么？与自己的专业方向结合，哪一种更适合？为什么？

二、实训

要求：

使用不同的方式（Eclipse 集成环境方式、命令行方式）创建一个新的 Android 工程，并部署运行。

使用不同的系统平台（Windows 系统平台、Linux 系统平台）创建一个新的 Android 工程，并部署运行。

分别使用 Android Studio、NDK、ADK 构建集成开发环境，并创建新的 Android 应用。

第 2 章

Android 项目及程序

学 习 目 标

本章主要介绍 AndroidManifest.xml 文件、gen 目录等 Android 项目构成文件及应用程序组成部分，对 Android 应用程序组件 Activity、Service、Intent 和 Intent Filter、BroadcastReceiver、ContentProvider 进行了简述，对 Fragement 及生命周期、Android 程序生命周期和组件生命周期进行了详细讲解。通过本章的学习，帮助读者完成以下知识要点的学习。

（1）Android 项目文件、应用程序组成部分。
（2）Android 系统应用程序组件功能及工作原理。
（3）Frgement 与 Activity 的关系、生命周期及应用。
（4）Android 程序生命周期、Activity 的生命周期中各状态的变化关系。
（5）Service 生命周期及其调用过程。

在 Android 的应用过程中，对于初学者而言，通常混淆 Android 项目和 Android 应用程序，它们之间既有区别又有联系。要创建 Android 应用，必须先创建 Android 项目，然后才能在项目中创建 Android 应用程序。下面就 Android 项目的构成和 Android 应用程序的组成进行介绍。

2.1 Android 项目构成

2.1.1 目录结构

在建立新项目的过程中，ADT 会自动建立一些目录和文件，这些目录和文件有其固定的作用，有的允许修改，有的不能修改。一个新创建的 Android 项目，项目结构包含 src 目录、gen 目录、assets 目录、res 目录、库文件 android4.2.jar、Android Dependencies（android-support-v4.jar）bin 目录、libs 目录以及三个项目工程文件 AndroidManifest.xml、project.properties、proguard-project.txt，如图 2-1 所示，下面逐一介绍。

src 目录：是源代码目录，所有允许用户修改的 Java 文件和用户自己添加的 Java 文件都保存在这个目录中。如建立 HelloWorld 工程，ADT 根据用户在工程向导中选择 Create Activity 选项，自动建立 HelloWorld.java 文件。

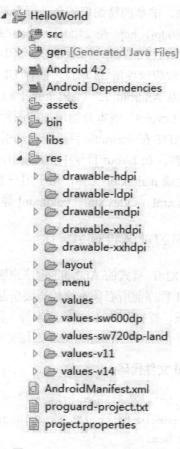

图 2-1　Android 工程目录结构

gen 目录：用来保存 ADT 自动生成的 R.java 文件，被程序代码用做资源引用映射。

Android 4.2.jar 文件：是 Android 程序所能引用的函数库文件，Android 通过平台所支持 API 都包含在这个文件中。

assets 目录：用来存放原始格式的文件，例如音频文件、视频文件等二进制格式文件。此目录中的资源不能被 R.java 文件索引，所以只能以资源流的形式读取。一般情况下为空。

res 目录：是资源目录，有三类 12 个子目录用来保存 Android 程序所有的资源。

project.properties、proguard-project.txt：在新版本的 ADT 创建项目后的工程中，混码的项目工程文件不再是 proguard.cfg 文件（Android 混淆器），proguard.cfg 可以用来防止程序被反编译，而在新的项目中实现全局混淆，需要将 project.properties 中 #proguard.config=${sdk.dir}/tools/proguard/proguard-android.txt:proguard-project.txt 前面的注释"#"去掉即可实现。

特别提醒：dpi 是 dot per inch 的简称，表示每英寸像素数。

在 Android 中，密度分类有 4 种，分别是 ldpi (low dpi)、mdpi (medium dpi)、hdpi (high dpi) 和 xhdpi（extra high dpi）。

对于屏幕来说，dpi 越大，屏幕的精细度越高，屏幕看起来就越清楚。一般情况下的 ldpi 是 120dpi，mdpi 是 160dpi，hdpi 是 240dpi，xhdpi 是 320dpi。dpi 是一种虚拟的像素单位，和具体像素值的对应公式是 dip/pixel=dip 值/160，也就是 px = dp * (dpi / 160)，当用户定义应用的布局的 UI 时应该使用 dp 单位，确保 UI 在不同的屏幕上正确显示。

需要注意的是：xhdpi 是从 Android 2.2（API Level 8）版本开始增加的图片分类。xlarge 是从 Android 2.3（API Level 9）版本开始增加的图片分类。

在 HelloWorld 工程中，ADT 在 drawable 目录中自动引入了 ic_laucher.png 文件，作为 HelloWorld 程序的图标文件；在 layout 目录中生成 activity_mail.xml 文件，用于描述用户界面，在 menu 目录下生成 main.xml 文件，用于用户界面 menu 菜单。在 values 目录下包括 strings.xml、dimens.xml、styles.xml、color.xml 等。

2.1.2 AndroidManifest.xml 文件简介

AndroidManifest.xml 是 XML 格式的 Android 程序声明文件，是全局描述文件，包含 Android 系统运行 Android 程序前所必须掌握的重要信息，这些信息包含应用程序名称、图标、包名称、模块组成、授权和 SDK 最低版本等。创建的每个 Android 项目应用程序必须在根目录下包含一个 AndroidManifest.xml 工程文件。

1. AndroidManifest.xml 文件代码

```
1.  <?xml version="1.0" encoding="utf-8"?>
2.  <manifest xmlns:android="http://schemas.android.com/apk/res/android"
3.      package="com.bcpl"
4.      android:versionCode="1"
5.      android:versionName="1.0">
6.      <uses-sdk android:minSdkVersion="10" />
7.      <application android:icon="@drawable/icon" android:label=
        "@string/app_name">
8.          <activity android:name=".HelloWorldActivity"
9.              android:label="@string/app_name">
10.             <intent-filter>
11.                 <action android:name="android.intent.action.MAIN" />
12.                 <category  android:name="android.intent.category.
                    LAUNCHER" />
13.             </intent-filter>
14.         </activity>
15.     </application>
16. </manifest>
```

AndroidManifest.xml 文件的根元素是 manifest，包含 xmlns:android、package、android:versionCode 和 android:versionName 共 4 个属性。

xmlns:android 定义了 Android 的命名空间，值为 http://schemas.android.com/apk/res/android。

package 定义了应用程序的包名称。

android:versionCode 定义了应用程序的版本号，是一个整数值，数值越大说明版本越新，但仅在程序内部使用，并不提供给应用程序的使用者。

android:versionName 定义了应用程序的版本名称，是一个字符串，仅限于为用户提供一个版本标识。

manifest 元素仅能包含一个 application 元素，application 元素中能够声明 Android 程序中最重要的 4 个组成部分，包括 Activity、Service、BroadcastReceiver 和 ContentProvider，所定义的属性将影响所有组成部分。

第 7 行属性 android:icon 定义了 Android 应用程序的图标。其中，@drawable/icon 是一种资源引用方式，表示资源类型是图像，资源名称为 icon，对应的资源文件为 res/drawable 目录下的 icon.png。

第 7 行属性 android:label 则定义了 Android 应用程序的标签名称。

activity 元素是对 Activity 子类的声明，必须在 AndroidManifest.xml 文件中声明的 Activity 才能在用户界面中显示。

第 8 行属性 android:name 定义了实现 Activity 类的名称，可以是完整的类名称，也可以是简化后的类名称。

第 9 行属性 android:label 则定义了 Activity 的标签名称，标签名称将在用户界面的 Activity 上部显示，@string/app_name 同样属于资源引用，表示资源类型是字符串，资源名称为 app_name，资源保存在 res/values 目录下的 strings.xml 文件中。

intent-filter 中声明了两个子元素 action 和 category，intent-filter 使 HelloAndroid 程序在启动时，将.HelloAndroid 这个 Activity 作为默认启动模块。

此外，Google 为了方便开发者对于各种分辨率机型的移植而在 AndroidManifest.xml 中增加了自动适配功能的标签设置，需要在<manifest>根元素中添加如下子元素。

```
1.    <supports-screens
2.    android:largeScreens="true"
3.    android:normalScreens="true"
4.    android:smallScreens="true"
5.    android:anyDensity="true"/>
```

上述属性 android:anyDensity="true"表示，应用程序安装在不同密度的终端上时，程序会分别加载 xxhdpi、xhdpi、hdpi、mdpi、ldpi 文件夹中的资源。相反，如果设为 false，即使在文件夹下拥有相同资源，应用也不会自动地去相应文件夹下寻找资源。

如果 drawable-hdpi、drawable-mdpi、drawable-ldpi 三个文件夹中有同一张图片资源的不同密度表示，那么系统会去加载 drawable-mdpi 文件夹中的资源；如果 drawable-hpdi

中有高密度图片，其他两个文件夹中没有对应图片资源，那么系统会去加载 drawable-hdpi 中的资源，其他同理；如果 drawable-hdpi，drawable-mdpi 中有图片资源，drawable-ldpi 中没有，系统会加载 drawable-mdpi 中的资源，其他同理，使用最接近的密度级别。

2．可视化编辑器

双击 AndroidManifest.xml 文件，直接进入可视化编辑器，如图 2-2 所示，用户可以直接编辑 Android 工程的应用程序名称、包名称、图标、标签和许可等相关属性。

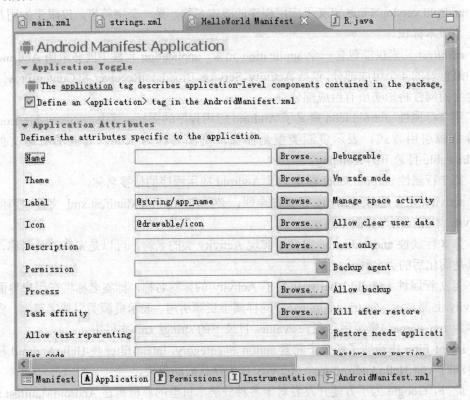

图 2-2　AndroidManifest.xml 文件可视化编辑器

2.1.3　gen 目录

在上述目录结构中已讲述，gen 目录下只存放一个由 ADT 自动生成，并不需要人工修改的 R.java 文件。

R.java 文件包含对 drawable、layout 和 values 目录内的资源的引用指针，Android 程序能够直接通过 R 类引用目录中的资源。

Android 系统中资源引用有两种方式：一种是在代码中引用资源；另一种是在资源中引用资源。

在代码中引用资源，需要使用资源的 ID，可以通过[R.resource_type.resource_name]或[android.R.resource_type.resource_name]获取资源 ID。

其中，resource_type 代表资源类型，也就是 R 类中的内部类名称；resource_name 代表资源名称，为对应资源的文件名或在 XML 文件中定义的资源名称属性。

在资源中引用资源，引用格式：

@[package:]type:name

其中，@表示对资源的引用；package 是包名称，如果在相同的包中，则 package 可以省略。

R.java 文件不能手动修改，如果向资源目录中增加或删除了资源文件，则需要在工程名称上右击，选择 Refresh 命令来更新 R.java 文件中的代码。

R 类包含的几个内部类，分别与资源类型相对应，资源 ID 便保存在这些内部类中，例如子类 drawable 表示图像资源，内部的静态变量 icon 表示资源名称，其资源 ID 为 0x7f020000。一般情况下，资源名称与资源文件名相同。

HelloWorld 工程生成的 R.java 文件的代码如下。

```java
package com.bcpl;

public final class R {
    public static final class attr {
    }
    public static final class drawable {
        public static final int icon=0x7f020000;
    }
    public static final class layout {
        public static final int main=0x7f030000;
    }
    public static final class string {
        public static final int app_name=0x7f040001;
        public static final int hello=0x7f040000;
    }
}
```

2.1.4　res 目录

res 目录中包含三大类 12 个子目录，分别如下。

（1）以 drawable 为前缀的目录。

drawable-hdpi 目录：里面主要存放高分辨率的图片，如 WVGA (480×800),FWVGA (480×854)，默认存放的是 icon.png 图片。

drawable-mdpi 目录：里面主要存放中等分辨率的图片，如 HVGA (320×480)，默认存放的是 icon.png 图片。

drawable-ldpi 目录：里面主要存放低分辨率的图片，如 QVGA (240×320)，默认存放的是 icon.png 图片。

drawable-xxhdpi 目录:里面存放高清 1280×720 分辨率的图片。
drawable-xhdpi 目录:里面存放标清 960×720 分辨率的图片。
系统会根据机器的分辨率来分别到这几个文件夹里面去找对应的图片。

(2) layout 目录:用来保存与用户界面相关的布局文件,这些布局文件都是 XML 文件,默认存放的是 main.xml 文件。

menu 目录:用来存放 menu 资源文件,文件名为 main.xml。

(3) 以 values 为前缀的目录:保存文件颜色、风格、主题和字符串等,默认存放的是 strings.xml 文件。

在 Android 3.2 以前,所有的资源文件都有相应的 xhdpi、hdpi、mdpi、ldpi 4 种文件来对应,Android 3.2 以后,Android 为了给开发者提供更精准的对布局文件的控制,可以通过为资源文件(res 目录下文件)增加后缀的方式来指定该文件夹里的 XML 布局文件或 color.xml、string.xml 是为哪种大小的屏幕使用。后缀的添加方式有三种,分别如下。

第一种后缀:sw<Num>dp,如 layout-sw600dp, values-sw600dp。sw 是 small width 的缩写,当所有屏幕的最小宽度都大于 600dp 时,屏幕就会自动到带 sw600dp 后缀的资源文件里去寻找相关资源文件,最小宽度是指屏幕宽高的较小值,每个屏幕都是固定的,不会随着屏幕横向纵向改变而改变。

第二种后缀:w<Num>dp,如 layout-w600dp, values-w600dp。设定了屏幕宽度大于<Num> dp 的情况下使用该资源文件。但它和 sw<Num>dp 不同的是,当屏幕横向纵向切换时,屏幕的宽度是变化的,以变化后的宽度来与 Num 相比,看是否使用此资源文件下的资源。

第三种后缀:h<Num>dp,如 layout-h600dp, values-h600dp。其使用方式和 w<Num>dp 一样,随着屏幕横纵向的变化,屏幕高度也会变化,根据变化后的高度值来判断是否使用 h<Num>dp。但这种方式不被推荐使用,因为屏幕在纵向上通常能够滚动导致长度变化,不像宽度那样基本固定,其灵活性不好。

2.1.5 layout 目录

layout 目录中存放 activity_main.xml 文件,是界面布局文件,利用 XML 描述的用户界面布局的相关内容将在后面详细介绍。activity_main.xml 文件代码如下:

```
<?xml version="1.0" encoding="utf-8"?>
<LinearLayout
xmlns:android="http://schemas.android.com/apk/res/android"
    android:orientation="vertical"
    android:layout_width="fill_parent"
    android:layout_height="fill_parent"
    >
<TextView
    android:layout_width="fill_parent"
    android:layout_height="wrap_content"
```

```
    android:text="@string/hello"
    />
</LinearLayout>
```

第 7 行的代码说明在界面中使用 TextView 控件，TextView 控件主要用来显示字符串文本。

第 10 行代码说明 TextView 控件需要显示的字符串，非常明显，@string/hello 是对资源的引用。

2.1.6 values 目录

values 目录中存放 strings.xml、styles.xml、dimens.xml 三个文件。其中，strings.xml 文件代码。

```
<?xml version="1.0" encoding="utf-8"?>
<resources>
    <string name="hello">Hello World, HelloWorldActivity!</string>
    <string name="app_name">HelloWorld</string>
</resources>
```

通过 strings.xml 文件的第 3 行代码分析，在 TextView 控件中显示的字符串应是"Hello World, HelloWorldActivity!"。

如果读者修改 strings.xml 文件的第 3 行代码的内容，重新编译、运行后，模拟器中显示的结果也应该随之更改。

2.1.7 project.properties 文件

project.properties 文件中的代码如下。

```
# This file is automatically generated by Android Tools.
# Do not modify this file -- YOUR CHANGES WILL BE ERASED!
#
# This file must be checked in Version Control Systems.
#
# To customize properties used by the Ant build system edit
# "ant.properties", and override values to adapt the script to your
# project structure.
#
# To enable ProGuard to shrink and obfuscate your code, uncomment this
(available properties: sdk.dir, user.home):
#proguard.config=${sdk.dir}/tools/proguard/proguard-android.txt:proguard-project.txt
# Project target.
target=android-17
```

project.properties 文件记录了 Android 工程的相关设置，该文件不能随意修改，需要右击工程名称，选择 Properties 进行修改。

在 project.properties 文件中只有第 12 行是有效代码，说明 Android 程序的编译目标。

2.2 Android 应用程序组成

2.2.1 Android 应用程序概述

Android 应用程序是在 Android 应用框架之上，由一些系统自带和用户创建的应用程序组成。组件是可以调用的基本功能模块，Android 应用程序就是由组件组成的。一个 Android 应用程序通常包含 4 个核心组件和一个 Intent。4 个核心组件分别是 Activity、Service、BroadcastReceiver 和 ContentProvider。Intent 是组件之间进行通信的载体，它不仅可以在同一个应用中起传递信息的作用，还可以在不同的应用间进行信息传递，如图 2-3 所示。

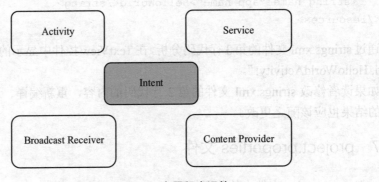

图 2-3 Android 应用程序组件

2.2.2 Activity 组件

Activity 是 Android 程序的表现层，显示可视化的用户界面，并接收与用户交互所产生的界面事件。一个 Android 应用程序可以包含一个或多个 Activity，其中一个作为 Main Activity 用于启动显示，一般在程序启动后会呈现一个 Activity，用于提示用户程序已经正常启动。

Activity 通过 View 管理用户界面 UI。View 绘制用户界面 UI 与处理用户界面事件（UI Event），View 可通过 XML 描述定义，也可在代码中生成。一般情况下，Android 建议将 UI 设计和逻辑分离，Android UI 设计类似 Swing，通过布局组织 UI 组件。

在应用程序中，每一个 Activity 都是一个单独的类，继承实现了 Activity 基础父类，这个类通过它的方法设置并显示由 Views 组成的用户界面 UI，并接受、响应与用户交互产生的界面事件，Activity 通过 startActivity 或 startActivityForResult 启动另外的 Activity。

在应用程序中，一个 Activity 在界面上的表现形式通常有：全屏窗体，非全屏悬浮窗体，对话框等。

2.2.3　Service 组件

Service 常用于没有用户界面，但需要长时间在后台运行的应用。与应用程序的其他模块（例如 Activity）一同运行于主线程中。一般通过 startService 或 bindService 方法创建 Service，通过 stopService 或 stopSelf 方法终止 Service。通常情况下，都在 Activity 中启动和终止 Service。

在 Android 应用中，Service 的典型应用是：音乐播放器。在一个媒体播放器程序中，大概要有一个或多个活动（Activity）来供用户选择歌曲并播放它。然而，音乐的回放就不能使用活动（Activity）了，因为用户希望能够切换到其他界面时音乐继续播放。这种情况下，媒体播放器活动（Activity）要用 Context.startService()启动一个服务来在后台运行保持音乐的播放。系统将保持这个音乐回放服务的运行直到它结束。需要注意，要用 Context.bindService()方法连接服务（如果它没有运行，要先启动它）。当连接到服务后，可以通过服务暴露的一个接口和它通信。对于音乐服务，它支持暂停、倒带、重放等功能。

2.2.4　Intent 和 Intent Filter 组件

1．Intent

Android 中提供了 Intent 机制来协助应用间的交互与通信，Intent 负责对应用中一次操作的动作、动作涉及数据、附加数据进行描述，Android 则根据此 Intent 的描述，负责找到对应的组件，将 Intent 传递给调用的组件，并完成组件的调用。Intent 不仅可用于应用程序之间，也可用于应用程序内部的 Activity/Service 之间的交互。因此，Intent 在这里起着一个媒体中介的作用，类似于消息、事件通知，它充当 Activity、Service、BroadcastReceiver 之间联系的桥梁，专门提供组件互相调用的相关信息，实现调用者与被调用者之间的解耦。具体详述见后续章节。

通常 Intent 分为显式和隐式两类。显式的 Intent，就是指定了组件名字的，是由程序指定具体的目标组件来处理，即在构造 Intent 对象时就指定接收者，指定了一个明确的组件(setComponent 或 setClass)来使用处理 Intent。

```
1.    Intent intent = new Intent(
2.        getApplicationContext(),
3.        Test.class
4.    );
5.    startActivity(intent);
```

特别注意：被启动的 Activity 需要在 AndroidManifest.xml 中进行定义。

隐式的 Intent，就是没有指定 Intent 的组件名字，没指定明确的组件来处理该 Intent。使用这种方式时，需要让 Intent 与应用中的 Intent Filter 描述表相匹配。需要 Android 根据 Intent 中的 Action、Data、Category 等来解析匹配。由系统接受调用并决定如何处理，即 Intent 的发送者在构造 Intent 对象时，并不知道也不关心接收者是谁，这有利于降低发送者和接收者之间的耦合。例如：startActivity(new Intent(Intent.ACTION_DIAL));。

```
1. Intent intent = new Intent();
2. intent.setAction("test.intent.IntentTest");
3. startActivity(intent);
```

目标组件（Activity、Service、BroadcastReceiver）是通过设置它们的 Intent Filter 来界定其处理的 Intent。如果一个组件没有定义 Intent Filter，那么它只能接收处理显式的 Intent，只有定义了 Intent Filter 的组件才能同时处理隐式和显式的 Intent。

一个 Intent 对象包含很多数据的信息，由以下 6 个部分组成。

（1）Action——要执行的动作。
（2）Data——执行动作要操作的数据。
（3）Category——被执行动作的附加信息。
（4）Extras——其他所有附加信息的集合。
（5）Type——显式指定 Intent 的数据类型（MIME）。
（6）Component——指定 Intent 的目标组件的类名称，比如要执行的动作、类别、数据、附加信息等。

2. Intent Filter

应用程序的组件为了告诉 Android 自己能响应、处理哪些隐式 Intent 请求，可以声明一个甚至多个 Intent Filter。每个 Intent Filter 描述该组件所能响应 Intent 请求的能力——组件希望接收什么类型的请求行为，什么类型的请求数据。比如请求网页浏览器这个例子中，网页浏览器程序的 Intent Filter 就应该声明它所希望接收的 Intent Action 是 WEB_SEARCH_ACTION，以及与之相关的请求数据是网页地址 URI 格式。如何为组件声明自己的 Intent Filter？常见的方法是在 AndroidManifest.xml 文件中用属性 <Intent-Filter>描述组件的 Intent Filter。

Intent 解析机制主要是通过查找已注册在 AndroidManifest.xml 中的所有 Intent Filter 及其中定义的 Intent，最终找到匹配的 Intent。在这个解析过程中，Android 是通过 Intent 的 action、type、category 这三个属性来进行判断的，判断方法如下。

（1）如果 Intent 指明定了 action，则目标组件的 Intent Filter 的 action 列表中就必须包含这个 action，否则不能匹配。

（2）如果 Intent 没有提供 type，系统将从 data 中得到数据类型。和 action 一样，目标组件的数据类型列表中必须包含 Intent 的数据类型，否则不能匹配。

（3）如果 Intent 中的数据不是 content: 类型的 URI，而且 Intent 也没有明确指定它的 type，将根据 Intent 中数据的 scheme（比如 http: 或者 mailto:）进行匹配。同上，Intent 的 scheme 必须出现在目标组件的 scheme 列表中。

（4）如果 Intent 指定了一个或多个 category，这些类别必须全部出现在组件的类别列表中。比如 Intent 中包含两个类别：LAUNCHER_CATEGORY 和 ALTERNATIVE_

CATEGORY，解析得到的目标组件必须至少包含这两个类别。

一个 Intent 对象只能指定一个 Action，而一个 Intent Filter 可以指定多个 Action，Action 的列表不能为空，否则它将组织所有的 Intent。

一个 Intent 对象的 Action 必须和 Intent Filter 中的某一个 Action 匹配，才能通过测试。如果 Intent Filter 的 Action 列表为空，则不通过。如果 Intent 对象不指定 Action，并且 Intent Filter 的 Action 列表不为空，则通过测试。

2.2.5　BroadcastReceiver 组件

在 Android 中，Broadcast 是一种广泛运用在应用程序之间传输信息的组件。而 BroadcastReceiver 是接收并响应广播消息的组件，对发送出来的 Broadcast 进行过滤接收并响应，它不包含任何用户界面，可以通过启动 Activity 或者 Notification 通知用户接收到重要信息，在 Notification 中有多种方法提示用户，如闪动背景灯、震动设备、发出声音或在状态栏上放置一个持久的图标。

BroadcastReceiver 过滤接收的过程如下。

在需要发送信息时，把要发送的信息和用于过滤的信息(如 Action、Category)装入一个 Intent 对象，然后通过调用 Context.sendBroadcast()、sendOrderBroadcast() 或 sendStickyBroadcast()方法，把 Intent 对象以广播方式发送出去。

当 Intent 发送后，所有已经注册的 BroadcastReceiver 会检查注册时的 Intent Filter 是否与发送的 Intent 相匹配，若匹配则调用 BroadcastReceiver 的 onReceive()方法。因此在定义一个 BroadcastReceiver 时，通常都需要实现 onReceive()方法。

BroadcastReceiver 注册有以下两种方式。

一种方式是，静态地在 AndroidManifest.xml 中用<receiver>标签声明注册，并在标签内用<intent- filter>标签设置过滤器。

另一种方式是，动态地在代码中先定义并设置好一个 Intent Filter 对象，然后在需要注册的地方调用 Context.registerReceiver() 方法，如果取消时就调用 Context.unregisterReceiver()方法。

不管是用 XML 注册的还是用代码注册的，在程序退出时，一般需要注销，否则下次启动程序时可能会有多个 BroadcastReceiver。另外，若在使用 sendBroadcast()的方法时指定了接收权限，则只有在 AndroidManifest.xml 中用<uses-permission>标签声明了拥有此权限的 BroascastReceiver 时才会有可能接收到发送来的 Broadcast。

同样，若在注册 BroadcastReceiver 时指定了可接收的 Broadcast 的权限，则只有在包内的 AndroidManifest.xml 中用<uses-permission>标签声明了拥有此权限的 Context 对象所发送的 Broadcast 时才能被这个 BroadcastReceiver 所接收。

2.2.6　ContentProvider 组件

ContentProvider 是 Android 系统提供的一种标准的共享数据的机制，在 Android 中每一个应用程序的资源都为私有，应用程序可以通过 ContentProvider 组件访问其他应用程序的私有数据（私有数据可以是存储在文件系统中的文件，或者是存放在 SQLite 中的数据库），如图 2-4 所示。

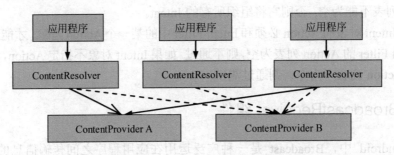

图 2-4　应用程序、ContentResolver 与 ContentProvider

对 ContentProvider 的使用，有以下两种方式。
（1）ContentResolver 访问。
（2）Context.getContentResolver()。

Android 系统内部也提供了一些内置的 ContentProvider，能够为应用程序提供重要的数据信息。使用 ContentProvider 对外共享数据的好处是统一了数据的访问方式。

2.3　Fragment 与 Activity

2.3.1　Fragment 简介

Fragment 在 Android 3.0 (API level 11)版本引入。可以把 Fragment 看作 Activity 中的模块，它有自己的布局、生命周期、单独处理自己的输入，在 Activity 运行的时候可以加载或者移除 Fragment 模块。通常把 Fragment 设计成可以在多个 Activity 中复用的模块，利用 Fragment 可以实现灵活的布局，改善用户体验。例如，可以在程序运行于大屏幕中时启动包含很多 Fragment 的 Activity，而在运行于小屏幕时启动一个包含少量 Fragment 的 Activity。如图 2-5 所示便笺簿的布局：一个 Activity 包含两个 Fragment。

例如，电视遥控器的布局，如图 2-6 所示。

Activity 与 Fragment 的区别是：Fragment 是 Activity 界面中的一部分，一个 Activity 可以包含多个 Fragment，也可以在多个 Activity 中重用一个 Fragment。Fragment 不能独立存在，它必须嵌入到 Activity 中，而且 Fragment 的生命周期受其所在的 Activity 的影响。

例如，当 Activity 暂停时，它拥有的所有的 Fragment 都暂停了，当 Activity 销毁时，它拥有的所有 Fragment 都被销毁。然而，当 Activity 运行时（在 onResume()之后，onPause()之前），可以单独地操作每个 Fragment，比如添加或删除它们。在执行上述针对 Fragment 的事务时，可以将事务添加到一个栈中（Activity 管理），栈中的每一条都是一个 Fragment 的一次事务，然后就可以反向执行 Fragment 的事务，在 Fragment 级支持"返回"键（向后导航）。

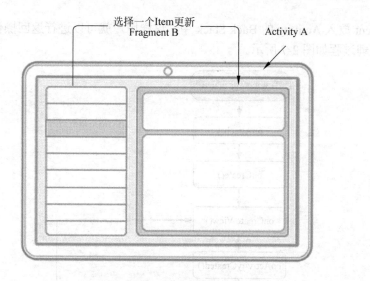

Activity A 包含Fragment A和Fragment B

图 2-5 便笺簿的布局

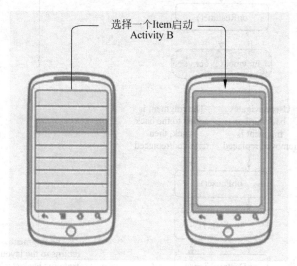

Activity A包含Fragment A　　　Activity B包含Fragment A

图 2-6 遥控器的布局

2.3.2 Fragment 的生命周期

　　Fragment 的生命周期和它所在的 Activity 是密切相关的。Fragment 必须嵌入在 Activity 中使用，如果 Activity 是暂停状态，其中所有的 Fragment 都是暂停状态；如果 Activity 是停止状态，这个 Activity 中所有的 Fragment 都不能被启动；如果 Activity 被销毁，那么它其中的所有 Fragment 都会被销毁。但是，当 Activity 在活动状态，可以独立控制 Fragment 的状态，如添加或者移除 Fragment。当这样进行 Fragment 转换时，可以把 Fragment 放入 Activity 的 Back Stack 中，这样用户就可以进行返回操作。Fragment 生命周期处理过程如图 2-7 所示。

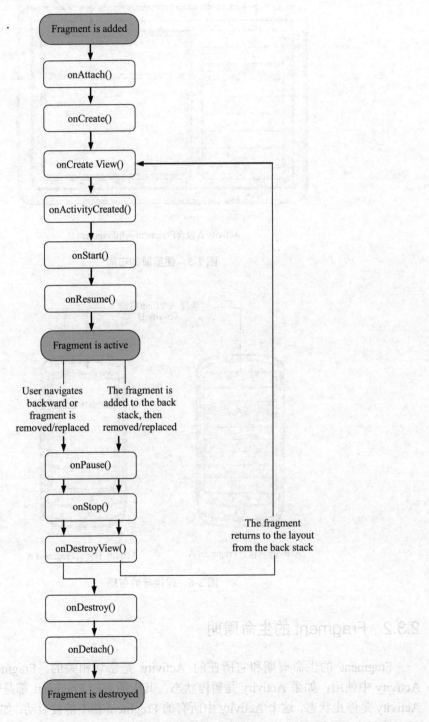

图 2-7 Fragment 生命周期处理

Fragment 有三种生存状态，分别是 Resumed、Paused、Stoped。

Resumed：Fragment 在一个运行的 Activity 中并且可见。

Paused：另一个 Activity 处于最顶层，但是 Fragment 所在的 Activity 并没有被完全覆盖（顶层的 Activity 是半透明的或不占据整个屏幕）。

Stoped：Fragment 不可见。可能是它所在的 Activity 处于 Stoped 状态或是 Fragment 被删除并添加到后退栈中了。此状态的 Fragment 仍然存在于内存中。

Fragment 的状态可以保存在一个 Bundle 中，在 Activity 被 Recreated 时需要，也可以在 onSaveInstanceState() 方法中保存状态并在 onCreate() 或 onCreateView() 或 onActivityCreated()中恢复。

2.3.3 Fragment 继承

使用 Fragment 时，需要继承 Fragment 或者 Fragment 的子类（DialogFragment, ListFragment, PreferenceFragment, WebViewFragment）。在 Android SDK 中 android-support-v4.jar 提供了 Fragment 的 APIs，在程序中应用时，只需使用导入如下包即可：

```
import android.support.v4.app.Fragment;
import android.support.v4.app.FragmentManager;
```

当创建包含 Fragment 的 Activity 时，如果使用 Support Library，继承的父类是 FragmentActivity 而不是 Activity。同时必须实现三个回调函数：onCreate()、onCreateView()、onPause()。

onCreate()方法：系统在创建 Fragment 时调用，它用来初始化相关的组件。

onCreateView()方法：第一次绘制 Fragment 的 UI 时系统调用，必须返回一个 View，如果 Fragment 不提供 UI 也可以返回 null。如果继承自 ListFragment，onCreateView()默认的实现会返回一个 ListView，无须单独实现。

onPause()方法：当用户离开 Fragment 时第一个调用，需要提交保存一些变化，以便进入后台的时候回调。

2.3.4 Fragment 创建方式

Fragment 的实现方式有三种，分为 UI 中实现 Fragment、布局中加入 Fragment 标签、程序代码中加入 Fragment。

1. UI 中实现 Fragment

在 UI 中实现 Fragment，必须实现 onCreateView()方法，实现代码如下。

```
1.  public static class TestFragment extends Fragment
2.  {
3.      @Override
```

```
4.    public View onCreateView(LayoutInflater inflater, ViewGroup
      container, Bundle savedInstanceState)
5.    {
6.        return inflater.inflate(R.layout.example_fragment,container,
   false);
7.    }
8. }
```

onCreateView()中，container 参数表示该 Fragment 在 Activity 中的父控件；savedInstanceState 提供了上一个实例的数据。

inflate()方法中的三个参数的含义为：第一个是 resource ID，表示当前的 Fragment 对应的资源文件；第二个参数是父容器控件；第三个布尔值参数表示是否连接该布局和其父容器控件，当前设置为 false，因系统已经插入这个布局到父控件，设置为 true 将会产生多余的一个 View Group。

注意：当 Fragment 被加入 Activity 中时，它会处在对应的 View Group 中。

2. 布局中加入 Fragment 标签声明

下述使用线性布局 LinearLayout 中加入 Fragment 标签声明，具体如下。

```
1.  <?xml version="1.0" encoding="utf-8"?>
2.  <LinearLayout xmlns:android="http://schemas.android.com/apk/res/android"
3.      android:orientation="horizontal"
4.      android:layout_width="match_parent"
5.      android:layout_height="match_parent">
6.      <fragment android:name="com.bcpl.ListFragment"
7.          android:id="@+id/list"
8.          android:layout_weight="1"
9.          android:layout_width="0dp"
10.         android:layout_height="match_parent" />
11.     <fragment android:name=" com.bcpl.ContentReaderFragment"
12.         android:id="@+id/viewer"
13.         android:layout_weight="2"
14.         android:layout_width="0dp"
15.         android:layout_height="match_parent" />
16. </LinearLayout>
```

当系统创建 Activity 的布局文件时，系统会实例化每一个 Fragment，并且调用它们的 onCreateView()方法，来获得相应 Fragment 的布局，并将返回值插入 Fragment 标签所在的位置。

其中，属性 android:name 添加自己创建的 Fragment 的完整类名。

为 Fragment 提供 ID 的方式有三种，分别如下。

（1）android:id 属性：唯一的 ID。

(2) android:tag 属性：唯一的字符串。
(3) 如果上面两个都没提供，系统使用容器 view 的 ID。

3. 程序代码中加入 Fragment

当 Activity 处于 Running 状态下的时候，可以在 Activity 的布局中动态地加入 Fragment，只需要指定加入 Fragment 的父 View Group 即可。

创建 FragmentTransaction 实例：

```
FragmentManager fragmentManager = getFragmentManager()
FragmentTransaction fragmentTransaction = fragmentManager.begin-
Transaction();
```

或

```
FragmentManager fragmentManager = getSupportFragmentManager();
// import android.support.v4.app.FragmentManager 时使用
```

创建 Fragment 对象：

```
TestFragment fragment = new TestFragment();
fragmentTransaction.add(R.id.fragment_container, fragment);
fragmentTransaction.commit();
```

其中，第一个参数是这个 Fragment 的容器，即父控件组，然后调用 commit()方法使得 FragmentTransaction 实例的改变生效。

2.3.5 Fragment 应用

写一个类继承自 Fragment 类，并且写好其布局文件（本例中是两个 TextView），在 Fragment 类的 onCreateView()方法中加入该布局。

之后用以下两种方法之一在 Activity 中加入这个 Fragment。

第一种是在 Activity 的布局文件中加入<fragment>标签；

第二种是在 Activity 的代码中使用 FragmentTransaction 的 add()方法加入 Fragment。

（1）定义 Fragment 类，代码如下。

```
1.  package com.bcpl.Testfragment;
2.  import android.os.Bundle;
3.  import android.support.v4.app.Fragment;
4.  import android.view.LayoutInflater;
5.  import android.view.View;
6.  import android.view.ViewGroup;
7.  public class TestFragment extends Fragment
8.  {
9.      @Override
```

```
10.     public void onCreate(Bundle savedInstanceState)
11.     {
12.         // TODO Auto-generated method stub
13.         super.onCreate(savedInstanceState);
14.         System.out.println("TestFragment--onCreate");
15.     }
16.
17.     @Override
18.     public View onCreateView(LayoutInflater inflater, ViewGroup container,
19.             Bundle savedInstanceState)
20.     {
21.         System.out.println("TestFragment--onCreateView");
22.         return   inflater.inflate(R.layout.example_fragment_layout,
                container, false);
23.
24.     }
25.
26.     @Override
27.     public void onPause()
28.     {
29.         // TODO Auto-generated method stub
30.         super.onPause();
31.         System.out.println("TestFragment--onPause");
32.     }
33.     @Override
34.     public void onResume()
35.     {
36.         // TODO Auto-generated method stub
37.         super.onResume();
38.         System.out.println("TestFragment--onResume");
39.     }
40.
41.     @Override
42.     public void onStop()
43.     {
44.         // TODO Auto-generated method stub
45.         super.onStop();
46.         System.out.println("TestFragment--onStop");
47.     }
48. }
```

(2) 创建 Fragment 的布局文件，代码如下。

```
1.  <?xml version="1.0" encoding="utf-8"?>
```

```
2.  <LinearLayout xmlns:android="http://schemas.android.com/apk/res/
    android"
3.      android:layout_width="match_parent"
4.      android:layout_height="match_parent"
5.      android:orientation="vertical" >
6.
7.      <TextView
8.          android:layout_width="match_parent"
9.          android:layout_height="wrap_content"
10.         android:text="@string/number1"
11.         />
12.     <TextView
13.         android:layout_width="match_parent"
14.         android:layout_height="wrap_content"
15.         android:text="@string/number2"
16.         />
17. </LinearLayout>
```

（3）创建 Activity 类，代码如下。

```
1.  package com.bcpl.Pfragment;
2.
3.  import android.os.Bundle;
4.  import android.support.v4.app.FragmentActivity;
5.  import android.support.v4.app.FragmentManager;
6.  import android.support.v4.app.FragmentTransaction;
7.
8.
9.  public class PFragment extends FragmentActivity
10. {
11.     @Override
12.     public void onCreate(Bundle savedInstanceState)
13.     {
14.         super.onCreate(savedInstanceState);
15.         setContentView(R.layout.activity_learn_fragment);
16.
17.         //在程序中加入 Fragment
18.         FragmentManager fragmentManager = getSupportFragmentManager
                ();
19.         FragmentTransaction fragmentTransaction = fragmentManager.
                beginTransaction();
20.
21.         TestFragment fragment = new TestFragment();
22.         fragmentTransaction.add(R.id.laylinear, fragment);
```

```
23.         fragmentTransaction.commit();
24.     }
25.
26. }
```

（4）创建 Activity 的布局文件 pfragment.xml，代码如下。

```
1.  <LinearLayout xmlns:android="http://schemas.android.com/apk/res/android"
2.      xmlns:tools="http://schemas.android.com/tools"
3.      android:layout_width="match_parent"
4.      android:layout_height="match_parent"
5.      android:orientation="vertical"
6.      >
7.      <Button
8.          android:id="@+id/btn1"
9.          android:layout_width="fill_parent"
10.         android:layout_height="wrap_content"
11.         android:text="@string/but1"
12.     />
13.     <fragment
14.         android:name="com.bcpl.pfragment.TestFragment"
15.         android:id="@+id/fragment1"
16.         android:layout_width="match_parent"
17.         android:layout_height="wrap_content"
18.
19.         />
20.     <Button
21.         android:id="@+id/btn2"
22.         android:layout_width="fill_parent"
23.         android:layout_height="wrap_content"
24.         android:text="@string/but2"
25.     />
26.     <LinearLayout
27.         xmlns:android="http://schemas.android.com/apk/res/android"
28.         android:id="@+id/laylinear"
29.         android:layout_width="fill_parent"
30.         android:layout_height="wrap_content"
31.         android:orientation="vertical"
32.         >
33.      <Button
34.          android:id="@+id/btn3"
35.          android:layout_width="fill_parent"
36.          android:layout_height="wrap_content"
37.          android:text="@string/but3"
```

```
38.         />
39.     </LinearLayout>
40.
41. </LinearLayout>
```

2.4 Android 生命周期

2.4.1 程序生命周期

程序的生命周期是指在 Android 系统中进程从启动到终止的所有阶段，也就是 Android 程序启动到停止的全过程。程序的生命周期是由 Android 系统进行调度和控制的。

Android 系统中的进程分为：前台进程、可见进程、服务进程、后台进程、空进程。Android 系统中的进程优先级由高到低，如图 2-8 所示。

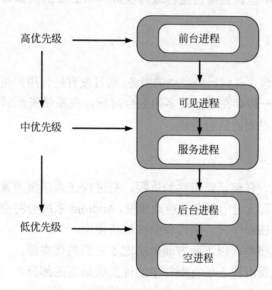

图 2-8　Android 系统的进程及优先级

1．前台进程

前台进程是 Android 系统中最重要的进程，是指与用户正在交互的进程，包含以下 4 种情况。

（1）进程中的 Activity 正在与用户进行交互。

（2）进程服务被 Activity 调用，而且这个 Activity 正在与用户进行交互。

（3）进程服务正在执行声明周期中的回调方法，如 onCreate()、onStart()或 onDestroy()。

（4）进程的 BroadcastReceiver 正在执行 onReceive()方法。

Android 系统在多个前台进程同时运行时，可能会出现资源不足的情况，此时会清除部分前台进程，保证主要的用户界面能够及时响应。

2. 可见进程

可见进程指部分程序界面能够被用户看见，但不在前台与用户交互，不响应界面事件的进程。如果一个进程包含服务，且这个服务正在被用户可见的 Activity 调用，此进程同样被视为可见进程。

Android 系统一般存在少量的可见进程，只有在特殊的情况下，Android 系统才会为保证前台进程的资源而清除可见进程。

3. 服务进程

服务进程是指包含已启动服务的进程，通常具有以下特点。
（1）没有用户界面。
（2）在后台长期运行。

Android 系统在不能保证前台进程或可视进程所必要的资源时，才会强行清除服务进程。

4. 后台进程

后台进程是指不包含任何已经启动的服务，而且没有任何用户可见的 Activity 的进程。

Android 系统中一般存在数量较多的后台进程，在系统资源紧张时，系统将优先清除用户较长时间没有见到的后台进程。

5. 空进程

空进程是指不包含任何活跃组件的进程，空进程在系统资源紧张时会被首先清除。但为了提高 Android 系统应用程序的启动速度，Android 系统会将空进程保存在系统内存中，在用户重新启动该程序时，空进程会被重新使用。

除了以上的优先级外，以下两方面也决定了它们的优先级。
（1）进程的优先级取决于所有组件中的优先级最高的部分。
（2）进程的优先级会根据与其他进程的依赖关系而变化。

2.4.2 组件生命周期

所有 Android 组件都具有自己的生命周期，是指从组件建立到组件销毁的整个过程。在生命周期中，组件会在可见、不可见、活动、非活动等状态中不断变化。下面就各个组件的生命周期逐一进行讲述。

1. Service 生命周期

Service 组件通常没有用户界面 UI，其启动后一直运行于后台；它与应用程序的其他模块（如 Activity）一同运行于程序的主线程中。

一个 Service 的生命周期通常包含：创建、启动、销毁这几个过程。

Service 只继承了 onCreate()，onStart()，onDestroy()三个方法，当第一次启动 Service 时，先后调用了 onCreate()，onStart()这两个方法，当停止 Service 时，则执行 onDestroy() 方法。需要注意的是，如果 Service 已经启动了，当再次启动 Service 时，不会再执行 onCreate()方法，而是直接执行 onStart()方法。

创建 Service 的方式有两种，一种是通过 startService 创建，另外一种是通过 bindService 创建。两种创建方式的区别在于，startService 是创建并启动 Service，而 bindService 只是创建了一个 Service 实例并取得了一个与该 Service 关联的 binder 对象，但没有启动它，如图 2-9 所示。

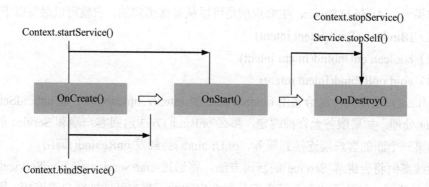

图 2-9　Service 生命周期

如果没有程序停止它或者它自己停止，Service 将一直运行。在这种模式下，Service 开始于调用 Context.startService()，停止于 Context.stopService()。Service 可以通过调用 stopService()或 Service.stopSelfResult()停止自己。不管调用多少次 startService()，只需要调用一次 stopService() 就可以停止 Service。一般在 Activity 中启动和终止 Service。它可以通过接口被外部程序调用。外部程序建立到 Service 的连接,通过连接来操作 Service。建立连接开始于 Context.bindService()，结束于 Context.unbindService()。多个客户端可以绑定到同一个 Service，如果 Service 没有启动，bindService()可以选择启动它。

上述两种方式不是完全分离的。通过 startService()启动的服务，如一个 Intent 想要播放音乐，通过 startService()方法启动后台播放音乐的 Service。然后，也许用户想要操作播放器或者获取当前正在播放的乐曲的信息，一个 Activity 就会通过 bindService()建立一个到这个 Service 的连接。这种情况下，stopService()在全部的连接关闭后才会真正停止 Service。

像 Activity 一样，Service 也有可以通过监视状态实现的生命周期。但是比 activity 要少，通常只有以下三个方法，而且是 public 的而不是 protected 的。

（1）void onCreate()

（2）void onStart(Intent intent)

（3）void onDestroy()

通过实现上述三个方法，可以监视 Service 生命周期的两个嵌套循环。

整个生命周期从 onCreate() 开始，从 onDestroy() 结束，像 Activity 一样，一个

Android Service 生命周期在 onCreate()中执行初始化操作,在 onDestroy()中释放所有用到的资源。例如,后台播放音乐的 Service 可能在 onCreate()中创建一个播放音乐的线程,在 onDestroy()中销毁这个线程。

活动生命周期开始于 onStart()。这个方法处理传入 startService()方法的 Intent。音乐服务会打开 Intent 查看要播放哪首歌曲,并开始播放。当服务停止的时候,没有方法检测到(没有 onStop()方法),onCreate()和 onDestroy() 用于所有通过 Context.startService()或 Context.bindService() 启动的 service。onStart() 只用于通过 startService()开始的 Service。

如果一个 Android Service 生命周期是可以从外部绑定的,它就可以触发以下的方法。

(1) IBinder onBind(Intent intent)

(2) boolean onUnbind(Intent intent)

(3) void onRebind(Intent intent)

onBind() 回调被传递给调用 bindService 的 Intent,onUnbind() 被 unbindService()中的 Intent 处理。如果服务允许被绑定,那么 onBind()方法返回客户端和 Service 的沟通通道。如果一个新的客户端连接到服务,onUnbind()会触发 onRebind()调用。

后续案例将会讲解 Sercice 的回调方法。将通过 startService 和 bindService()启动的 Service 分开了,但是要注意不管它们是怎么启动的,都有可能被客户端连接,因此都有可能触发到 onBind()和 onUnbind()方法。

2. Service 生命周期应用案例

具体实现步骤如下。

(1) 在 Eclipse 中选择 File→New→Android Project 命令,创建一个新的 Android 工程,项目名为 ServiceTestDemo,目标 API 选择 17(即 Android 4.2 版本),应用程序名为 ServiceTestDemo,包名为 com.bcpl.cn,创建的 Activity 的名字为 ServiceTestDemoActivity,最小 SDK 版本根据选择的目标 API 会自动添加为 17。

(2) 修改 res 目录下 layout 文件夹中的 activity_main.xml 代码,添加 4 个按钮,代码如下。

```
1.  <?xml version="1.0" encoding="utf-8"?>
2.  <LinearLayout xmlns:android="http://schemas.android.com/apk/res/
    android"
3.     android:orientation="vertical"
4.     android:layout_width="fill_parent"
5.     android:layout_height="fill_parent"
6.     >
7.     <TextView
8.        android:id="@+id/text"
9.        android:layout_width="fill_parent"
10.       android:layout_height="wrap_content"
```

```
11.         android:text="@string/hello"
12.     />
13.     <Button
14.         android:id="@+id/startservice"
15.         android:layout_width="fill_parent"
16.         android:layout_height="wrap_content"
17.         android:text="启动 Service"
18.     />
19.     <Button
20.         android:id="@+id/stopservice"
21.         android:layout_width="fill_parent"
22.         android:layout_height="wrap_content"
23.         android:text="停止 Service"
24.     />
25.     <Button
26.         android:id="@+id/bindservice"
27.         android:layout_width="fill_parent"
28.         android:layout_height="wrap_content"
29.         android:text="绑定 Service"
30.     />
31.     <Button
32.         android:id="@+id/unbindservice"
33.         android:layout_width="fill_parent"
34.         android:layout_height="wrap_content"
35.         android:text="解除 Service"
36.     />
37. </LinearLayout>
```

（3）在上述包下，新建一个 Service，命名为 MyService.java，代码如下。

```
1.  package com.bcpl.service;
2.  import android.app.Service;
3.  import android.content.Intent;
4.  import android.os.Binder;
5.  import android.os.IBinder;
6.  import android.text.format.Time;
7.  import android.util.Log;
8.  public class MyService extends Service {
9.  //定义一个Tag标签
10. private static final String TAG = "TestService";
11. //这里定义一个Binder类，用在onBind()方法里，这样Activity那边可以获取到
12. private MyBinder mBinder = new MyBinder();
13. @Override
14. public IBinder onBind(Intent intent) {
15. Log.e(TAG, " -----start IBinder------~~~");
```

```
16.     return mBinder;
17. }
18. @Override
19. public void onCreate() {
20.     Log.e(TAG, "-----start onCreate-----");
21.     super.onCreate();
22. }
23.
24. @Override
25. public void onStart(Intent intent, int startId) {
26.     Log.e(TAG, "-----start onStart-----");
27.     super.onStart(intent, startId);
28. }
29.
30. @Override
31. public void onDestroy() {
32.     Log.e(TAG, "-----start onDestroy-----");
33.     super.onDestroy();
34. }
35.
36.
37. @Override
38. public boolean onUnbind(Intent intent) {
39.     Log.e(TAG, "-----start onUnbind-----");
40.     return super.onUnbind(intent);
41. }
42.
43. //这里写了一个获取当前时间的函数,不过没有格式化
44. public String getSystemTime(){
45.
46.     Time t = new Time();
47.     t.setToNow();
48.     return t.toString();
49. }
50.
51. public class MyBinder extends Binder{
52.     MyService getService()
53.     {
54.         return MyService.this;
55.     }
56. }
57. }
```

（4）修改 com.bcpl.cn 包下的 ServiceTestDemoActivity.java 文件，代码如下。

```
1.  package com.bcpl.cn;
2.  import android.app.Activity;
3.  import android.content.ComponentName;
4.  import android.content.Context;
5.  import android.content.Intent;
6.  import android.content.ServiceConnection;
7.  import android.os.Bundle;
8.  import android.os.IBinder;
9.  import android.view.View;
10. import android.view.View.OnClickListener;
11. import android.widget.Button;
12. import android.widget.TextView;
13. public class ServiceTestDemoActivity extends Activity implements OnClickListener{
14.     private MyService mMyService;
15.     private TextView mTextView;
16.     private Button startServiceButton;
17.     private Button stopServiceButton;
18.     private Button bindServiceButton;
19.     private Button unbindServiceButton;
20.     private Context mContext;
21.     //这里需要用到ServiceConnection,在Context.bindService和context.
        //unBindService()里用到
22.     private ServiceConnection mServiceConnection = new ServiceConnection() {
23.         //bindService时,让TextView显示MyService里getSystemTime()
            //方法的返回值
24.         public void onServiceConnected(ComponentName name, IBinder service) {
25.             // TODO Auto-generated method stub
26.             //mMyService = ((MyService.MyBinder)service).
                    getService();
27.         }
28.
29.         public void onServiceDisconnected(ComponentName name) {
30.             // TODO Auto-generated method stub
31.
32.         }
33.     };
34.     public void onCreate(Bundle savedInstanceState) {
35.         super.onCreate(savedInstanceState);
36.         setContentView(R.layout.main);
```

```
37.        setupViews();
38.    }
39.
40.    public void setupViews(){
41.
42.     mContext = ServiceDemo.this;
43.     mTextView = (TextView)findViewById(R.id.text);
44.
45.     startServiceButton = (Button)findViewById(R.id.startservice);
46.     stopServiceButton = (Button)findViewById(R.id.stopservice);
47.     bindServiceButton = (Button)findViewById(R.id.bindservice);
48.     unbindServiceButton = (Button)findViewById(R.id.unbindservice);
49.
50.     startServiceButton.setOnClickListener(this);
51.     stopServiceButton.setOnClickListener(this);
52.     bindServiceButton.setOnClickListener(this);
53.     unbindServiceButton.setOnClickListener(this);
54.    }
55.
56.    public void onClick(View v) {
57.        // TODO Auto-generated method stub
58.        if(v == startServiceButton){
59.            Intent i = new Intent();
60.            i.setClass(ServiceTestDemoActivity.this,MyService.class);
61.            mContext.startService(i);
62.        }else if(v == stopServiceButton){
63.            Intent i = new Intent();
64.            i.setClass(ServiceTestDemoActivity.this,
                MyService.class);
65.            mContext.stopService(i);
66.        }else if(v == bindServiceButton){
67.            Intent i = new Intent();
68.            i.setClass(ServiceTestDemoActivity.this, MyService.class);
69.            mContext.bindService(i, mServiceConnection, BIND_AUTO_
            CREATE);
70.        }else{
71.            mContext.unbindService(mServiceConnection);
72.        }
73.    }
74.
75. }
```

（5）修改 AndroidManifest.xml 代码，在<application>标签下根目录下，添加注册新创建的 MyService，如以下代码第 14 行所示。

```
1.  <?xml version="1.0" encoding="utf-8"?>
2.  <manifest xmlns:android="http://schemas.android.com/apk/res/android"
3.  package="com.bcpl.cn "
4.  android:versionCode="1"
5.  android:versionName="1.0">
6.  <application android:icon="@drawable/icon" android:label="@string/
    app_name">
7.  <activity android:name=". ServiceTestDemoActivity"
8.  android:label="@string/app_name">
9.  <intent-filter>
10. <action android:name="android.intent.action.MAIN" />
11. <category android:name="android.intent.category.LAUNCHER" />
12. </intent-filter>
13. </activity>
14. <service android:name=".MyService" android:exported="true"></service>
15. </application>
16. <uses-sdk android:minSdkVersion="17" />
17. </manifest>
```

（6）部署工程，并执行上述工程，运行结果如图 2-10 所示。

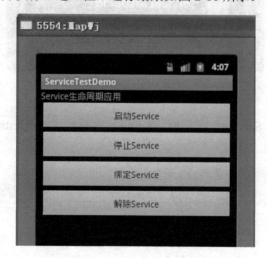

图 2-10　Service 生命周期运行界面

程序执行运行过程如下：

（1）单击"启动 Service"按钮时，程序先后执行了 Service 中 onCreate()→onStart() 这两个方法，打开日志界面 LogCat 视窗，如图 2-11 所示。

```
Log (64)  w
Time                        pid    Message
09-15 16:09:43.306     E    610    -----start onCreate-----
09-15 16:09:43.306     E    610    -----start onStart-----
```

图 2-11　启动 Service 调用顺序

（2）然后按 Home 键进入 Settings→Applications→Running Services 查看刚才新启动的服务，如图 2-12 所示。

图 2-12　新启动的服务

（3）单击"停止 Service"按钮时，Service 则执行了 onDestroy()方法，如图 2-13 所示。

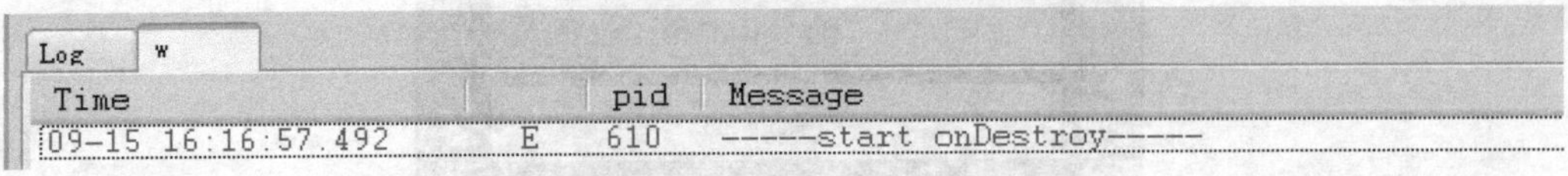

```
Log    W
Time                          pid    Message
09-15 16:16:57.492         E   610    -----start onDestroy-----
```

图 2-13　停止 Service

（4）再次单击"启动 Service"按钮，然后再单击"绑定 Service"按钮（通常 bind Service 都是 bind 已经启动的 Service），查看 Service 的 IBinder()方法执行情况，如图 2-14 所示。

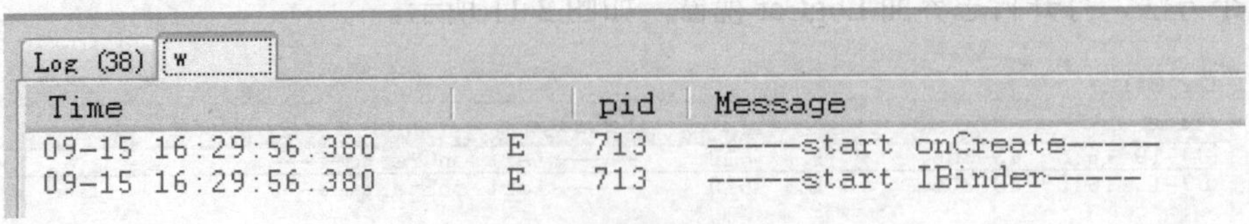

```
Log (38)    W
Time                          pid    Message
09-15 16:29:56.380         E   713    -----start onCreate-----
09-15 16:29:56.380         E   713    -----start IBinder-----
```

图 2-14　Service 的 IBinder()方法执行

（5）最后单击"解除 Service"按钮，则 Service 执行了 onUnbind()方法，如图 2-15 所示。

图 2-15　Service 的 onUnbind()方法执行

3．BroadcastReceiver 生命周期

Android 在接收到一个广播 Intent 之后，找到了处理该 Intent 的 BroadcastReceiver，创建一个对象来处理 Intent。然后，调用被创建的 BroadcastReceiver 对象的 onReceive 方法进行处理，然后就撤销这个对象，如图 2-16 所示。只有在执行这个方法时，BroadcastReceiver 才是活动的。当 onReceive()方法执行完，BroadcastReceiver 才成为非活动的。

```
Create Object  →  onReceive  →  Destroy Object
```

图 2-16　BroadcastReceiver 处理过程

BroadcastReceiver 活动时，它的进程不能被杀掉，而当它的进程中只包含不活动组件时，可能会被系统随时杀掉（其他进程需要消耗它所占用的内存）。解决这个问题的办法是 onReceive() 方法启动一个 Android Service 生命周期，让 Service 去做耗时的工作，这样系统就知道此进程中还有活动的工作。

需要注意的是，对象在 onReceive 方法返回之后就被撤销，所以在 onReceive 方法中不宜处理异步的过程。例如，弹出对话框与用户交互，可使用消息栏替代。

4．Activity 生命周期

1）Activity 状态

在 Activity 生命周期中，其表现状态有 4 种，分别是：活动状态、暂停状态、停止状态和非活动状态。

（1）Active（活动状态）：是指 Activity 通过 onCreate 被创建，Activity 在用户界面中处于最上层，完全能让用户看到，能够与用户进行交互。

（2）Pause（暂停状态）：是指当一个 Activity 失去焦点，该 Activity 将进入 Pause 状态，Activity 在界面上被部分遮挡，该 Activity 不再处于用户界面的最上层，且不能够与用户进行交互，系统在内存不足时会将其终止。

（3）Stop（停止状态）：是指当一个 Activity 被另一个 Activity 覆盖，该 Activity 将

进入 Stop 状态，Activity 在界面上完全不能被用户看到，也就是说这个 Activity 被其他 Activity 全部遮挡，系统在需要内存的时候会将其终止。

（4）非活动状态：不在以上三种状态中的 Activity 则处于非活动状态。

当 Activity 处于 Pause 或者 Stop 状态时，都可能被系统终止并回收。因此，有必要在 onPause 和 onStop 方法中将应用程序运行过程中的一些状态，例如用户输入等，保存到持久存储中。如果程序中启动了其他后台线程，也需要注意在这些方法中进行一些处理，例如在线程中打开了一个进度条对话框，如果不在 Pause 或 Stop 中取消线程，则当线程运行完取消掉对话框时就会抛出异常，如图 2-17 所示。

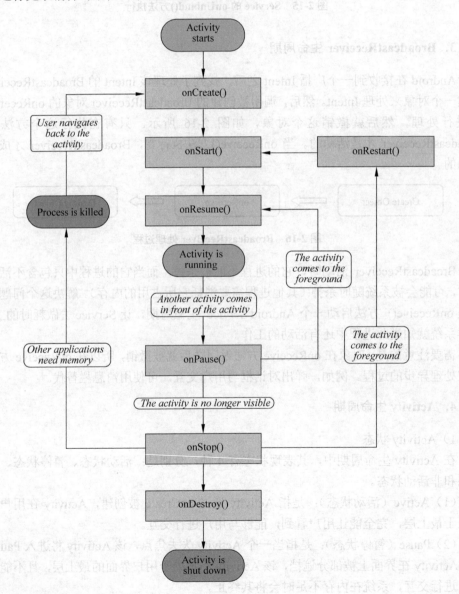

图 2-17　Activity 生命周期调用流程

在 Activity 生命周期中，其事件的回调方法有 7 个，以及 Activity 状态保存/恢复的事件回调方法有两个，如下所示。

```
public class MyActivity extends Activity {
    protected void onCreate(Bundle savedInstanceState);
    public void onRestoreInstanceState(Bundle savedInstanceState);
    public void onSaveInstanceState(Bundle savedInstanceState);
    protected void onStart();
    protected void onRestart();
    protected void onResume();
    protected void onPause();
    protected void onStop();
    protected void onDestroy();
}
```

具体说明如表 2-1 和表 2-2 所示。

表 2-1 Activity 生命周期的事件回调方法

方　　法	是否可终止	说　　明
onCreate()	否	Activity 启动后第一个被调用的函数，常用来进行 Activity 的初始化，例如创建 View、绑定数据或恢复信息等
onStart()	否	当 Activity 显示在屏幕上时，该函数被调用
onRestart()	否	当 Activity 从停止状态进入活动状态前，调用该函数
onResume()	否	当 Activity 能够与用户交互，接收用户输入时，该函数被调用。此时的 Activity 位于 Activity 栈的栈顶
onPause()	是	当 Activity 进入暂停状态时，该函数被调用。一般用来保存持久的数据或释放占用的资源
onStop()	是	当 Activity 进入停止状态时，该函数被调用
onDestroy()	是	在 Activity 被终止前，即进入非活动状态前，该函数被调用

表 2-2 Activity 状态保存/恢复的事件回调方法

方　　法	是否可终止	说　　明
onSaveInstanceState()	否	Android 系统因资源不足终止 Activity 前调用该函数，用以保存 Activity 的状态信息，供 onRestoreInstanceState() 或 onCreate() 恢复之用
onRestoreInstanceState()	否	恢复 onSaveInstanceState() 保存的 Activity 状态信息，在 onStart() 和 onResume() 之间被调用

2）Activity 生命周期分类

Activity 生命周期指 Activity 从启动到销毁的过程。Activity 的生命周期可分为全生命周期、可视生命周期和活动生命周期。在 Activity 的每个生命周期中包含不同的事件回调方法。

（1）全生命周期。

全生命周期是从 Activity 建立到销毁的全部过程，始于 onCreate()，结束于

onDestroy()。使用者通常在 onCreate()中初始化 Activity 所能使用的全局资源和状态，并在 onDestroy()中释放这些资源。在特殊的情况下，Android 系统会不调用 onDestroy()函数，而直接终止进程。

（2）可视生命周期。

可视生命周期是 Activity 在界面上从可见到不可见的过程，开始于 onStart()，结束于 onStop()。

在可视生命周期中，onStart()方法一般用来初始化或启动与更新界面相关的资源，onStop()一般用来暂停或停止一切与更新用户界面相关的线程、计时器和服务。onRestart()方法在 onSart()前被调用，用来在 Activity 从不可见变为可见的过程中，进行一些特定的处理过程。在可视生命周期中，onStart()和 onStop()一般会被多次调用。此外，onStart()和 onStop()也经常被用来注册和注销 BroadcastReceiver。

（3）活动生命周期。

活动生命周期是指 Activity 在屏幕的最上层，并能够与用户交互的阶段，开始于 onResume()，结束于 onPause()。在 Activity 的状态变换过程中，onResume()和 onPause()经常被调用，因此应简捷、高效地实现这两个方法。onPause()是第一个被标识为"可终止"的方法，在 onPause()返回后，onStop()和 onDestroy()随时能被 Android 系统终止，onPause()常用来保存持久数据，如界面上用户的输入信息等。

具体 Activity 事件的生命周期划分及回调方法的调用顺序如图 2-18 所示。

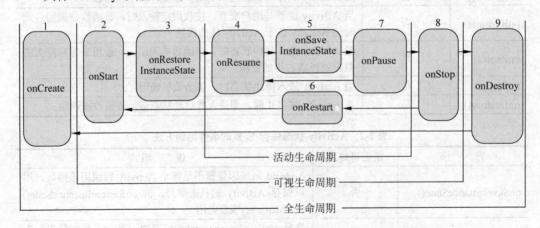

图 2-18　Activity 生命周期划分及事件回调方法的调用顺序

在活动生命周期中，关于 onPause()和 onSaveInstanceState()方法，它们之间的异同主要是：相同之处是这两个方法都可以用来保存界面中的用户输入数据。区别在于：onPause()一般用于保存持久性数据，并将数据保存在存储设备上的文件系统或数据库系统中。onSaveInstanceState()主要用来保存动态的状态信息，信息一般保存在 Bundle（保存多种格式数据的对象）中，系统在调用 onRestoreInstanceState()和 onCreate()时，会同样利用 Bundle 将数据传递给方法。

2.5 项目案例

学习目标：学习 Activity 生命周期中的事件调用顺序及上述介绍的 Activity 中方法的应用过程，掌握它们的测试及转换过程。

案例描述：使用 Activity 的 onCreate()、onStart()、onRestoreInstanceState()、onResume()、InstanceState()、onPause()、onStop()、onDestroy()方法，在不同生命周期中，方法的调用，并在日志 logcat 中输出其相关调用顺序。

案例要点：采用不同生命周期分类并就上述方法的实现及测试步骤进行介绍。

案例步骤：

（1）创建一个新的 Android 工程，工程名称为 ActivityLifeCycle，包名称为 com.bcpl.ActivityLifeCycle，Activity 名称为 ActivityLifeCycleActivity，使用 Android 4.2（API Level 17）作为目标平台，创建工程。

（2）修改 ActivityLifeCycleActivity.java 文件，代码如下。

```
1.  package com.bcpl.ActivityLifeCycle;
2.  import android.app.Activity;
3.  import android.os.Bundle;
4.  import android.util.Log;
5.  public class ActivityLifeCycleActivity extends Activity {
6.     private static String TAG = "LIFTCYCLE";
7.     @Override   //完全生命周期开始时被调用，初始化 Activity
8.     public void onCreate(Bundle savedInstanceState) {
9.        super.onCreate(savedInstanceState);
10.       setContentView(R.layout.main);
11.       Log.i(TAG, "(1) onCreate()");
12.    }
13.    @Override   //可视生命周期开始时被调用，对用户界面进行必要的更改
14.    public void onStart() {
15.       super.onStart();
16.       Log.i(TAG, "(2) onStart()");
17.    }
18.    @Override
19.    //在 onStart()后被调用，用于恢复 onSaveInstanceState()保存的用户界面信息
20.    public void onRestoreInstanceState(Bundle savedInstanceState) {
21.       super.onRestoreInstanceState(savedInstanceState);
22.       Log.i(TAG, "(3) onRestoreInstanceState()");
23.    }
24.    @Override
25.    //在活动生命周期开始时被调用，恢复被 onPause()停止的用于界面更新的资源
26.    public void onResume() {
```

```
27.    super.onResume();
28.    Log.i(TAG, "(4) onResume()");
29. }
30. @Override
31. // 在 onResume()后被调用,保存界面信息
32. public void onSaveInstanceState(Bundle savedInstanceState) {
33.    super.onSaveInstanceState(savedInstanceState);
34.    Log.i(TAG, "(5) onSaveInstanceState()");
35. }
36. @Override
37. //在重新进入可视生命周期前被调用,载入界面所需要的更改信息
38. public void onRestart() {
39.    super.onRestart();
40.    Log.i(TAG, "(6) onRestart()");
41. }
42. @Override
43. //在活动生命周期结束时被调用,用来保存持久的数据或释放占用的资源
44. public void onPause() {
45.    super.onPause();
46.    Log.i(TAG, " (7) onPause()");
47. }
48. @Override//在可视生命周期结束时被调用,一般用来保存持久的数据或释放占用的资源
49. public void onStop() {
50.    super.onStop();
51.    Log.i(TAG, "(8) onStop()");
52. }
53. @Override //在完全生命周期结束时被调用,释放资源,包括线程、数据连接等
54. public void onDestroy() {
55.    super.onDestroy();
56.    Log.i(TAG, "(9) onDestroy()");
57. }
58. }
```

上面的程序主要通过在生命周期函数中添加"日志点"的方法进行调试,程序的运行结果将会显示在 LogCat 中。为了观察和分析程序运行结果,在 LogCat 中设置过滤器 LifeCycleFilter,过滤方法选择 by Log Tag,过滤关键字为 LIFTCYCLE。

1. 全生命周期

启动和关闭 ActivityLifeCycleActivity 的 LogCat 输出。

启动 ActivityLifeCycleActivity,按下模拟器上的"返回"键,关闭 ActivityLifeCycleActivity。

LogCat 输出结果如图 2-19 所示。

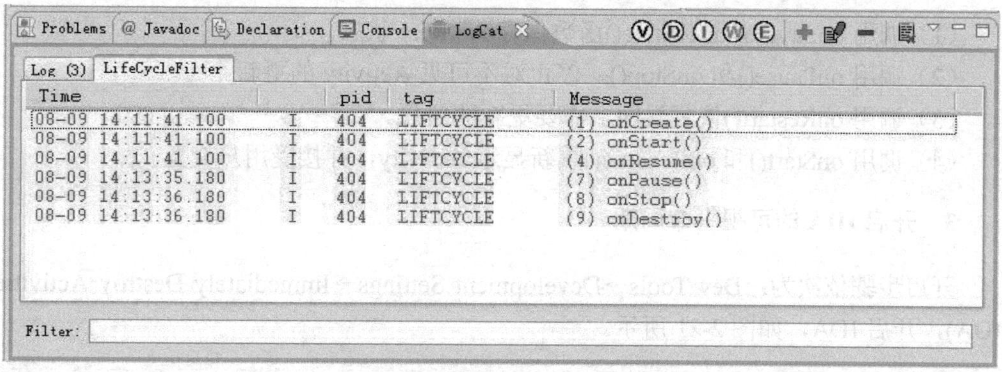

图 2-19 Activity 的全生命周期

从图 2-19 可以看出，方法的调用顺序为：onCreate()→onStart()→onResume()→onPause()→onStop()→onDestroy()。

（1）调用 onCreate()函数分配资源。
（2）调用 onStart()将 Activity 显示在屏幕上。
（3）调用 onResume()获取屏幕焦点。
（4）调用 onPause()、onStop()和 onDestroy()，释放资源并销毁进程。

2．可视生命周期

可视生命周期中的状态转换及测试步骤如下。
（1）启动 ActivityLifeCycle。
（2）按"呼出/接听"键启动内置的拨号程序。
（3）再通过"返回"键退出拨号程序。
（4）ActivityLifeCycle 重新显示在屏幕中。

可视生命周期的 LogCat 输出结果，如图 2-20 所示。

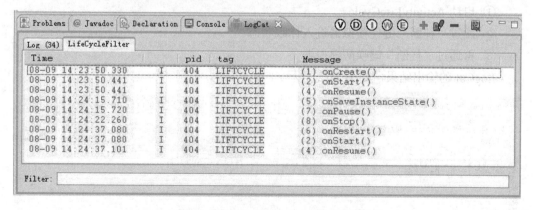

图 2-20 可视生命周期中方法调用过程

其方法的调用顺序为：onSaveInstanceState()→onPause()→onStop()→onRestart()→onStart()→onResume()。

(1)调用 onSaveInstanceState()函数保存 Activity 状态。
(2)调用 onPause()和 onStop(),停止对不可见 Activity 的更新。
(3)调用 onRestart()恢复界面上需要更新的信息。
(4)调用 onStart()和 onResume()重新显示 Activity,并接受用户交互。

3.开启 IDA 的可视生命周期

开启步骤依次为:Dev Tools→Development Settings→Immediately Destroy Activities (IDA),开启 IDA,如图 2-21 所示。

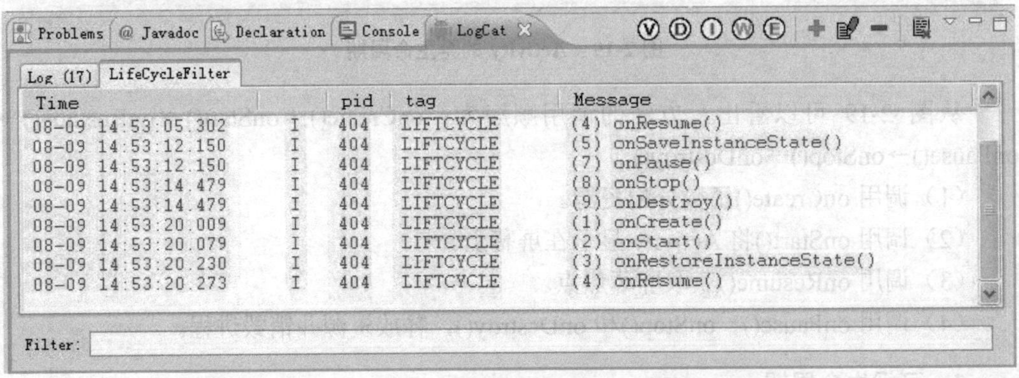

图 2-21 开启 IDA 可视生命周期的调用顺序

开启 IDA 的可视生命周期的方法调用顺序依次为:onSaveInstanceState()→onPause()→onStop()→onDestroy()→onCreate()→onStart()→onRestoreInstanceState()→onResume()。

(1)调用 onRestoreInstanceState()恢复 Activity 销毁前的状态。
(2)其他的函数调用顺序与程序启动过程的调用顺序相同。

活动生命周期的测试步骤如下。

① 启动 ActivityLifeCycle。
② 通过"挂断"键使模拟器进入休眠状态。
③ 再通过"挂断"键唤醒模拟器。

LogCat 的输出结果,如图 2-22 所示。

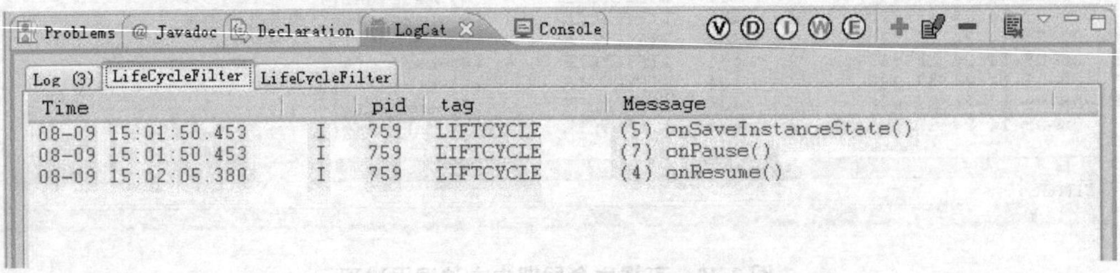

图 2-22 保存、停止及恢复调用过程

方法调用顺序依次为:onSaveInstanceState()→onPause()→onResume()。

（1）调用 onSaveInstanceState ()保存 Activity 的状态。
（2）调用 onPause()停止与用户交互。
（3）调用 onResume()恢复与用户的交互。

习 题

一、简答题

1. 在 AndroidManefiest.xml 中如何修改应用程序名称及图标？
2. Android 4.0 之后应用程序目录中增加了哪些内容？
3. 简述 Fragment 的作用及其生命周期。
4. 简述组件生命周期中 Service 生命周期过程。
5. 项目案例描述的 Activity 生命周期中表现状态分为哪些？它们与组件生命周期之间有什么关系？

二、实训

要求：
使用不同方式，创建 Fragment 实例，并完成 Fragment 生命周期过程的测试。

第 3 章

Android 界面设计基础

学习目标

本章主要介绍了 Android 移动和触摸设备设计原则、Android 界面框架及结构、常用的 Android 界面布局。同时,就不同的布局 LinearLayout、RelativeLayout、TableLayout、FrameLayout、AbsoluteLayout、GridLayout 应用进行了讲解。读者通过本章的学习,能够完成以下知识要点的学习。

(1) Android 移动和触摸设备的设计原则、界面框架及 MVC 设计。
(2) Android 界面控件类的分类。
(3) Android 各种界面布局应用及其特点。

在实际的 Android 项目开发中,不论是 Android 应用,还是基于 Web 项目或基于物联网的 Android 客户端应用开发,其开发设计都必须先进行项目的策划方案设计,然后按照软件开发的流程来进行项目的设计与开发。一个完整的软件开发流程通常都必须经过如下几个阶段:项目计划、软件需求分析、软件概要设计、软件详细设计、数据库设计、软件开发、软件测试等。

贯穿本书知识点的案例项目是妈咪宝贝系统开发项目,因本书重点介绍基于 Android 移动端的应用,所以在系统的设计和开发流程中,对 Web 服务器端和数据库存储部分不再进行介绍,移动端具体应用详见后续各章项目案例。

3.1 Android 界面设计简介

用户界面(User Interface,UI)是 Android 系统和用户之间进行信息交换的媒介,它实现了信息的内部表示形式与用户可以接受的形式之间的转换。在计算机出现的早期,批处理界面(1945—1968)和命令行界面(1969—1983)就已经广泛应用。

通常用户界面的设计需要先进行原型设计、数字工具实现、用户测试等阶段。UI 设计工具有 Adobe、OmniGraffle(Mac 系统使用)、WireframeSketcher(可作为 Eclipse 插件)、Balsamiq、图形用户界面(Graphical User Interface,GUI,基于 Eclipse 的插件)、Pencil(火狐浏览器插件)、Android Design Preview、Android Asset Studio 等。采用图形方式与用户进行交互,是当前用户界面比较流行的一种设计和应用。未来的用户界面的设计发展趋势是,将虚拟现实技术应用到界面设计中,使用户能够摆脱键盘与鼠标的交互方式，而通过动作、语言，甚至是脑电波来控制计算机。

3.1.1 移动和触摸设备设计原则

在实践中，移动设备使用的环境非常嘈杂，其上的应用关系用户注意力的刺激和吸引，其用户的界面设计必须直观，所有的控件必须显而易见，需要考虑移动设备的使用场景和移动多任务的处理。此外，还需考虑设备的限制，因不同移动终端设备厂商生产的屏幕尺寸大小也有可能不同。所以，针对 Android UI 的设计与 PC 终端的 UI 设计有很大的不同，在设计手机用户界面时，应全面考虑手持移动终端的特点，并坚持界面设计与程序逻辑松散耦合的理念。因此，在 Android UI 设计中，应解决的问题和坚持的原则有以下几点。

（1）需要界面设计与程序逻辑完全分离，这样不仅有利于它们的并行开发，而且在后期修改界面时，也不用再次修改程序的逻辑代码。

（2）根据不同型号手机的屏幕解析度、尺寸和纵横比各不相同，自动调整界面上部分控件的位置和尺寸，避免因为屏幕信息的变化而出现显示错误。

（3）能够合理利用较小的屏幕显示空间，构造出符合人机交互规律的用户界面，避免出现凌乱、拥挤的用户界面。

（4）界面的设计必须直观、易见、易用，便于用户依靠手势功能控制触摸界面。

Android 系统已经解决了前两个问题，使用 XML 文件描述用户界面；资源中的资源文件分为不同的类别独立保存在不同类别的资源文件夹中（如第 2 章讲述的 res 目录下 drawable-hdpi 等）；对用户界面描述非常灵活，允许不明确定义界面元素的位置和尺寸，仅声明界面元素的相对位置和粗略尺寸。

3.1.2 触摸屏与物理按键设计

Android 系统设备主要以手机和平板电脑为主，它们都把触摸屏作为主要的系统控制设备。目前，触摸屏设备主要分为电容式触摸屏和电阻式触摸屏。电容式触摸是利用人体的电流感应原理进行工作，通过增加互电容的电极，实现多点触控。电阻式触摸屏是通过用户用力按压屏幕来引发屏幕的响应，它更廉价，主要应用在低端和老的设备上。屏幕技术的变革，将会对 UI 设计带来革命性的变化。

一般 Android 手机上都会有三个物理按键，分别为 Menu（菜单）、Home（主界面）和 Back（后退）键。在 Android 3.0 版本之后，物理按键变为可选，可以通过软件渲染的按钮来进行代替。这也给用户界面的操作栏设计模式、软键盘设计、操作按钮设计带来变化。

3.2 Android 界面框架及部件

框架（Framework）的发展已经有三十多年历史了，最早的是 1980 年的 Smalltalk 语言的 MVC，在框架发展过程中，典型的有：MVC Framework（20 世纪 80 年代初期）、MacApp Framework（20 世纪 80 年代中期）、MFC Framework（20 世纪 90 年代初期）、San Francisco Framework（20 世纪 90 年代中期）、.NET Framework（2000 年）、Android 框

架（2007 年）。Android 用户界面框架（Android UI Framework），采用的是比较流行的 MVC（Model-View-Controller）框架模型。MVC 模型提供了处理用户输入的控制器（Controller），显示用户界面和图像的视图（View），以及保存数据和代码的模型（Model），如图 3-1 所示。

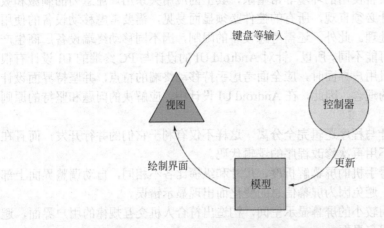

图 3-1 Android MVC 框架模型

3.2.1 Android 用户界面结构

Android 用户界面通常包含活动（Activity）、片段（Fragment）、布局（Layout）、小部件（Widget）几部分，如图 3-2 所示。其用户界面组件有许多，包含文本标签、图库小部件、列表小部件、标签页容器等。

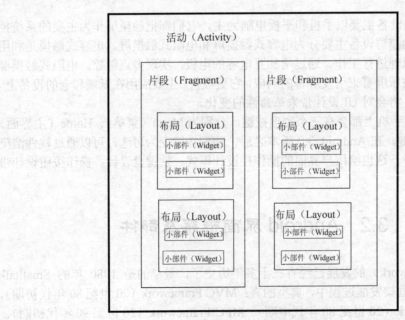

图 3-2 Android 用户界面结构

活动（Activity）如前面所述，它是 Android 应用的核心组件，通常用来表示一个界面，可以添加多个界面到一个活动中，它是负责界面显示什么的控制实例，可以用来移除或添加新的组件，也可以通过 Intent（触发意图）来启动触发新的活动。

片段（Fragment）是界面上独立的一个部分，可以和其他片段放在一起，也可以单独放置，通常把它作为一个子活动。它是 Android 3.0 版本之后引进的新特性，其目的是为了让用户应用程序具有更强的跨设备扩展能力（在智能手机和平板电脑之间）。

布局（Layout）是对用户界面中小部件排列设置的容器，如同规定房间中的家具如何放置及其放置的位置。后续章节中会详细讲解。

小部件（Widget）是 Android 中独立的组件，包含按钮、文本框、编辑框等，这些在后续章节中都会讲解。

3.2.2　Android 与 MVC 设计

MVC（Model-View-Controller），即把一个应用的输入、处理、输出流程，按照 Model、View、Controller 的方式进行分离，这样一个应用被分成三个层——模型、视图、控制器。

在 Android 系统中，视图（View）代表用户交互界面。一个应用可能有很多不同的视图，MVC 设计模式对于视图的处理仅限于视图上数据的采集和处理，以及用户的请求，而不包括在视图上的业务流程的处理。业务流程的处理交予模型（Model）处理。比如一个订单的视图只接受来自模型的数据并显示给用户，以及将用户界面的输入数据和请求传递给控制和模型。

模型（Model）：代表业务逻辑 Bean，就是业务流程/状态的处理以及业务规则的制定。业务流程的处理过程对其他层来说是黑箱操作，模型接收视图请求的数据，并返回最终的处理结果。业务模型的设计可以说是 MVC 最主要的核心。

控制器（Controller）：可以理解为从用户接收请求，将模型与视图匹配在一起，共同完成用户的请求。划分控制层的作用明确了它就是一个分发器，选择什么样的模型，选择什么样的视图，可以完成什么样的用户请求。控制层并不做任何的数据处理。例如，用户单击一个链接，控制层接受请求后，并不处理业务信息，它只把用户的信息传递给模型，告诉模型做什么，选择符合要求的视图返回给用户。因此，一个模型可能对应多个视图，一个视图可能对应多个模型。在 Android 系统中，由 Activity 充当控制器的角色。

从开发者的角度，MVC 把应用程序的逻辑层与视图层（界面）完全分开，最大的好处是：界面设计人员可以直接参与到界面开发中，程序员可以把精力放在逻辑层处理上，有利于提高效率和明确任务分工。

MVC 模型中的控制器（Controller）能够接受并响应程序的外部动作，如按键动作或触摸屏动作等，控制器使用队列处理外部动作，每个外部动作作为一个对立的事件被加入队列中,然后 Android 用户界面框架按照"先进先出"的规则从队列中获取事件,并将这个事件分配给所对应的事件处理函数。

3.2.3 视图树模型

Android 用户界面框架(Android UI Framework)采用视图树(View Tree)模型。即在 Android 用户界面框架中界面元素是以一种树状结构组织在一起,并称为视图树。视图树由 View 和 ViewGroup 构成,如图 3-2 所示。

Android 系统会依据视图树的结构从上至下绘制每一个界面元素。每个元素负责对自身的绘制,如果元素包含子元素,该元素会通知其下所有子元素进行绘制。

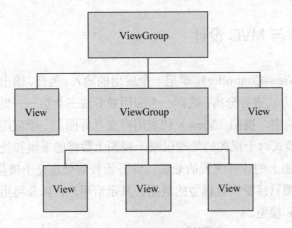

图 3-3 Android 视图树模型

3.3 Android 界面控件类简介

在 Android 中使用各种控件可以实现 UI 的外观,而 View 是各类控件的基类,是创建交互式的图形用户界面的基础。ViewGroup 是布局管理器(Layout)及 View 容器的基类。下面就 View 类和 ViewGroup 类进行简述。

3.3.1 View 类

1. View 类

View 是界面中最基本的可视单元,呈现了最基本的 UI 构造块。一个视图占据屏幕上的一个方形区域,存储了屏幕上特定矩形区域内所显示内容的数据结构,并能够实现所占据区域的界面绘制、焦点变化、用户输入和界面事件处理等功能。

View 是 Android 中最基础的类之一,所有在界面上的可见元素都是 View 的子类,类的视图结构是 android.view.View，如 Button、RadioButton、CheckBox 等，都是通过继承 View 的方法来实现的。通过继承 View，可以很方便地定制出有个性的控件出来。

View 有众多的扩展者，它们大部分是在 android.widget 包中，这些继承者实际上就是 Android 系统中的"控件"。View 的直接继承者包括文本视图(TextView)、图像视图(ImageView)、进度条(ProgressBar)等。它们各自又有众多的继承者。每个控件除了继承父类的功能之外，一般还具有自己的公有方法、保护方法、XML 属性等。

在 Android 中一般情况是在布局文件中，通过使用各种控件实现 UI 的外观，然后在 Java 文件中实现对各种控件的控制动作。控件类的名称也是它们在布局文件 XML 中使用的标签名称。

2．View 类通用行为和属性

View 是 Android 中所有控件类的基类，因此 View 中一些内容是所有控件类都具有的通用行为和属性。

View 作为各种控件的基类，其 XML 属性被所有控件通用，几个重要的 XML 属性如表 3-1 所示。

表 3-1 View 中几个重要 XML 属性及其对应的方法

XML 属性名称	Java 中对应方法	描　　述
android:visibility	setVisibility(int)	描述 View 的可见性
android:id	setId(int)	设置 View 的标识符，一般通过 findViewById 方法获取
android:background	setBackgroundResource(int)	设置背景
android:clickable	setClickable(boolean)	设置 View 响应单击事件

注意：由于 Java 语言不支持多重继承，因此 Android 控件不可能以基本功能的"排列组合"的方式实现。在这种情况下，为了实现功能的复用，基类的功能比较多，作为控件的父类，View 所实现的功能也较多。

3.3.2 ViewGroup 类

ViewGroup 是个特殊的 View，它继承于 Android.view.View。它的功能就是装载和管理下一层的 View 对象或 ViewGroup 对象，也就说它是一个容纳其他元素的容器。ViewGroup 是布局管理器（Layout）及 View 容器的基类，如图 3-3 所示。

ViewGroup 中，还定义了一个嵌套类 ViewGroup.LayoutParams。这个类定义了一个显示对象的位置、大小等属性，View 通过 LayoutParams 中的这些属性值来告诉父级，它们将如何放置。

ViewGroup 是一个抽象类，所以真正充当容器的是它的子类们，也就是下面要重点讲述的 LinearLayout、AbsoluteLayout、FrameLayout 等布局管理器。

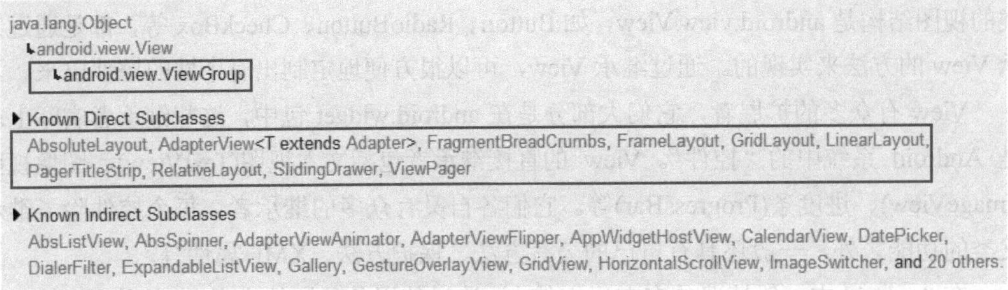

```
java.lang.Object
   ↳android.view.View
      ↳android.view.ViewGroup
```

Known Direct Subclasses
AbsoluteLayout, AdapterView<T extends Adapter>, FragmentBreadCrumbs, FrameLayout, GridLayout, LinearLayout, PagerTitleStrip, RelativeLayout, SlidingDrawer, ViewPager

Known Indirect Subclasses
AbsListView, AbsSpinner, AdapterViewAnimator, AdapterViewFlipper, AppWidgetHostView, CalendarView, DatePicker, DialerFilter, ExpandableListView, Gallery, GestureOverlayView, GridView, HorizontalScrollView, ImageSwitcher, and 20 others.

图 3-4 ViewGroup 类继承结构

ViewGroup 功能：一个是承载界面布局，另一个是承载具有原子特性的重构模块。

3.3.3 界面控件

Android 系统的界面控件分为定制控件和系统控件两种。

定制控件是用户独立开发的控件，或通过继承并修改系统控件后所产生的新控件。能够为用户提供特殊的功能或与众不同的显示需求方式。

系统控件是 Android 系统提供给用户已经封装的界面控件，提供在应用程序开发过程中的常见功能控件。系统控件更有利于帮助用户进行快速开发，同时能够使 Android 系统中应用程序的界面保持一致性。

常见的系统控件包括文本控件（TextView、EditText）、按钮控件（Button、ImageButton）、单选和复选按钮控件（Checkbox、RadioButton）、Spinner、ListView 和 TabHost 等。界面控件的介绍将在后续的章节中进行详述，此处不再赘述。

3.4 Android 界面布局

3.4.1 Android 布局策略

UI（用户界面）布局（Layout）是用户界面结构的描述，定义了界面中所有的元素、结构及它们之间的相互关系。在 Android 中布局的缩放，通过定义固定区域和可调整大小的区域来告诉操作系统如何缩放，也可以通过布局嵌套布局的方式实现。

Android 下创建界面布局的方法有以下三种。

（1）XML 文件布局方式，使用 XML 文件描述界面布局。
（2）程序代码创建，在程序运行时动态添加或修改界面布局。
（3）XML 文件和程序代码创建相结合。

一般都采用 XML 布局文件创建用户界面，当然，用户既可以独立使用任何一种声明界面布局的方式，也可以同时使用两种方式。与使用代码方式相比较，使用 XML 文件声明界面布局的优势主要有以下几点。

（1）将程序的表现层和控制层分离。

（2）在后期修改用户界面时，无须更改程序的源代码。

（3）用户还能够通过可视化工具直接看到所设计的用户界面，有利于加快界面设计的过程，并且为界面设计与开发带来极大的便利性。

在 Android 系统中，布局管理器是控件的容器，每个控件在窗体中都有具体的位置和大小，在窗体中摆放各种控件时，很难判断其具体位置和大小。不过，使用 Android 布局管理器可以很方便地控制各控件的位置和大小。Android 提供了 5 种布局管理器用来管理控件，它们是线性布局管理器（LinearLayout）、表格布局管理器（TableLayout）、帧布局管理器（FrameLayout）、相对布局管理器（RelativeLayout）和绝对布局管理器（AbsoluteLayout）。布局管理器的主要作用有以下几个。

（1）适应不同的移动设备屏幕分辨率。

（2）方便横屏和竖屏之间相互切换。

（3）管理每个控件的大小以及位置。

3.4.2 线性布局 LinearLayout 及应用

1．线性布局简介

线性布局（LinearLayout）是一种常用的界面布局，也是 RadioGroup, TabWidget, TableLayout, TableRow, ZoomControls 类的父类。在线性布局中，LinearLayout 可以让它的子元素以垂直或水平的方式排成一行（不设置方向的时候默认按照垂直方向排列），通过将 android:orientation 属性设置为 Horizontal（水平）或 Vertical（垂直）来达到设置线性布局的目的，线性布局不支持元素的自动浮动。

如果是垂直排列，则每行仅包含一个界面元素，如图 3-5 所示。

如果是水平排列，则每列仅包含一个界面元素，如图 3-6 所示。

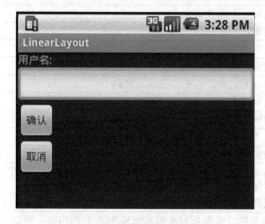
图 3-5　垂直排列

图 3-6　水平排列

表 3-2　LinearLayout 常用属性及对应方法

XML 属性名	对应的方法	描　述
Android:id		设置组件的名称
android:divider	setDividerDrawable(Drawable)	设置用于在按钮间垂直分割的可绘制对象
android:gravity	setGravity(int)	指定在对象内部，横纵方向上如何放置对象的内容
android:orientation	setOrientation(int)	设置线性布局管理器内组件的排列方式，可以设置为 Horizontal（水平排列）、Vertical（垂直排列、默认值）两个值的其中之一
android:weightSum		定义最大的权值和
android:layout_height		设置组件的基本高度，可选：fill_parent、match_parent、wrap_content
android:layout_width		设置组件的基本宽度，可选：fill_parent、match_parent、wrap_content
android:background		设置组件的背景，可以是背景图片、也可以是背景颜色

注明：fill_parent 表示组件的宽度与父容器的宽度相同，match_parent 的作用一样。wrap_content 表示该组件的高度恰能包裹它的内容。

LinearLayout 常用属性及对应方法如表 3-2 所示。在 setGravity(int)方法中，可以设置参数，设置线性布局 LinearLayout 中放置的对象元素的排列对齐方式。如果需要设置多个或者组合设置对齐方式，属性常量由"|"分隔。具体属性及描述如表 3-3 所示。

表 3-3　setGravity(int)方法可取的属性常量及描述

属性常量	值	描　述
top	0x30	将对象放在其容器的顶部，不改变其大小
bottom	0x50	将对象放在其容器的底部，不改变其大小
left	0x03	将对象放在其容器的左侧，不改变其大小
right	0x05	将对象放在其容器的右侧，不改变其大小
center_vertical	0x10	将对象纵向居中，不改变其大小
fill_vertical	0x70	必要的时候增加对象的纵向大小，以完全充满其容器
center_horizontal	0x01	将对象横向居中，不改变其大小
fill_horizontal	0x07	必要的时候增加对象的横向大小，以完全充满其容器
center	0x11	将对象横纵居中，不改变其大小
fill	0x77	必要的时候增加对象的横纵向大小，以完全充满其容器
clip_vertical	0x80	附加选项，用于按照容器的边来剪切对象的顶部和/或底部的内容。剪切基于其纵向对齐设置：顶部对齐时，剪切底部；底部对齐时，剪切顶部；除此之外剪切顶部和底部
clip_horizontal	0x08	附加选项，用于按照容器的边来剪切对象的左侧和/或右侧的内容。剪切基于其横向对齐设置：左侧对齐时，剪切右侧；右侧对齐时，剪切左侧；除此之外剪切左侧和右侧

2. 线性布局应用

上述内容中对 LinearLayout 线性布局知识进行了简要介绍，本节下面将就线性布局在实际程序中如何使用和基本的实现流程进行详述。使用线性布局垂直排列实现了 5 个按钮排列的功能，具体实现步骤如下。

（1）在 Eclipse 中选择 File→New→Android Project 命令，创建一个新的 Android 工程，项目名为 LinearLayoutDemo，目标 API 选择 17（即 Android 4.2 版本），应用程序名为 LinearLayoutDemo，包名为 com.bcpl.activity，创建的 Activity 的名字为 LinearLayoutActivity，最小 SDK 版本根据选择的目标 API 会自动添加为 8。

（2）打开项目工程中 res→layout 目录下的 activity_main.xml 文件，设置布局，添加 5 个按钮，按钮分别显示 button1、button2 等，代码如下所示。

```
1.  <?xml version="1.0" encoding="utf-8"?>
2.  <LinearLayout xmlns:android="http://schemas.android.com/apk/res/android"
3.      android:orientation="vertical"
4.      android:layout_width="fill_parent"
5.      android:layout_height="fill_parent"
6.      >
7.      <Button
8.      android:id="@+id/but1"
9.      android:layout_width="wrap_content"
10.     android:layout_height="wrap_content"
11.     android:text="button1"
12.     />
13.     <Button
14.     android:id="@+id/but2"
15.     android:layout_width="wrap_content"
16.     android:layout_height="wrap_content"
17.     android:text="button2"
18.     />
19.     <Button
20.     android:id="@+id/but3"
21.     android:layout_width="wrap_content"
22.     android:layout_height="wrap_content"
23.     android:text="button3"
24.     />
25.     <Button
26.     android:id="@+id/but4"
27.     android:layout_width="wrap_content"
28.     android:layout_height="wrap_content"
29.     android:text="button4"
```

```
30.        />
31.        <Button
32.            android:id="@+id/b5"
33.            android:layout_width="wrap_content"
34.            android:layout_height="wrap_content"
35.            android:text="button5"
36.        />
37.    </LinearLayout>
```

第 2-6 行声明了一个线性布局，第 2 行代码是声明 XML 文件的根元素为线性布局，第 3 行设置了线性布局的元素排列方式是垂直排列。

第 4、5 行设置了线性布局在所属的父容器中的布局方式为横向和纵向填充父容器，表示线性布局宽度等于父控件的宽度，就是将线性布局在横向和纵向上占据父控件的所有空间。

第 7～12 行声明了一个按钮控件，第 8 行设置了 ID 为 but1，第 11 行设置了按钮显示为 "button1"。

第 9 行设置了按钮控件在父容器中的布局方式为只占据自身大小的空间，表示线性布局宽度等于所有子控件的宽度总和，也就是线性布局的宽度会刚好将所有子控件包含其中。

第 10 行设置了按钮控件在父容器中的布局方式为只占据自身大小的空间，表示线性布局高度等于所有子控件的高度总和，也就是线性布局的高度会刚好将所有子控件包含其中。

（3）src 目录下 com.hisoft.activity 包下的 LinearLayoutActivity.java 文件和 res→values 目录下的 strings.xml 文件，都暂不做修改。部署运行项目工程，项目运行效果如图 3-7 所示。

图 3-7　线性布局运行结果

注意：LinearLayout 布局中控件按顺序从左到右或从上到下依次排列。

建立横向线性布局与纵向线性布局相似，只需注意将线性布局的 Orientation 属性的值设置为 Horizontal 即可。

3.4.3 相对布局 RelativeLayout 及应用

1. 相对布局简介

相对布局（RelativeLayout）是一种非常灵活的布局方式，按照控件之间所指定的相对位置参数来自动对控件进行排列，确定界面中所有元素的布局位置。让子元素指定它们相对于其他元素的位置（通过 ID 来指定）或相对于父布局对象，与 AbsoluteLayout 这个绝对坐标布局是相反的。实际开发中，一般推荐使用这种布局。

相对布局的特点：能够最大程度地保证在各种屏幕类型的手机上正确显示界面布局。

在 RelativeLayout 布局里的控件包含丰富的排列属性，总地可以分为以下三类。

（1）以 parent（父控件）为参照物的 XML 属性，属性取值可以为 true 或者 false，如表 3-4 所示。

表 3-4 以 parent（父控件）为参照物的 XML 属性及描述

XML 属性名称	描 述
android:layout_alignParentTop	如果为 true，将该控件的顶部与其父控件的顶部对齐
android:layout_alignParentBottom	如果为 true，将该控件的底部与其父控件的底部对齐
android:layout_alignParentLeft	如果为 true，将该控件的左部与其父控件的左部对齐
android:layout_alignParentRight	如果为 true，将该控件的右部与其父控件的右部对齐
android:layout_centerHorizontal	如果为 true，将该控件的置于父控件的水平居中位置
android:layout_centerVertical	如果为 true，将该控件的置于父控件的垂直居中位置
android:layout_centerInParent	如果为 true，将该控件的置于父控件的中央

（2）指定参照物的 XML 属性，layout_alignBottom, layout_toLeftOf, layout_above, layout_alignBaseline 系列和其他控件 ID，如表 3-5 所示。

表 3-5 参照物的 XML 属性及描述

XML 属性名称	描 述
android:layout_alignBaseline	将该控件的 baseline 与给定 ID 的 baseline 对齐
android:layout_alignTop	将该控件的顶部边缘与给定 ID 的顶部边缘对齐
android:layout_alignBottom	将该控件的底部边缘与给定 ID 的底部边缘对齐
android:layout_alignLeft	将该控件的左边缘与给定 ID 的左边缘对齐
android:layout_alignRight	将该控件的右边缘与给定 ID 的右边缘对齐
android:layout_above	将该控件的底部置于给定 ID 的控件之上
android:layout_below	将该控件的底部置于给定 ID 的控件之下
android:layout_toLeftOf	将该控件的右边缘与给定 ID 的控件左边缘对齐
android:layout_toRightOf	将该控件的左边缘与给定 ID 的控件右边缘对齐

（3）指定移动像素的 XML 属性，如表 3-6 所示。

表 3-6 移动像素的 XML 属性及描述

XML 属性名称	描 述
android:layout_marginTop	上偏移的值
android:layout_marginBottom	下偏移的值
android:layout_marginLeft	左偏移的值
android:layout_marginRight	右偏移的值

注意事项：

（1）使用 RelativeLayout 布局的时候，避免程序运行时做控件布局的更改，因为 RelativeLayout 布局里面的属性之间很容易冲突。

（2）在相对布局 RelativeLayout 的大小和它的子控件位置之间要避免出现循环依赖，如设置相对布局 RelativeLayout 高度属性为 WRAP_CONTENT，就不能再设置它的子控件高度属性为 ALIGN_PARENT_BOTTOM。

2．相对布局应用

在前面对 RelativeLayout 布局及属性进行了简要介绍后，本节将就 RelativeLayout 布局在实际程序中的使用和基本的实现进行详述。使用文本控件 TextView 和 EditView 及两个按钮控件实现相对排列的功能，具体实现步骤如下。

（1）如前所述，在 Eclipse 中选择 File→New→Android Project 命令，创建一个新的 Android 工程，项目名为 RelativeLayoutDemo，目标 API 选择 17（即 Android 4.2 版本），应用程序名为 RelativeLayoutDemo，包名为 com.bcpl.activity，创建的 Activity 的名字为 RelativeLayoutActivity。

（2）打开项目工程中 res→layout 目录下的 activity_main.xml 文件，设置布局，添加 TextView、EditText 控件和两个按钮控件，代码如下所示。

```
1.   <?xml version="1.0" encoding="utf-8"?>
2.   <RelativeLayout
     xmlns:android="http://schemas.android.com/apk/res/android"
3.       android:layout_width="fill_parent"
4.       android:layout_height="fill_parent">
5.      <TextView
6.          android:id="@+id/label"
7.          android:layout_width="fill_parent"
8.          android:layout_height="wrap_content"
9.          android:text="请输入:"/>
10.     <EditText
11.         android:id="@+id/entry"
12.         android:layout_width="fill_parent"
13.         android:layout_height="wrap_content"
14.           android:background="@android:drawable/editbox_background"
15.         android:layout_below="@id/label"/>
16.     <Button
```

```
17.         android:id="@+id/ok"
18.         android:layout_width="wrap_content"
19.         android:layout_height="wrap_content"
20.         android:layout_below="@id/entry"
21.         android:layout_alignParentRight="true"
22.         android:layout_marginLeft="10dip"
23.         android:text="OK" />
24.     <Button
25.         android:layout_width="wrap_content"
26.         android:layout_height="wrap_content"
27.         android:layout_toLeftOf="@id/ok"
28.         android:layout_alignTop="@id/ok"
29.         android:text="Cancel" />
30. </RelativeLayout>
```

第 2～4 行声明了一个相对布局，第 2 行代码是声明 XML 文件的根元素为相对布局，第 3、4 行设置了相对布局在所属的父容器中的布局方式为横向和纵向填充父容器。

第 5～9 行声明了一个 TextView 控件，第 6 行设置了 ID 为 label，第 7 行设置宽度为填充父容器，第 8 行设置了高度为控件自身内容，第 9 行设置了 TextView 显示的文字内容。

第 10～15 设置了一个 EditView 控件，第 11 行设置了 ID 为 entry，第 12 行设置宽度为填充父容器，第 13 行设置了高度为控件自身内容，第 14 行设置了背景，第 15 行设置了 ID 为 entry 位于 ID 为 label 的下面。

第 16～29 行设置了两个 Button 按钮控件，第 20 行设置了按钮位于 ID 为 entry 的控件下方，第 21 行设置了该控件的右部与其父控件的右部对齐；第 22 行设置了按钮从右边框左偏移 10 个 dip；第 23 行设置了按钮显示文字为 OK。

第 27 行设置了 Cancel 按钮控件的右边缘与给定 ID 为 OK 的控件左边缘对齐。

第 28 行设置了 Cancel 按钮控件的顶部边缘与给定 ID 为 OK 的控件的顶部边缘对齐。

第 29 行设置了按钮显示文字为 Cancel。

（3）src 目录下 com.bcpl.activity 包下的 RelativeLayoutActivity.java 文件和 res→values 目录下的 strings.xml 文件，都暂不做修改。部署运行 RelativeLayoutDemo 项目工程，项目运行效果如图 3-8 所示。

图 3-8　RelativeLayoutDemo 项目运行结果

3.4.4 表格布局 TableLayout 及应用

1. 表格布局简介

表格布局（TableLayout）也是一种常用的界面布局，采用行、列的形式来管理 UI 组件，它是将屏幕划分成网格单元（网格的边界对用户是不可见的），然后通过指定行和列的方式，将界面元素添加到网格中。它并不需要明确地声明包含多少行、列，而是通过添加 TableRow、其他组件来控制表格的行数和列数。每次向 TableLayout 中添加一个 TableRow，该 TableRow 就是一个表格行，TableRow 也是容器，因此它可以不断地添加其他组件，每添加一个子组件该表格就增加一列。每一行可以有 0 个或多个单元格，每个单元格就是一个 View，一个 Table 中可以有空的单元格，单元格可以像在 HTML 中使用的方式一样，合并多个单元格，跨越多列。这些 TableRow 中，单元格不能设置 layout_width，宽度属性默认是 fill_parent，只有高度 layout_height 可以自定义，默认值是 wrap_content。

在表格布局中，一个列的宽度由该列中最宽的单元格决定。表格布局支持嵌套，可以将另一个表格布局放置在前一个表格布局的网格中，也可以在表格布局中添加其他界面布局，如线性布局、相对布局等。常用具体属性及方法如表 3-7 所示。

表 3-7 表格布局 TableLayout 常用属性及相关方法

属性名称	相关方法	描述
android:collapseColumns	setColumnCollapsed(int,boolean)	设置指定列为 collapse，列索引从 0 开始
android:shrinkColumns	setShrinkAllColumns(boolean)	设置指定列为 shrink，列索引从 0 开始
android:stretchColumns	setStretchAllColumns(boolean)	设置指定列为 stretch，列索引从 0 开始

在表格布局 TableLayout 中，如果一个列通过 setColumnShrinkable()方法设置为 shrinkable，则该列的宽度可以进行收缩，使表格能够适应它父容器的大小。

如果一个列通过 setColumnStretchable()方法设置为 stretchable，则该列的宽度可以进行拉伸，扩展它的宽度填充空余的空间。

建立表格布局时要注意以下几点。

（1）向界面中添加一个线性布局，无须修改布局的属性值。其中，Id 属性为 TableLayout01，Layout width 和 Layout height 属性都为 wrap_content。

（2）向 TableLayout01 中添加两个 TableRow。TableRow 代表一个单独的行，每行被划分为几个小的单元，单元中可以添加一个界面控件。其中，Id 属性分别为 TableRow01 和 TableRow02，Layout width 和 Layout height 属性都为 wrap_content，表格布局示意和布局效果如图 3-9 和图 3-10 所示。

第 3 章　Android 界面设计基础

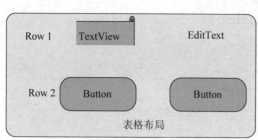

图 3-9　表格布局　　　　　　　　　图 3-10　表格布局效果图

2．表格布局应用

本节将就 TableLayout 布局的使用和应用进行详述。本案例使用 TableRow 和 View 实现 Table 的功能，具体实现步骤如下。

（1）在 Eclipse 中选择 File→New→Android Project 命令，创建一个新的 Android 工程，项目名为 TableLayoutDemo，目标 API 选择 17（即 Android 4.2 版本），应用程序名为 TableLayoutDemo，包名为 com.bcpl.activity，创建的 Activity 的名字为 TableLayoutActivity。

（2）打开项目工程中 res→layout 目录下的 activity_main.xml 文件，设置布局，添加 6 个 TableRow 和两个 View，代码如下所示。

```
1.   <?xml version="1.0" encoding="utf-8"?>
2.   <TableLayout xmlns:android="http://schemas.android.com/apk/res/android"
3.       android:layout_width="fill_parent"
4.       android:layout_height="fill_parent"
5.       android:stretchColumns="1">
6.
7.       <TableRow>
8.           <TextView
9.               android:layout_column="1"
10.              android:text="姓名："
11.              android:padding="3dip" />
12.          <TextView
13.              android:text="张三"
14.              android:gravity="right"
15.              android:padding="3dip" />
16.      </TableRow>
17.
18.      <TableRow>
19.          <TextView
```

```
20.         android:layout_column="1"
21.         android:text="年龄: "
22.         android:padding="3dip" />
23.     <TextView
24.         android:text="10"
25.         android:gravity="right"
26.         android:padding="3dip" />
27.   </TableRow>
28.
29.   <TableRow>
30.     <TextView
31.         android:layout_column="1"
32.         android:text="性别: "
33.         android:padding="3dip" />
34.     <TextView
35.         android:text="男"
36.         android:gravity="right"
37.         android:padding="3dip" />
38.   </TableRow>
39.
40.   <View
41.       android:layout_height="2dip"
42.       android:background="#FF909090" />
43.
44.   <TableRow>
45.     <TextView
46.         android:layout_column="1"
47.         android:text="所在城市: "
48.         android:padding="3dip" />
49.     <TextView
50.         android:text="北京"
51.         android:gravity="right"
52.         android:padding="3dip" />
53.   </TableRow>
54.
55.   <TableRow>
56.     <TextView
57.         android:layout_column="1"
58.         android:text="国籍: "
59.         android:padding="3dip" />
60.     <TextView
61.         android:text="中国"
62.         android:gravity="right"
63.         android:padding="3dip" />
64.   </TableRow>
65.
```

```
66.        <View
67.            android:layout_height="2dip"
68.            android:background="#FF909090" />
69.
70.        <TableRow>
71.            <TextView
72.                android:layout_column="1"
73.                android:text="附加信息："
74.                android:padding="3dip" />
75.        </TableRow>
76. </TableLayout>
```

第 2~5 行声明了一个表格布局，第 2 行代码是声明 XML 文件的根元素为表格布局，第 3、4 行设置了表格布局在所属的父容器中的布局方式为横向和纵向填充父容器。第 5 行是设置 TableLayout 所有行的第二列为扩展列，剩余的空间由第二列补齐。

第 7~16 行声明了一个 TableRow。

第 8~11 行声明了一个 TextView，第 9 行设置了从第二列开始填写（0 是起始列），第 11 行设置了字符四周到 TextView 的空白边的大小。

第 12~15 行声明了第二个 TextView，第 14 行设置了 TextView 内字符的对齐方式，此为右对齐。

第 40~42 行声明了一个 View，加一个分割线，View 是 TextView 的父类；第 41 行设置了线的高度为 2；第 42 行设置了背景颜色。

（3）src 目录下 com.bcpl.activity 包下的 TableLayoutActivity.java 文件和 res→values 目录下的 strings.xml 文件，都暂不做修改。部署运行 TableLayoutDemo 项目工程，项目运行效果如图 3-11 所示。

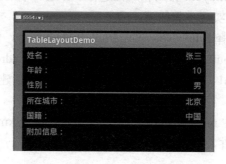

图 3-11　TableLayoutDemo 运行结果

3.4.5　帧布局 FrameLayout 及应用

1. 帧布局简介

帧布局（FrameLayout）是 Android 布局系统中最简单的界面布局，是用来存放一个元素的空白空间，且子元素的位置是不能够指定的，只能够放置在空白空间的左上角。在帧布局中，如果先后存放多个子元素，后放置的子元素将遮挡先放置的子元素。

帧布局由 FrameLayout 所代表，帧布局容器为每个加入其中的组件创建一个空白的区域（成为一帧），所有每个子组件占据一帧，这些帧都会根据 gravity 属性执行自动对齐。也就是说，把组件一个一个地叠加在一起。

FrameLayout 控件继承自 ViewGroup，它在 ViewGroup 的基础上，定义了自己的三个属性，对应的 XML Attributes 分别为 android:foreground，android:foregroundGravity，android:measureAllChildren。第一个属性是设置前景色；第二个属性是控制前景色的重心，前两个属性其实是对 android:backgroud 的重写，其目的是可以控制背景的重心；第三个属性如果为 true，则在测量时测量所有的子元素（即使该子元素为 gone）。帧布局常用属性及方法如表 3-8 所示。

表 3-8 帧布局 FrameLayout 常用属性及相关方法

属 性 名 称	相 关 方 法	描 述
android:foreground	setForeground(Drawable)	设置绘制在子控件之上的内容，设置前景色
android:foregroundGravity	setForegroundGravity(int)	设置应用于绘制在子控件之上内容的 gravity 属性，控制前景色的重心
android:measureAllChildren	setMeasureAllChildren(boolean)	根据参数值，决定是设置测试所有的元素还是仅测量状态是 VISIBLE 或 INVISIBLE 的元素

2. 帧布局应用

前面章节已就 FrameLayout 布局和属性进行了详述。本案例使用 TableRow 和 View 实现 Table 的功能，具体实现步骤如下：

（1）在 Eclipse 中选择 File→New→Android Project 命令，创建一个新的 Android 工程，项目名为 FrameLayoutDemo1，目标 API 选择 17（即 Android 4.2 版本），应用程序名为 FrameLayoutDemo1，包名为 com.bcpl.activity，创建的 Activity 的名字为 MainActivity。

（2）打开项目工程中 res→layout 目录下的 activity_main.xml 文件，设置线性布局，添加一个 Button 按钮控件，代码如下所示。

```
1.  <?xml version="1.0" encoding="utf-8"?>
2.  <LinearLayout
3.    xmlns:android="http://schemas.android.com/apk/res/android"
4.    android:orientation="vertical"
5.    android:layout_width="match_parent"
6.    android:layout_height="match_parent">
7.    <Button android:text="Click to Framelayout"
8.        android:id="@+id/button1"
9.        android:layout_width="wrap_content"
10.       android:layout_height="wrap_content"
```

```
11.        </Button>
12. </LinearLayout>
```

（3）在 src 目录下 com.bcpl.activity 包下的 MainActivity.java 文件中，声明创建按钮，为按钮添加监听器，并通过使用 Intent 实现从 MainActivity 到 FrameLayoutActivity 的跳转，代码如下：

```
1.  package com.bcpl.activity;

2.  import android.app.Activity;
3.  import android.content.Intent;
4.  import android.os.Bundle;
5.  import android.view.View;
6.  import android.view.View.OnClickListener;
7.  import android.widget.Button;

8.  public class MainActivity extends Activity {

9.      private Button button1;

10.     @Override
11.     protected void onCreate(Bundle savedInstanceState) {
12.         super.onCreate(savedInstanceState);
13.         setContentView(R.layout.main);

14.         button1 = (Button)findViewById(R.id.button1);
15.         //为按钮绑定一个单击事件的监听器
16.         button1.setOnClickListener(new OnClickListener(){
17.             public void onClick(View v)
18.             {
19.                 //通过 Intent 跳转到 FrameLayoutActivity
20.                 Intent intent = new Intent();
21.                 intent.setClass(MainActivity.this,
FrameLayoutActivity.class);
22.                 startActivity(intent);
23.             }
24.         });
25.     }
26. }
```

（4）在项目工程 res→layout 目录下新建 framelayout.xml 文件，声明 FrameLayout 布局，并添加两个 ImageView 控件，代码如下。

```
1.  <?xml version="1.0" encoding="utf-8"?>
```

```
2.  <FrameLayout
    xmlns:android="http://schemas.android.com/apk/res/android"
3.      android:id="@+id/frameLayout1"
4.      android:layout_width="fill_parent"
5.      android:layout_height="fill_parent">
6.      <ImageView android:src="@drawable/frame"
7.          android:id="@+id/imageView1"
8.          android:layout_width="wrap_content"
9.          android:layout_height="wrap_content">
10.     </ImageView>
11.     <ImageView android:src="@drawable/icon"
12.         android:id="@+id/imageView2"
13.         android:layout_width="wrap_content"
14.         android:layout_height="wrap_content">
15.     </ImageView>
16. </FrameLayout>
```

（5）在 res 目录下，把 frame.jpg 图片添加到 drawable-hdpi、drawable-mdpi、drawable-ldpi 文件夹中，以供步骤（4）第 6 行控件 ImageView 使用，第 11 行第二个 ImageView 控件用系统自带的图片。

（6）在 src 目录下 com.hisoft.activity 包下，创建 FrameLayoutActivity.java 文件，调用上面创建的帧布局文件 framelayout.xml，代码如下。

```
1.  package com.hisoft.activity;
2.  import android.app.Activity;
3.  import android.os.Bundle;
4.  public class FrameLayoutActivity extends Activity
5.  {
6.      @Override
7.      public void onCreate(Bundle savedInstanceState)
8.      {
9.          super.onCreate(savedInstanceState);
10.         setContentView(R.layout.framelayout);
11.     }
12. }
```

（7）在 AndroidManifest.xml 文件中在<application>标签节点下，添加新创建的 FrameLayout Activity 注册声明，代码如下：

```
1.  <activity    android:name=".FrameLayoutActivity"    android:label=
    "@string/app_name" />
```

（8）部署运行 FrameLayoutDemo1 项目工程，项目运行效果如图 3-12 所示。然后单击 Click to Framelayout 按钮，出现如图 3-13 所示界面。所有的图片都显示在帧布局的左上角,并且第二个覆盖第一个,如果把 framelayout.xml 中的两个 ImageView 控件的描述互换位置,则大图片完全覆盖住小图片。

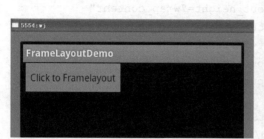

图 3-12　项目运行结果　　　　图 3-13　帧布局图片叠加效果

3.4.6　绝对布局 AbsoluteLayout 及应用

绝对布局（AbsoluteLayout）能通过指定界面元素的坐标位置,来确定用户界面的整体布局。绝对布局是一种不推荐使用的界面布局,因为通过 X 轴和 Y 轴确定界面元素位置后,Android 系统不能够根据不同屏幕对界面元素的位置进行调整,降低了界面布局对不同类型和尺寸屏幕的适应能力。

上述已就 AbsoluteLayout 布局进行了简介。下面将通过一个应用的开发来介绍 AbsoluteLayout 的使用方法,具体实现步骤如下。

（1）在 Eclipse 中选择 File→New→Android Project 命令,创建一个新的 Android 工程,项目名为 AbsoluteLayoutDemo,目标 API 选择 17（即 Android 4.2 版本）,应用程序名为 AbsoluteLayoutDemo,包名为 com.bcpl.activity,创建的 Activity 的名字为 AbsoluteLayoutActivity。

（2）打开项目工程中 res→layout 目录下的 activity_main.xml 文件,设置绝对布局,添加三个 TextView 控件,代码如下所示。

```
1.    <?xml version="1.0" encoding="utf-8"?>
2.    <AbsoluteLayout
         xmlns:android="http://schemas.android.com/apk/res/android"
3.                android:id="@+id/absoluteLayout1"
4.                android:layout_width="fill_parent"
5.                android:layout_height="fill_parent">
6.       <TextView android:textSize="18pt"
```

```
7.         android:id="@+id/tv1"
8.         android:layout_height="wrap_content"
9.         android:layout_width="wrap_content"
10.        android:text="@string/tv1"
11.        android:layout_x="37dp"
12.        android:layout_y="37dp">
13.    </TextView>
14.    <TextView android:textSize="18pt"
15.        android:id="@+id/tv2"
16.        android:layout_height="wrap_content"
17.        android:layout_width="wrap_content"
18.        android:text="@string/tv2"
19.        android:layout_x="186dp"
20.        android:layout_y="104dp">
21.    </TextView>
22.    <TextView android:textSize="18pt"
23.        android:id="@+id/tv3"
24.        android:layout_height="wrap_content"
25.        android:layout_width="wrap_content"
26.        android:text="@string/tv3"
27.        android:layout_x="106dp"
28.        android:layout_y="188dp">
29.    </TextView>
30. </AbsoluteLayout>
```

第2～5行声明了一个绝对布局，第2行代码是声明XML文件的根元素为绝对布局，第3行设置了ID，第4、5行设置了绝对布局在所属的父容器中的布局方式为横向和纵向填充父容器。

第6～13行声明了一个TextView控件，第6行设置了文本的大小，第7行设置了ID的名称，第8、9行设置了控件在父容器中的布局方式为只占据自身内容大小的空间，第10行设置了TextView控件显示的文本内容为资源文件strings.xml中设置的名称为tv1的值，第11、12行设置了控件显示的起始坐标位置。

第14～21行声明了第二个TextView控件。

第22～29行声明了第三个TextView控件。

（3）打开res目录下values文件中的strings.xml文件，修改文件代码如下。

```
1.  <?xml version="1.0" encoding="utf-8"?>
2.  <resources>
3.      <string name="app_name">AbsoluteLayoutDemo</string>
4.      <string name="tv1">文本1</string>
5.      <string name="tv2">文本2</string>
6.      <string name="tv3">文本3</string>
```

7. `</resources>`

(4)部署运行 AbsoluteLayoutDemo 项目工程,项目运行效果如图 3-14 所示。

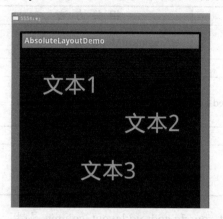

图 3-14 项目运行结果

上述使用的是默认屏幕大小为 432 的模拟器或屏幕,如果使用不同屏幕大小的模拟器或设备,如 800 大小,则显示效果有可能发生差异及变形。

3.4.7 网格布局 GridLayout 及应用

1. GridLayout 网络布局简介

在 Android 4.0 之前,网格布局通常通过 LiearLayout 的嵌套或 TableLayout 布局来实现,但无论采用哪一种方式都存在明显的缺陷。如采用 LinearLayout 布局实现,存在着X、Y 轴方向上不能同时对控件进行对齐、多层布局嵌套存在性能等问题;采用 TableLayout 布局可能会出现控件不能占据多个行或列等问题。为了克服上述问题,Android 4.0 之后的版本,新增加了网格布局管理器。网格布局管理器将布局划分为行、列和单元格,类似于 HTML 中的 Table 标签,把整个容器划分为行×列个网格,每个网格可以放置一个组件。网格布局 GridLayout 类的继承结构如图 3-15 所示。

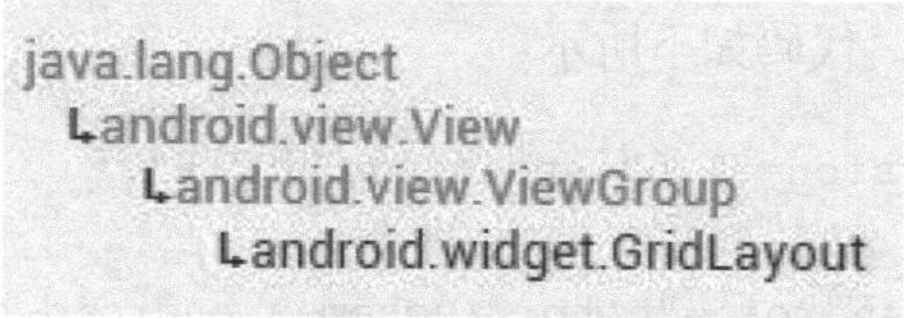

```
java.lang.Object
  └android.view.View
      └android.view.ViewGroup
          └android.widget.GridLayout
```

图 3-15 GridLayout 类的继承结构

网格布局实质上有点类似 LinearLayout 布局的 API。它与 LinearLayout 布局一样,也分为水平和垂直两种方式,默认是水平布局,一个控件紧靠一个控件从左到右依次排列,通过指定布局属性 android:columnCount 设置列数后,控件会自动换行进行排列,同时对于 GridLayout 布局中的子控件,默认按照 wrap_content 的方式设置其显示。

GridLayout 布局的常用属性及相关方法如表 3-9 所示。

表 3-9 GridLayout 布局的常用属性及相关方法

属性名称	相关方法	描述
android:alignmentMode	setAlignmentMode(int)	设置布局外部边界的对齐方式
android:columnCount	setColumnCount(int)	设置最大列数量
android:columnOrderPreserved	setColumnOrderPreserved(boolean)	设置是否保留列序号，为 true 时，必须保留与列索引一样的升序
android:rowCount	setRowCount(int)	设置最大行数量
android:rowOrderPreserved	setRowOrderPreserved(boolean)	设置是否保留行序号，为 true 时，必须保留与行索引一样的升序

注明：android:layout_row 和 android:layout_column 属性，指定某控件显示在固定的行或列，android:layout_row="0"表示从第一行开始，android:layout_column="0"表示从第一列开始，android:layout_rowSpan 或者 layout_columnSpan 属性设置控件跨越多行或多列，同时设置其 layout_gravity 属性为 fill（表明该控件填满所跨越的整行或整列）。

在 GridLayout 中，设置控件大小、外边距及对齐、位置属性（Alignment/gravity），与 LinearLayout 一样，如 gravity 在 LinearLayout 中取值一样，也选取 left、top、right、bottom、center_horizontal、center_vertical、center、fill_horizontal、fill_vertical 和 fill 这些常量值。

2．GridLayout 布局应用

下面通过一个 Email 邮件登录界面，来讲述 GridLayout 布局的应用，具体步骤如下：

（1）在 Eclipse 中选择 File→New→Android Project 命令，创建一个新的 Android 工程，项目名为 GridLayoutDemo，目标 API 选择 17（即 Android 4.2 版本），应用程序名为 GridLayoutDemo，包名为 com.bcpl.activity，创建的 Activity 的名字为 GridLayoutActivity。

（2）打开项目工程中 res→layout 目录下的 activity_main.xml 文件，设置绝对布局，添加三个 TextView 控件，代码如下所示。

```
1.      <?xml version="1.0" encoding="utf-8"?>
2.      <GridLayout
3.          xmlns:android="http://schemas.android.com/apk/res/android"
4.          android:layout_width="match_parent"
5.          android:layout_height="match_parent"
6.          android:useDefaultMargins="true"
7.          android:alignmentMode="alignBounds"
8.          android:columnOrderPreserved="false"
9.          android:columnCount="4"
```

```
10.    >
11.    <TextView
12.            android:text="Email Configure"
13.            android:textSize="32dip"
14.            android:layout_columnSpan="4"
15.            android:layout_gravity="center_horizontal"
16.            />
17.    <TextView
18.            android:text="you can configure email in next steps: "
19.            android:textSize="16dip"
20.            android:layout_columnSpan="4"
21.            android:layout_gravity="left"
22.            />
23.    <TextView
24.            android:text="Email address:"
25.            android:layout_gravity="right"
26.            />
27.    <EditText
28.            android:ems="10"
29.            />
30.    <TextView
31.            android:text="Password:"
32.            android:layout_column="0"
33.            android:layout_gravity="right"
34.            />
35.    <EditText
36.            android:ems="8"
37.            />
38.    <Space
39.            android:layout_row="4"
40.            android:layout_column="0"
41.            android:layout_columnSpan="3"
42.            android:layout_gravity="fill"
43.            />
44.    <Button
45.            android:text="Next"
46.            android:layout_row="5"
47.            android:layout_column="3"
48.            />
49. </GridLayout>
```

注意：GridLayout 中每个子元素控件的布局参数属性中有行与列的索引，定义了控件存放位置，但当其中一个参数或这两个参数都没有指定的时候，GridLayout 会提供默认的值。此外，新加了 Space 标签，用于跨行、列填充空白。

在 GridLayout 当子元素控件被加入一个 GridLayout 的时候，GridLayout 会维护这一个位置指针，以及一个所谓的"高水位标志"，用来将控件摆放到那些还闲置着的单元格里去，进行自动索引的分配。

（3）部署运行 GridLayoutDemo 项目工程，项目运行效果如图 3-16 所示。

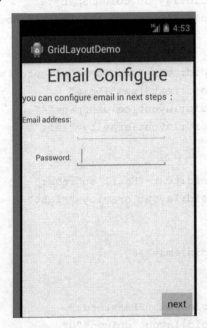

图 3-16　Email 配置运行界面

此外，关于 GridLayout 布局编写的简易计算器工程中的 XML 文件中布局代码如下。

```
1.    <?xml version="1.0" encoding="utf-8"?>
2.    <GridLayout xmlns:android="http://schemas.android.com/apk/res/android"
3.        android:layout_width="wrap_content"
4.        android:layout_height="wrap_content"
5.        android:orientation="horizontal"
6.        android:rowCount="5"
7.        android:columnCount="4" >
8.        <Button
9.         android:id="@+id/one"
10.        android:text="1"/>
11.       <Button
12.        android:id="@+id/two"
13.        android:text="2"/>
14.        <Button
15.        android:id="@+id/three"
16.        android:text="3"/>
17.       <Button
```

```
18.        android:id="@+id/devide"
19.        android:text="/"/>
20.    <Button
21.        android:id="@+id/four"
22.        android:text="4"/>
23.    <Button
24.        android:id="@+id/five"
25.        android:text="5"/>
26.    <Button
27.        android:id="@+id/six"
28.        android:text="6"/>
29.    <Button
30.        android:id="@+id/multiply"
31.        android:text="×"/>
32.    <Button
33.        android:id="@+id/seven"
34.        android:text="7"/>
35.    <Button
36.        android:id="@+id/eight"
37.        android:text="8"/>
38.    <Button
39.        android:id="@+id/nine"
40.        android:text="9"/>
41.    <Button
42.        android:id="@+id/minus"
43.        android:text="-"/>
44.    <Button
45.        android:id="@+id/zero"
46.            android:layout_columnSpan="2"
47.            android:layout_gravity="fill"
48.        android:text="0"/>
49.    <Button
50.        android:id="@+id/point"
51.        android:text="."/>
52.    <Button
53.        android:id="@+id/plus"
54.            android:layout_rowSpan="2"
55.            android:layout_gravity="fill"
56.        android:text="+"/>
57.    <Button
58.        android:id="@+id/equal"
59.            android:layout_columnSpan="3"
60.            android:layout_gravity="fill"
61.        android:text="="/>
```

62.   </GridLayout>

运行结果如图 3-17 所示。

图 3-17 简易计算器结果

3.5 项目案例

学习目标：学习在 Fragment 中综合应用各种布局，Android UI 布局的不同使用方法及应用，尤其是线性布局 LinearLayout、表格布局 TableLayout 的使用，以及它们的混合使用方法、属性的设置，修改 AndroidManifest.XML 文件、activity_main.xml 文件的方法。

案例描述：Fragment 综合应用各种布局，使用 Android UI 布局中的线性布局 LinearLayout，并在其中使用若干 TableRow，在每一个 TableRow 中添加 TextView、EditText 控件，然后设置每一个布局及控件的属性，实现用户注册界面。

案例要点：Fragment 中添加各种布局文件、TableRow 及属性设置。

案例步骤：

（1）创建新的项目工程 Project_Chapter_3，选择目标平台 Android4.2，在 src 目录下创建包 com.bcpl.activity 和 MainActivity。

（2）在 res 目录下 layout 文件夹中修改 activity_main.xml 文件，修改其布局为线性布局，并进行线性布局嵌套，然后在第二个线性布局中添加 4 个 Button 按钮，代码如下。

```
1.    <LinearLayout    xmlns:android="http://schemas.android.com/apk/res/
android"
2.        xmlns:tools="http://schemas.android.com/tools"
3.        android:layout_width="match_parent"
4.        android:layout_height="match_parent"
5.        android:orientation="vertical"
```

```xml
6.     tools:context=".MainActivity" >
7.
8.     <LinearLayout
9.         android:id="@+id/layoutCenter"
10.        android:layout_width="match_parent"
11.        android:layout_height="0dip"
12.        android:layout_weight="1"
13.        android:orientation="vertical" >
14.    </LinearLayout>
15.
16.    <LinearLayout
17.        android:layout_width="match_parent"
18.        android:layout_height="wrap_content" >
19.
20.        <Button
21.            android:id="@+id/button1"
22.            android:layout_width="wrap_content"
23.            android:layout_height="wrap_content"
24.            android:onClick="btnClickA"
25.            android:text="MS" />
26.
27.        <Button
28.            android:id="@+id/button2"
29.            android:layout_width="wrap_content"
30.            android:layout_height="wrap_content"
31.            android:onClick="btnClickB"
32.            android:text="ShowMS" />
33.
34.        <Button
35.            android:id="@+id/button3"
36.            android:layout_width="wrap_content"
37.            android:layout_height="wrap_content"
38.            android:onClick="btnClickC"
39.            android:text="C" />
40.        <Button
41.            android:id="@+id/button4"
42.            android:layout_width="wrap_content"
43.            android:layout_height="wrap_content"
44.            android:onClick="btnClickD"
45.            android:text="Login" />
46.    </LinearLayout>
47. </LinearLayout>>
```

（3）创建 register.xml 和 a.xml、b.xml、c.xml 文件，这里重点介绍 register.xml 文件，

其他的都与其类似，register.xml 文件代码如下。

```xml
1.  <?xml version="1.0" encoding="utf-8"?>
2.  <LinearLayoutxmlns:android="http://schemas.android.com/apk/res/android"
3.      android:orientation="vertical" android:layout_width="fill_parent"
4.      android:layout_height="fill_parent">
5.  <!--使用线性布局垂直排列,加入表格4行2列 -->
6.      <TableRow android:id="@+id/TableRow01"
7.          android:layout_width="fill_parent"
8.          android:layout_height="wrap_content">
9.          <TextView android:text="@string/name"
10.             android:layout_width="70px"
11.             android:layout_height="40px"
12.             >
13.         </TextView>
14.         <EditText android:layout_width="fill_parent"
15.             android:layout_height="40px"
16.             android:id="@+id/ename"
17.             android:singleLine="true"
18.             android:inputType="textPersonName"
19.             android:background="@android:drawable/editbox_background"
20.             >
21.         </EditText>
22.     </TableRow>
23.     <TableRow android:id="@+id/TableRow02"
24.         android:layout_width="fill_parent"
25.         android:layout_height="wrap_content">
26.         <TextView android:text="@string/phone_number"
27.             android:layout_width="70px"
28.             android:layout_height="40px"
29.             >
30.         </TextView>
31.         <EditText android:layout_width="fill_parent"
32.             android:layout_height="40px"
33.             android:id="@+id/etel"
34.             android:singleLine="true"
35.             android:inputType="phone"
36.             android:background="@android:drawable/editbox_background"
37.
38.             >
39.         </EditText>
40.     </TableRow>
```

```
41.    <TableRow android:id="@+id/TableRow03"
42.        android:layout_width="fill_parent"
43.        android:layout_height="wrap_content">
44.        <TextView android:text="@string/email"
45.            android:layout_width="70px"
46.            android:layout_height="40px"
47.            >
48.        </TextView>
49.        <EditText android:layout_width="fill_parent"
50.            android:layout_height="40px"
51.            android:id="@+id/email"
52.            android:singleLine="true"
53.            android:inputType="textEmailAddress"
54.            android:background="@android:drawable/editbox_back
                ground"
55.            >
56.        </EditText>
57.    </TableRow>
58.    <TableRow android:id="@+id/TableRow04"
59.        android:layout_width="fill_parent"
60.        android:layout_height="wrap_content">
61.        <TextView android:text="@string/addr"
62.            android:layout_width="70px"
63.            android:layout_height="40px"
64.            >
65.        </TextView>
66.        <EditText android:layout_width="fill_parent"
67.            android:layout_height="40px"
68.            android:id="@+id/eaddress"
69.            android:singleLine="true"
70.            android:inputType="textPostalAddress"
71.            android:background="@android:drawable/editbox_back
                ground">
72.        </EditText>
73.    </TableRow>
74. </LinearLayout>
```

（4）在 src 目录 com.bcpl.activity 包下，修改 MainActivity.java 文件，调用 register.xml，代码如下。

```
1. public class MainActivity extends FragmentActivity {
2.
3.     @Override
4.     protected void onCreate(Bundle savedInstanceState) {
```

```
5.         super.onCreate(savedInstanceState);
6.         setContentView(R.layout.register);
7.
8.     }
9.     public void changeView(int param){
10.
11.    }
12.    @Override
13.    public boolean onCreateOptionsMenu(Menu menu) {
14.        // Inflate the menu; this adds items to the action bar if it
           is present.
15.        getMenuInflater().inflate(R.menu.main, menu);
16.        return true;
17.    }
18.
19. }
```

（5）部署工程，程序运行结果如图 3-18 所示。

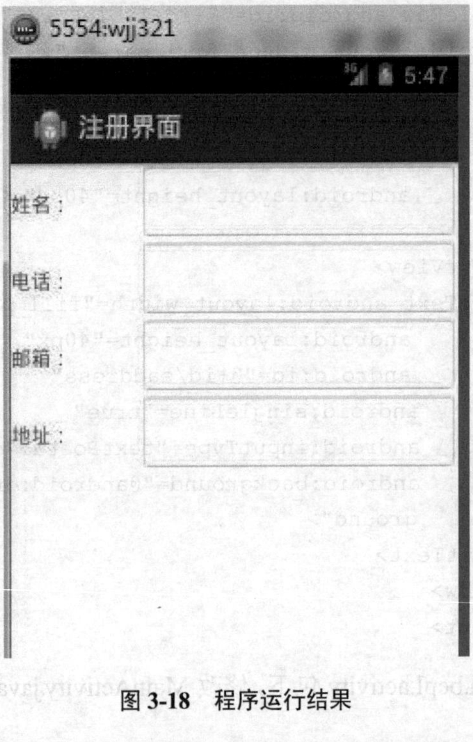

图 3-18 程序运行结果

注意：上述只是实现了注册登录界面和主 activity_main.xml 文件，Fragment 调用不同的界面功能并没有在此写出，需要作为拓展由读者自己完成（但程序代码中已经给出详细实现）。具体实现如图 3-19 所示(点单击 MS 按钮)；单击 Save 按钮后，再单击 ShouMS 按钮，显示如图 3-20 所示；单击 Login 按钮，如图 3-21 所示。

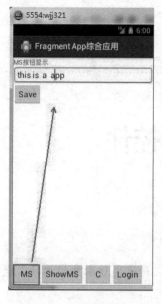  

图 3-19　单击 MS 按钮　　图 3-20　单击 ShowMS 按钮　　图 3-21　单击 Login 按钮

习　题

一、简答题

1. Android 移动终端界面的设计原则包含哪些？其设计与 iOS 移动终端界面设计有何差异？

2. 简述 Android 界面框架在 4.0 版本之后添加了什么特性，在 Android Studio 中与 SDK 有何不同？

3. GridLayout 布局是在哪一个版本之后才添加的？为什么增加它呢？

4. 界面布局嵌套已经成为常态，界面布局与 UI 设计之间有什么关系？有人常说移动 UI 工程师必须有美术背景，你是如何看待这一现象和技术之间的关联呢？

二、实训

要求：

（1）完成上述 Fragment 的综合布局应用功能。

（2）使用不同尺寸大小的移动设备，完成注册界面设计，并要求在不同屏幕尺寸大小下（如 800、432、400 等），不发生变形且能正常显示。

第 4 章

Android 界面基础控件

学习目标

本章主要介绍了 Android 界面系统控件中的基础控件。读者通过本章的学习,能够深入了解 Android UI 基础控件,学习以下知识要点。

(1) 文本控件 TextView、EditText 的属性及方法、应用。

(2) 按钮控件 Button、图片按钮控件 ImageButton 通常用法及属性。

(3) 单选按钮 RadioButton、复选按钮 CheckBox 常用方法、引用处理。

(4) 时间选择器 TimePicker、日期选择器 DatePicker 使用方法及属性设置。

(5) 图片按钮 ImageView 常用方法、引用处理及设置。

(6) 模拟时钟 AnalogClock 和数字时钟 DiditalClock 的应用。

Android 系统提供了许多控件给开发者使用,开发者通过对这些控件编码与控件组合能够实现系统设想的模型和相应的功能。Android 系统的界面控件分为定制控件和系统控件,系统控件是 Android 系统提供给用户已经封装的界面控件,是在应用程序开发过程中常见功能控件。系统控件更有利于帮助用户进行快速开发,同时能够帮助开发者在使用 Android 系统进行开发过程中,保持应用程序的界面一致性。

在开发应用中,经常使用的系统控件有 TextView、EditText、Button、ImageButton、Checkbox、RadioButton、Spinner、ListView 和 TabHost 等。

Android UI 系统控件的使用,除了在传统的程序代码中直接声明、创建之外,其最能体现设计思想的是,在 XML 文件中来描述控件。在 XML 中可以描述控件的宽度、长度、控件上的文本、控件的背景、控件的填充、设置源等。

4.1 文本控件简介

在 Android 系统中,文本控件包含 TextView 和 EditText 控件,它们都继承 android.view.View,在 android.widget 包中。本节就文本控件的属性及使用方法进行详述。

4.1.1 文本框 TextView 及应用

1. 文本框 TextView 创建方式

android.widget 包中的 TextView 是文本表示控件，一般用来展示文本，是一种用于显示字符串的控件。其主要功能是向用户展示文本的内容，可以作为应用程序的标签或者邮件正文的显示，默认情况下不允许用户直接编辑。

在程序设计和开发中，使用 TextView 可以采用的方式有以下两种。

（1）在程序中以创建控件的对象方式来使用 TextView 控件。

如 TextView 控件，可以通过编写如下代码完成控件使用。

```
TextView tv=new TextView(this);
tv.setText("大家好");
setContentView(tv);
```

（2）使用 XML 描述控件，并在程序中引用和使用。

① 在 res/layout 文件下的 XML 文件中描述控件。

```
<TextView
Android:id="@+id/text_view"
Android:layout_width="fill_parent"
Android:layout_height="wrap_content"
Android:textSize="16sp"
Android:padding="10dip"
Android:background="#00f0d0"
Android:text="大家好，这里是 TextView"/>
```

② 在程序中引用 XML 描述的 TextView。

```
TextView text_view = (TextView) findViewById(R.id.text_view);
```

上述两种方式的使用，各有优缺点，根据不同的需要，采用相应的方法。相比而言，采用第二种方法更好，其主要优势：方便代码的维护，编码灵活，利于分工协作。

TextView 控件常用的方法有：getText()、setText()。

TextView 控件有着与之相应的属性，通过选择不同的属性，给予其值，能够实现不同的效果。TextView 控件属性的设置既可以在 XML 文件中通过属性名称进行设定赋值，也可以采用对应的方法，在程序代码中设定。其常用属性及对应方法如表 4-1 所示

表 4-1 TextView 控件常用 XML 属性及对应方法

属 性 名 称	对 应 方 法	说　明
android:text	setText(CharSequence)	设置 TextView 控件文字显示
android:autoLink	setAutoLinkMask(int)	设置是否当文本为 URL 链接/email/电话号码/map 时，文本显示为可单击的链接。可选值(none/web/email/phone/map/all)

续表

属性名称	对应方法	说明
android:hint	setHint(int)	当 TextView 中显示的内容为空时，显示该文本
android:textColor	setTextColor(ColorStateList)	设置字体颜色
android:textSize	setTextSize(float)	设置字体大小
android:typeface	setTypeface(Typeface)	设置文本字体，必须是以下常量值之一：normal 0、sans 1、serif 2、monospace(等宽字体) 3
android:ellipsize	setEllipsize(TextUtils.TruncateAt)	如果设置了该属性，当 TextView 中要显示的内容超过了 TextView 的长度时，会对内容进行省略。可取的值有 start、middle、end 和 marquee
android:gravity	setGravity(int)	定义 TextView 在 x 轴和 y 轴方向上的显示方式
android:height	setHeight(int)	设置文本区域的高度，支持度量单位：px/dp/sp/in/mm
android:minHeight	setMinHeight(int)	设置文本区域的最小高度
android:maxHeight	setMaxHeight(int)	设置文本区域的最大高度
android:width	setWidth(int)	设置文本区域的宽度，支持度量单位：px/dp/sp/in/mm
android:minWidth	setMinWidth(int)	设置文本区域的最小宽度
android:maxWidth	setMaxWidth(int)	设置文本区域的最大宽度

2. TextView 应用

本节在上述 TextView 控件讲解的基础之上，通过案例熟悉 TextView 控件的属性和用法，具体步骤如下。

（1）创建一个新的 Android 工程，工程名为 TextViewDemo，目标 API 选择 17（即 Android 4.2 版本），应用程序名为 TextViewDemo，包名为 com.bcpl.cn，创建的 Activity 的名字为 MainActivity。

（2）打开项目工程中 res→layout 目录下的 activity_main.xml 文件，设置线性布局，添加 4 个 TextView 控件，并设置属性，代码如下所示。

```
1.   <?xml version="1.0" encoding="utf-8"?>
2.   <?xml version="1.0" encoding="utf-8"?>
3.   <LinearLayout
xmlns:android="http://schemas.android.com/apk/res/android"
4.      android:orientation="vertical"
5.      android:layout_width="fill_parent"
6.      android:layout_height="fill_parent"
7.      >
8.      <TextView
9.      android:layout_width="fill_parent"
10.     android:layout_height="wrap_content"
11.     android:text="字体大小为14的文本"
```

```
12.     android:textSize="14pt"
13.     />

14.     <TextView
15.     android:layout_width="fill_parent"
16.     android:layout_height="wrap_content"
17.     android:singleLine="true"
18.     android:text="TextView 示例"
19.     android:ellipsize="middle"
20.     />

21.     <TextView
22.     android:layout_width="fill_parent"
23.     android:layout_height="wrap_content"
24.     android:singleLine="true"
25.     android:text="访问: http://www.zfjsjx.cn"
26.     android:autoLink="web"
27.     />

28.     <TextView
29.     android:layout_width="fill_parent"
30.     android:layout_height="wrap_content"
31.     android:text="红色并带阴影的文本"
32.     android:shadowColor="#0000ff"
33.     android:shadowDx="15.0"
34.     android:shadowDy="20.0"
35.     android:shadowRadius="45.0"
36.     android:textColor="#ff0000"
37.     android:textSize="20pt"
38.     />

39. </LinearLayout>
```

（3）src 目录下 com.bcpl.cn 包下的 TextViewActivity.java 文件和 res→values 目录下的 strings.xml 文件，都暂不做修改。部署运行项目工程，项目运行效果如图 4-1 所示。

图 4-1 TextViewDemo 运行结果

4.1.2 编辑框 EditText 及应用

1．编辑框 EditText 简介

EditText 控件继承自 android.widget.TextView，在 android.widget 包中。EditText 为输入框，是编辑文本控件，主要功能是让用户输入文本的内容，它是可编辑的，用来输入和编辑字符串。

利用控件 EditText 不仅可以实现输入信息，还可以根据需要对输入信息进行限制约束。例如，限制控件 EditText 输入信息：

```
<EditText
 android:layout_width="fill_parent"
 android:layout_height="wrap_content"
 android:inputType="numeber"/>
```

与 TextView 一样，EditText 控件的使用方法也有两种，一种是以在程序中创建控件的对象方式来使用 EditText 控件；另外一种是在 res/layout 文件下的 XML 文件中描述控件，程序中使用 EditText 控件。

例如：

（1）用 XML 描述一个 EditText。

```
<EditText Android:id="@+id/edit_text"
Android:layout_width="fill_parent"
Android:layout_height="wrap_content"
Android:text="这里可以输入文字" />
```

（2）在程序中引用 XML 描述的 EditText。

EditText editText = (EditText) findViewById(R.id.editText);

EditText 常用方法：getText()。它也有着与之相应的属性，通过选择不同的属性，给予其值，能够实现其不同的效果。其常用属性及对应方法如表 4-2 所示。

表 4-2 EditText 控件常用 XML 属性及对应方法

属性名称	对应方法	说 明
android:hint		输入框的提示文字
android:password	setTransformationMethod (TransformationMethod)	设置文本框中的内容是否显示为密码，当为 true 时，以小点"."显示文本
android:phoneNumber	setKeyListener(KeyListner)	设置文本框中的内容只能是电话号码，当为 true 时，表示电话框
android:digits	setKeyListener(KeyListener)，可使用此方法监听键盘来实现	设置允许输入哪些字符。如"1234567890.+-*/%\n()"
android:numeric	setKeyListener(KeyListener)，可使用此方法监听键盘来实现	设置只能输入数字，并且置顶可输入的数字格式，可选值有 integer| signed| decimal。integer 正整数，signed 整数(可带负号)，decimal 浮点数

续表

属性名称	对应方法	说明
android:singleLine	setTransformationMethod(TransformationMethod)	设置文本框的单行模式
android:maxLength	setFilters(InputFilter)	设置最大显示长度
android:cursorVisible	setCursorVisible(boolean)	设置光标是否可见，默认可见
android:lines	setLines(int)	通过设置固定的行数来决定 EditText 的高度
android:maxLines	setMaxLines(int)	设置最大的行数
android:minLines	setMinLines(int)	设置最小的行数
android:scrollHorizontally	setHorizontallyScrolling(boolean)	设置文本框是否可以进行水平滚动
android:selectAllOnFocus	setSelectAllOnFocus(boolean)	如果文本内容可选中，当文本框获得焦点时自动选中全部文本内容
android:shadowColor	setShadowLayer(float,float,float,int)	为文本框设置指定颜色的阴影，需要与 shadowRadius 一起使用
android:shadowDx	setShadowLayer(float,float,float,int)	设置阴影横向坐标开始位置，为浮点数
android:shadowDy	setShadowLayer(float,float,float,int)	设置阴影纵向坐标开始位置，为浮点数
android:shadowRadius	setShadowLayer(float,float,float,int)	为文本框设置阴影的半径，为浮点数

2．EditText 应用

在上述 EditText 控件讲解的基础之上，通过案例熟悉 EditText 控件的属性和用法，具体步骤如下。

（1）创建一个新的 Android 工程，工程名为 EditTextDemo，目标 API 选择 17（即 Android 4.2 版本），应用程序名为 EditTextDemo，包名为 com.bcpl.cn，创建的 Activity 的名字为 MainActivity。

（2）打开项目工程中 res→layout 目录下的 activity_main.xml 文件，设置线性布局，添加一个 TextView 控件和一个 EditText 控件，并设置相关属性，代码如下所示。

```
1.   <?xml version="1.0" encoding="utf-8"?>
2.   <LinearLayout
xmlns:android="http://schemas.android.com/apk/res/android"
3.       android:orientation="vertical"
4.       android:layout_width="fill_parent"
5.       android:layout_height="fill_parent"
6.       >
7.       <TextView android:text="请输入："
8.           android:id="@+id/textView1"
9.           android:layout_width="wrap_content"
```

```
10.            android:layout_height="wrap_content">
11.        </TextView>
12.        <EditText android:layout_height="wrap_content"
13.            android:layout_width="match_parent"
14.            android:id="@+id/editText1"
15.            android:hint="这里键入输入内容">
16.            <requestFocus></requestFocus>
17.        </EditText>
```

(3) src 目录下 com.bcpl.cn 包下的 MainActivity.java 文件和 res→values 目录下的 strings.xml 文件，都暂不做修改。部署运行项目工程，项目运行效果如图 4-2 所示。

图 4-2　EditTextDemo 工程运行结果

4.2　按钮控件简介

4.2.1　按钮 Button 及应用

1．按钮 Button 创建方式

Button 是一种常用的按钮控件，继承自 android.widget.TextView，在 android.widget 包中。用户能够在该控件上单击，然后引发相应的事件处理函数。它的常用子类有 CheckBox, RadioButton, ToggleButton 等，在后续章节中会讲到。

Button 控件的通常用法是：在程序中通过 super.findViewById(id)得到在 Layout 中 XML 文件中声明的 Button 的引用，然后使用 setOnClickListener(View.OnClickListener) 添加监听，再在 View.OnClickListener 监听器中使用 v.equals(View)方法判断是哪一个按钮被单击了，调用不同方法进行分别处理。例如：

1）用 XML 描述一个 Button

```
<Button Android:id="@+id/button"
```

```
Android:layout_width="wrap_content"
Android:layout_height="wrap_content"
Android:text="这是一个button" />
```

2）在程序代码中引用用 XML 描述的 Button

```
Button button = (Button) findViewById(R.id.button);
```

3）给 Button 设置事件响应

```
button.setOnClickListener(button_listener);
```

4）生成一个按钮事件监听器

```
private Button.OnClickListener button_listener = new
Button.OnClickListener() {
public void onClick(View v) {
switch(v.getId()){
        case R.id.Button:
            textView.setText("Button 按钮 1");
            return;
        case R.id.Button01:
            textView.setText("Button 按钮 2");
            return;
        }

    }
};
```

此外，也可以采用在 Layout 中 XML 文件中声明分配一个方法给 Button 按钮，使用 android:onClick 属性，例如：

```
1.      <Button
2.        android:layout_height="wrap_content"
3.        android:layout_width="wrap_content"
4.        android:text="@string/self_destruct"
5.        android:onClick="selfDestruct" />
```

当用户单击 Button 按钮时，Android 系统会自动调用 Activity 中的 selfDestruct(View) 方法，但 selfDestruct(View) 方法必须声明为 public，并只能接受 View 作为其唯一的参数。传递给这个方法的 View 是被单击的控件的一个引用，如下：

```
1.     public void selfDestruct(View view) {
2.       // Kabloey
3.     }
```

2. Button 应用

在开发应用中,Button 按钮的使用较为常见,下面通过一个单击 Button 按钮修改标题的案例来介绍 Button 的应用。具体步骤如下。

(1)创建一个新的 Android 工程,工程名为 ButtonDemo,目标 API 选择 17(即 Android4.2 版本),应用程序名为 ButtonDemo,包名为 com.hisoft.activity,创建的 Activity 的名字为 ButtonActivity。

(2)打开项目工程中 res→layout 目录下的 activity_button.xml 文件,设置线性布局,添加一个 Button 按钮控件,并设置相关属性,代码如下所示。

```xml
1.  <?xml version="1.0" encoding="utf-8"?>
2.  <LinearLayout xmlns:android="http://schemas.android.com/apk/res/android"
3.      android:orientation="vertical"
4.      android:layout_width="fill_parent"
5.      android:layout_height="fill_parent"
6.      >
7.  <Button android:text="按钮 1"
8.      android:id="@+id/button1"
9.      android:layout_width="wrap_content"
10.     android:layout_height="wrap_content">
11. </Button>

12. </LinearLayout>
```

(3)打开 src 目录下 com.bcpl.activity 包下的 ButtonActivity.java 文件,声明 Button 按钮,并获取引用,然后添加监听器,代码如下。

```java
1.  package com.bcpl.activity;

2.  import android.app.Activity;
3.  import android.os.Bundle;
4.  import android.view.View;
5.  import android.view.View.OnClickListener;
6.  import android.widget.Button;
7.  public class ButtonActivity extends Activity
8.  {
9.      private Button button1;
10.
11.     @Override
12.     public void onCreate(Bundle savedInstanceState)
13.     {
14.         super.onCreate(savedInstanceState);
```

```
15.        setContentView(R.layout.main);
16.
17.        button1 = (Button) this.findViewById(R.id.button1);
18.
19.        //给 button1 设置监听
20.        button1.setOnClickListener(new OnClickListener() {
21.
22.            public void onClick(View v) {
23.
24.                setTitle("按钮被点击了!!!");
25.            }
26.        });
27.    }
28. }
```

（4）部署运行 ButtonDemo 项目工程，项目运行效果如图 4-3 所示。单击"按钮 1"按钮后，效果如图 4-4 所示。

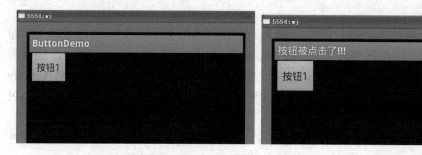

图 4-3 ButtonDemo 程序运行结果　　　　　图 4-4 单击"按钮 1"按钮运行结果

4.2.2 图片按钮 ImageButton 及应用

1. ImageButton 简介

ImageButton 继承自 ImageView 类，是用以实现能够显示图像功能的控件按钮，既可以显示图片又可以作为 Button 使用。

ImageButton 与 Button 之间的区别：ImageButton 中没有 text 属性。

ImageButton 控件中设置按钮中显示的图片可以通过 android：src 属性来设置。也可以通过 setImageResource(int)来设置。默认情况下，ImageButton 与 Button 具有一样的背景色，当按钮处于不同的状态时，背景色会发生变化，一般将 ImageButton 控件背景色设置为图片或者透明，以避免控件显示的图片不能完全覆盖背景色时，影响显示效果。

下面通过例子说明使用 XML 描述 ImageButton 控件，并在程序中引用和使用的简要过程。

（1）在 res/layout 文件下的 XML 文件中描述 ImageButton 控件。

```
    <ImageButton android:id="@+id/ImageButton01"
        android:layout_width="wrap_content"
        android:layout_height="wrap_content">
    </ImageButton>
```

（2）在程序中引用 XML 描述的 ImageButton。

```
    ImageButton imageButton = (ImageButton)findViewById(R. id.
    ImageButton01);
```

（3）利用 setImageResource()函数，将新加入的 png 文件 R.drawable.download 传递给 ImageButton。

```
imageButton.setImageResource(R.drawable.download);
```

2．ImageButton 应用

下面通过单击一个 Button 按钮显示 ImageButton 的案例来介绍 ImageButton 的应用。具体步骤如下。

（1）创建一个新的 Android 工程，工程名为 ImageButtonDemo，目标 API 选择 17（即 Android 4.2 版本），应用程序名为 ImageButtonDemo，包名为 com.hisoft.activity，创建的 Activity 的名字为 ImageButtonActivity，布局文件 activity_image_button.xml 命名与 Activity 的名称 ImageButtonActivity 中 ImageButton 相映射，最小 SDK 版本根据选择的目标 API 会自动添加为 8。

（2）打开项目工程中 res→layout 目录下的 activity_image_button.xml 文件，设置线性布局，添加一个 Button 按钮控件，并设置相关属性，代码如下所示。

```
1.    <?xml version="1.0" encoding="utf-8"?>
2.    <LinearLayout xmlns:android="http://schemas.android.com/apk/res/android"
3.        android:orientation="vertical"
4.        android:layout_width="fill_parent"
5.        android:layout_height="fill_parent"
6.        >
7.        <Button android:text="普通按钮"
8.            android:id="@+id/button1"
9.            android:layout_width="wrap_content"
10.           android:layout_height="wrap_content">
11.       </Button>
12.   </LinearLayout>
```

（3）在项目工程 res→layout 目录下创建 imagebutton.xml 文件，设置线性布局，添加一个 TextView 控件和一个 ImageButton 按钮，代码如下。

```
1.  <?xml version="1.0" encoding="utf-8"?>
2.  <LinearLayout xmlns:android="http://schemas.android.com/apk/res/android"
3.      android:orientation="vertical" android:layout_width="fill_parent"
4.      android:layout_height="wrap_content">

5.  <TextView
6.      android:layout_width="wrap_content"
7.      android:layout_height="wrap_content"
8.      android:text="图片按钮:" />
9.  <ImageButton android:src="@drawable/icon"
10.             android:layout_height="wrap_content"
11.             android:layout_width="wrap_content"
12.             android:id="@+id/imageButton1">
13. </ImageButton>
14. </LinearLayout>
```

（4）在 src 目录下 com.bcpl.activity 包下，打开 ImageButtonActivity.java 文件，设置界面显示 imagebutton.xml 文件内容。

```
1.  package com.bcpl.activity;

2.  import android.app.Activity;
3.  import android.os.Bundle;
4.  import android.view.View;
5.  import android.view.View.OnClickListener;
6.  import android.widget.Button;


7.  public class ImageButtonActivity extends Activity
8.  {
9.      private Button button1;

10.     @Override
11.     public void onCreate(Bundle savedInstanceState)
12.     {
13.         super.onCreate(savedInstanceState);
14.         setContentView(R.layout.imagebutton);

15.     }
16. }
```

（5）在 src 目录下 com.bcpl.activity 包下，新建 MainActivity.java 文件，代码如下。

```
1.   package com.bcpl.activity;

2.   import android.app.Activity;
3.   import android.content.Intent;
4.   import android.os.Bundle;
5.   import android.view.View;
6.   import android.view.View.OnClickListener;
7.   import android.widget.Button;

8.   public class MainActivity extends Activity {

9.       private Button button1;

10.      @Override
11.      protected void onCreate(Bundle savedInstanceState) {
12.          super.onCreate(savedInstanceState);
13.          setContentView(R.layout.main);

14.          button1 = (Button) this.findViewById(R.id.button1);
15.          //给 button1 设置监听
16.          button1.setOnClickListener(new OnClickListener() {

17.              public void onClick(View v) {
18.                  //通过 Intent 跳转到 ImageButtonActivity
19.                  Intent intent = new Intent();
20.                  intent.setClass(MainActivity.this,
ImageButtonActivity.class);
21.                  startActivity(intent);
22.              }
23.          });
24.      }

25.  }
```

（6）部署运行 ImageButtonDemo 项目工程，项目运行效果如图 4-5 所示。单击"普通按钮"按钮后，效果如图 4-6 所示。

图 4-5　ImageButtonDemo 运行结果

图 4-6　图片按钮运行效果

4.3　单选与复选按钮简介

4.3.1　单选按钮 RadioButton

单选按钮 RadioButton 是仅可以选择一个选项的控件，继承自 android.widget.CompoundButton，在 android.widget 包中。

单选按钮要声明在 RadioGroup 中，RadioGroup 是 RadioButton 的承载体，程序运行时不可见，应用程序中可能包含一个或多个 RadioGroup，RadioGroup 是线性布局 LinearLayout 的子类。其类的继承结构如图 4-7 所示，一个 RadioGroup 包含多个 RadioButton，RadioGroup 用于对单选框进行分组，在每个 RadioGroup 中（相同组内的单选框），用户仅能够选择其中一个 RadioButton。

```
java.lang.Object
  ↳android.view.View
    ↳android.view.ViewGroup
      ↳android.widget.LinearLayout
        ↳android.widget.RadioGroup
```

图 4-7　RadioGroup 的类继承图

单选按钮状态更改的监听，是要给它的 RadioGroup 添加 setOnCheckedChangeListener(RadioGroup.OnCheckedChangeListener)监听器。注意监听器类型和复选按钮 CheckBox 是不相同的。

单选按钮的通常用法如下。

1．用 XML 描述的 RadioGroup 和 RadioButton 应用的界面设计

```
<?xml version="1.0" encoding="utf-8"?>
<LinearLayout
```

```xml
    xmlns:android="http://schemas.android.com/apk/res/android"
        android:orientation="vertical"
        android:layout_width="fill_parent"
        android:layout_height="fill_parent"
        >
<RadioGroup android:id="@+id/radioGroup"
   xmlns:android="http://schemas.android.com/apk/res/android"
   android:layout_width="wrap_content"
   android:layout_height="wrap_content">
<RadioButton android:id="@+id/java"
        android:layout_width="wrap_content"
        android:layout_height="wrap_content"
        android:text="java" />
    <RadioButton android:id="@+id/dotNet"
        android:layout_width="wrap_content"
        android:layout_height="wrap_content"
        android:text="dotNet" />
    <RadioButton android:id="@+id/php"
        android:layout_width="wrap_content"
        android:layout_height="wrap_content"
        android:text="PHP" />
</RadioGroup>
</LinearLayout>
```

2. 引用处理程序

```
public void onCreate(Bundle savedInstanceState) {
    ...
    RadioGroup radioGroup = (RadioGroup) findViewById (R.id.radio
    Group);
    radioGroup.setOnCheckedChangeListener(new  RadioGroup.OnChecked
        ChangeListener() {
      public void onCheckedChanged(RadioGroup group, int checkedId)
{
            RadioButton radioButton = (RadioButton) findView ById
            (checkedId);
            Log.i(TAG, String.valueOf(radioButton.getText()));
        }
    });
}
```

RadioButton 和 RadioGroup 常用的方法及说明，如表 4-3 所示。

表 4-3 RadioButton 和 RadioGroup 常用的方法及描述

方 法 名 称	描 述
RadioGroup.check (int id)	通过传递的参数，设置 RadioButton 单选框
RadioGroup.clearCheck ()	清空选中的项
RadioGroup.setOnCheckedChangeListener()	处理单选框 RadioButton 被选择事件，把 RadioGroup.OnCheckedChangeListener 实例作为参数传入
RadioButton.getText( )	获取单选框的值

如下代码：

```
RadioGroup.check(R.id.dotNet);   //将 id 名为 dotNet 的单选框设置成选中状态。
(RadioButton) findViewById(radioGroup.getCheckedRadioButtonId());
                                //获取被选中的单选框
RadioButton.getText( );         //获取单选框的值
```

4.3.2 复选按钮 CheckBox

复选按钮 CheckBox 是一个同时可以选择多个选项的控件，继承自 android.widget.CompoundButton，在 android.widget 包中，如图 4-8 所示。

```
java.lang.Object
  ↳android.view.View
    ↳android.widget.TextView
      ↳android.widget.Button
        ↳android.widget.CompoundButton
          ↳android.widget.CheckBox
```

图 4-8 CheckBox 类继承结构

每个多选框都是独立的，可以通过迭代所有多选框，然后根据其状态是否被选中再获取其值。CheckBox 常用方法如表 4-4 所示。

表 4-4 CheckBox 常用方法及描述

方 法 名 称	描 述
isChecked()	检查是否被选中
setChecked(boolean)	如为 true，设置成选中状态
setOnCheckedChangeListener()	处理多选框 CheckBox 被选择事件，监听按钮状态是否更改，把 CompoundButton.OnChecked Change Listener 实例作为参数传入
getText()	获取多选框的值

复选按钮 CheckBox 的通常用法如下。

1. 用 XML 描述的 CheckBox 应用界面设计

```xml
<?xml version="1.0" encoding="utf-8"?>
<LinearLayout
  xmlns:android="http://schemas.android.com/apk/res/android"
  android:layout_width="wrap_content"
  android:layout_height="fill_parent">
  <CheckBox android:id="@+id/checkboxjava"
    android:layout_width="wrap_content"
    android:layout_height="wrap_content"
    android:text="java" />
  <CheckBox android:id="@+id/checkboxdotNet"
    android:layout_width="wrap_content"
    android:layout_height="wrap_content"
    android:text="dotNet" />
  <CheckBox android:id="@+id/checkboxphp"
    android:layout_width="wrap_content"
    android:layout_height="wrap_content"
    android:text="PHP" />

  <Button android:id="@+id/checkboxButton"
    android:layout_width="fill_parent"
    android:layout_height="wrap_content"
    android:text="获取值" />
</LinearLayout>
```

2. 引用 XML 描述的代码处理

```java
public class CheckBoxActivity extends Activity {
private static final String TAG = "CheckBoxActivity";
private List<CheckBox> checkboxs = new ArrayList<CheckBox>();

    @Override
    public void onCreate(Bundle savedInstanceState) {
        super.onCreate(savedInstanceState);
        setContentView(R.layout.checkbox);
        checkboxs.add((CheckBox) findViewById(R.id.checkboxdotNet));
        checkboxs.add((CheckBox) findViewById(R.id.checkboxjava));
        checkboxs.add((CheckBox) findViewById(R.id.checkboxphp));
        checkboxs.get(1).setChecked(true);//设置成选中状态
        for(CheckBox box : checkboxs){
          box.setOnCheckedChangeListener(listener);
```

```
        }
        Button button = (Button)findViewById(R.id.checkboxButton);
        button.setOnClickListener(new View.OnClickListener() {

    @Override
    public void onClick(View v) {
        List<String> values = new ArrayList<String>();
        for(CheckBox box : checkboxs){
            if(box.isChecked()){
                values.add(box.getText().toString());
            }
        }
        Toast.makeText(CheckBoxActivity.this, values.toString(), 1).
            show();
    }
        });
    }
    CompoundButton.OnCheckedChangeListener listener = new Compound Button.
    OnCheckedChangeListener() { @Override
    public void onCheckedChanged(CompoundButton buttonView, boolean isChecked) {
    CheckBox checkBox = (CheckBox) buttonView;
    Log.i(TAG, "isChecked="+ isChecked +",value="+ checkBox.getText());
    //输出单选框的值
    }
    };
}
```

4.3.3 RadioButton 和 CheckBox 综合应用

通过上述的 RadioButton 和 CheckBox 基本介绍，下面通过一个应用，加深读者对 RadioButton 和 CheckBox 的用法和应用的熟悉和掌握。具体步骤如下。

（1）创建一个新的 Android 工程，工程名为 RadioButtonAndCheckboxDemo，目标 API 选择 17（即 Android 4.2 版本），应用程序名为 RadioButtonAndCheckboxDemo，包名为 com.bcpl.activity，创建的 Activity 的名字为 RadioButtonAndCheckboxActivity，最小 SDK 版本根据选择的目标 API 会自动添加为 8。

（2）打开项目工程中 res→layout 目录下的 activity_radio_button_and_checkbox.xml 文件，设置线性布局，添加两个 TextView 控件、一个 RadioGroup、两个 RadioButton 和三个 CheckBox，并设置相关属性，代码如下所示：

```
1.  <?xml version="1.0" encoding="utf-8"?>
2.  <LinearLayout    xmlns:android="http://schemas.android.com/apk/res/
```

```
          android"
3.        android:orientation="vertical"
4.        android:layout_width="fill_parent"
5.        android:layout_height="fill_parent"
6.        >
7.        <TextView android:text="性别："
8.            android:id="@+id/textView1"
9.            android:layout_width="wrap_content"
10.           android:layout_height="wrap_content">
11.       </TextView>
12.       <RadioGroup
13.          android:layout_width="fill_parent"
14.          android:layout_height="wrap_content"
15.          android:orientation="vertical"
16.          android:checkedButton="@+id/radioButton1"
17.          android:id="@+id/rg">
18.          <RadioButton android:text="男"
19.              android:id="@+id/radioButton1"
20.              android:layout_width="wrap_content"
21.              android:layout_height="wrap_content">
22.          </RadioButton>
23.          <RadioButton android:text="女"
24.              android:id="@+id/radioButton2"
25.              android:layout_width="wrap_content"
26.              android:layout_height="wrap_content">
27.          </RadioButton>
28.       </RadioGroup>

29.       <TextView android:text="爱好："
30.           android:id="@+id/textView2"
31.           android:layout_width="wrap_content"
32.           android:layout_height="wrap_content">
33.       </TextView>
34.       <CheckBox android:text="音乐"
35.           android:id="@+id/checkBox1"
36.           android:layout_width="wrap_content"
37.           android:layout_height="wrap_content">
38.       </CheckBox>
39.       <CheckBox android:text="体育"
40.           android:id="@+id/checkBox1"
41.           android:layout_width="wrap_content"
42.           android:layout_height="wrap_content">
43.       </CheckBox>
44.       <CheckBox android:text="收藏"
```

```
45.            android:id="@+id/checkBox1"
46.            android:layout_width="wrap_content"
47.            android:layout_height="wrap_content">
48.     </CheckBox>

49. </LinearLayout>
```

(3)src 目录下 com.bcpl.activity 包下的 RadioButtonAndCheckboxActivity.java 文件和 res→values 目录下的 strings.xml 文件,都暂不做修改。部署运行项目工程,项目运行效果如图 4-9 所示。

图 4-9 **RadioButtonAndCheckboxDemo** 运行结果

4.4 时间与日期控件简介

4.4.1 时间选择器 TimePicker

时间选择器 TimePicker,是 Android 的时间设置控件,继承自 android.widget.FrameLayout,在 android.widget 包中。TimePicker 类的继承图如图 4-10 所示。

```
java.lang.Object
   ↳android.view.View
      ↳android.view.ViewGroup
         ↳android.widget.FrameLayout
            ↳android.widget.TimePicker
```

图 4-10 **TimePicker** 类继承图

TimePicker 控件向用户显示时间，并允许用户选择（24 小时制或 AM/PM 制），改变时间，会触发 OnTimeChanged 事件，可以通过添加 OnTimeChangedListener 监听器，监听事件。

时间选择器 TimePicker 的通常用法如下。

1. 用 XML 描述一个 TimePicker

```
<TimePicker android:id="@+id/time_picker"
    android:layout_width="wrap_content"
    android:layout_height="wrap_content"/>
```

2. 程序中引用 XML 描述的 TimePicker

```
TimePicker tp = (TimePicker)this.findViewById(R.id.time_picker);
```

然后，在使用的时候可以初始化时间。

时间选择器 TimePicker 常用的方法如表 4-5 所示。

表 4-5 时间选择器 TimePicker 常用的方法及描述

方法名称	描述
setCurrentMinute(Integer currentMinute)	设置当前时间的分钟
setCurrentHour(Integer currentHour)	设置当前时间的小时
setIs24HourView(boolean)	设置为 24 小时制，如为 true，则显示
setEnabled(boolean enabled)	设置当前视图是否可以编辑
setOnTimeChangedListener(TimePicker.OnTimeChangedListener onTimeChangedListener)	为 OnTimeChangedListener 设置监听器，当时间改变时调用
getCurrentMinute()	获取时间控件的当前分钟，返回为 Integer 类型对象
getCurrentHour()	获取时间控件的当前小时，返回为 Integer 类型对象

相关类包：TimePickerDialog、DatePickerDialog，以对话框形式显示日期时间视图。Calendar 日历是设定年度日期对象和一个整数字段之间转换的抽象基类，如月，日，小时等。

4.4.2 日期选择器 DatePicker

日期选择器 DatePicker，是 Android 的日期设置控件，也继承自 android.widget.FrameLayout，在 android.widget 包中。DatePicker 类的继承图如图 4-11 所示。

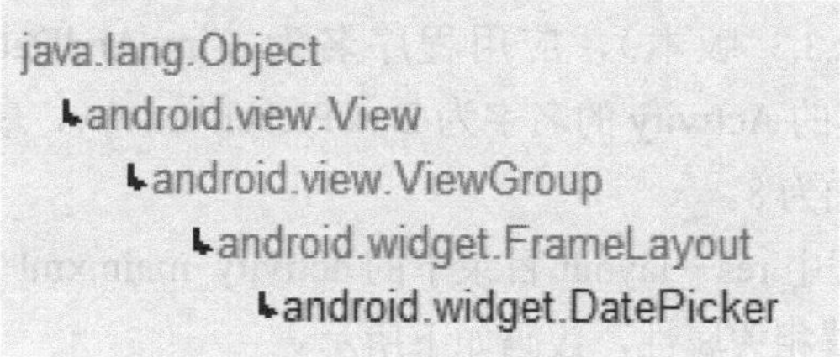

```
java.lang.Object
   ↳android.view.View
      ↳android.view.ViewGroup
         ↳android.widget.FrameLayout
            ↳android.widget.DatePicker
```

图 4-11　DatePicker 类继承图

DatePicker 控件提供年、月、日的日期数据，并允许用户进行选择。改变日期，会触发 onDateChanged 事件，通过添加 onDateChangedListener 监听器可以监听捕获事件。日期选择器 DatePicker 通常用法如下。

1．用 XML 描述一个 DatePicker

```
<DatePicker
android:id="@+id/date_picker"
android:layout_width="wrap_content"
android:layout_height="wrap_content" />
```

2．程序中引用 XML 描述的 DatePicker

```
DatePicker dp = (DatePicker)this.findViewById(R.id.date_picker);
dp.init(2012, 8, 17, null);//使用的时候可以初始化时间
```

日期选择器 DatePicker 常用的方法如表 4-6 所示

表 4-6　日期选择器 DatePicker 常用的方法及说明

方 法 名 称	描　　述
getDayOfMonth()	获取当前 Day
getMonth()	获取当前月
getYear()	获取当前年
updateDate(int year, int monthOfYear, int dayOfMonth)	更新日期
setEnabled(boolean enabled)	根据参数设置日期选择器控件是否可用或编辑
init(int year,int monthOfYear,int dayOfMonth, DatePicker.OnDateChangedListeneronDate Change Listener)	初始化日期选择器控件的属性，参数 onDateChangedListener 为监听器对象，监听日期数据变化

4.4.3　时间与日期控件综合应用

前面介绍了 TimePicker 和 DatePicker 的基本用法和方法，下面通过一个案例应用，加深读者对 TimePicker 和 DatePicker 应用的熟悉和掌握。具体步骤如下。

（1）创建一个新的 Android 工程，工程名为 TimeAndDatePickerDemo，目标 API 选

择 17（即 Android 2.3.3 版本），应用程序名为 TimeAndDatePickerDemo，包名为 com.bcpl.activity，创建的 Activity 的名字为 DatePickerActivity，最小 SDK 版本根据选择的目标 API 会自动添加为 8。

（2）打开项目工程中 res→layout 目录下的 activity_main.xml 文件，设置线性布局，添加两个 Button，并设置相关属性，代码如下所示。

```
1.    <?xml version="1.0" encoding="utf-8"?>
2.    <LinearLayout
xmlns:android="http://schemas.android.com/apk/res/android"
3.        android:orientation="vertical"
4.        android:layout_width="fill_parent"
5.        android:layout_height="fill_parent"
6.        >
7.        <Button android:text="to TimePicker"
8.            android:id="@+id/bn_time"
9.            android:layout_width="wrap_content"
10.           android:layout_height="wrap_content">
11.       </Button>
12.       <Button android:text="to DatePicker"
13.           android:id="@+id/bn_date"
14.           android:layout_width="wrap_content"
15.           android:layout_height="wrap_content">
16.       </Button>
17.   </LinearLayout>
```

（3）在项目工程中 res→layout 目录下，创建 timepicker.xml 文件，设置线性布局，添加一个 TimePicker 控件描述，并设置相关属性，代码如下所示。

```
1.    <?xml version="1.0" encoding="utf-8"?>
2.    <LinearLayout
3.      xmlns:android="http://schemas.android.com/apk/res/android"
4.      android:orientation="vertical"
5.      android:layout_width="match_parent"
6.      android:layout_height="match_parent">
7.      <TimePicker android:id="@+id/timePicker1"
8.          android:layout_width="wrap_content"
9.          android:layout_height="wrap_content">
10.     </TimePicker>
11.   </LinearLayout>
```

（4）在项目工程中 res→layout 目录下，创建 datepicker.xml 文件，设置线性布局，添加一个 DatePicker 控件描述，并设置相关属性，代码如下所示。

```
1.  <?xml version="1.0" encoding="utf-8"?>
2.  <LinearLayout
3.    xmlns:android="http://schemas.android.com/apk/res/android"
4.    android:orientation="vertical"
5.    android:layout_width="match_parent"
6.    android:layout_height="match_parent">
7.      <DatePicker android:id="@+id/datePicker1"
8.          android:layout_width="wrap_content"
9.          android:layout_height="wrap_content">
10.     </DatePicker>
11. </LinearLayout>
```

（5）修改 src 目录中 com.bcpl.activity 包下的 DatePickerActivity.java 文件，代码如下。

```
1.  package com.bcpl.activity;
2.  import android.app.Activity;
3.  import android.os.Bundle;
4.  public class DatePickerActivity extends Activity {

5.      @Override
6.      protected void onCreate(Bundle savedInstanceState) {
7.          super.onCreate(savedInstanceState);
8.          this.setContentView(R.layout.datepicker);
9.      }

10. }
```

（6）在 src 目录中 com.bcpl.activity 包下创建的 TimePickerActivity.java 文件，代码如下。

```
1.  package com.bcpl.activity;

2.  import android.app.Activity;
3.  import android.os.Bundle;

4.  public class TimePickerActivity extends Activity {

5.      @Override
6.      protected void onCreate(Bundle savedInstanceState) {
7.          super.onCreate(savedInstanceState);
8.          this.setContentView(R.layout.timepicker);
9.      }

10. }
```

（7）在 src 目录中 com.bcpl.activity 包下创建的 MainActivity.java 文件，代码如下。

```
1.  package com.bcpl.activity;
2.  import android.app.Activity;
3.  import android.content.Intent;
4.  import android.os.Bundle;
5.  import android.view.View;
6.  import android.view.View.OnClickListener;
7.  import android.widget.Button;
8.  public class MainActivity extends Activity {
9.      private Button bn_time, bn_date;
10.     @Override
11.     protected void onCreate(Bundle savedInstanceState) {
12.         super.onCreate(savedInstanceState);
13.         setContentView(R.layout.main);
14.         this.bn_time = (Button) this.findViewById(R.id.bn_time);
15.         this.bn_date = (Button) this.findViewById(R.id.bn_date);
16.         MyListener ml = new MyListener();
17.         this.bn_time.setOnClickListener(ml);
18.         this.bn_date.setOnClickListener(ml);
19.     }
20.     class MyListener implements OnClickListener{
21.         private Intent intent = new Intent();
22.         public void onClick(View v) {
23.             if(v == bn_time){
24.                 intent.setClass(MainActivity.this, TimePickerActivity.class);
25.             }
26.             if(v == bn_date){
27.                 intent.setClass(MainActivity.this, DatePickerActivity.class);
28.             }
29.             startActivity(intent);
30.         }
31.     }
32. }
```

（8）部署运行 TimeAndDatePickerDemo 项目工程，项目运行效果如图 4-12 所示。

图 4-12　TimeAndDatePickerDemo 运行结果图

单击 to TimePicker 按钮，显示时间界面，如图 4-13 所示。
单击 to DatePicker 按钮，显示日期界面，如图 4-14 所示。

 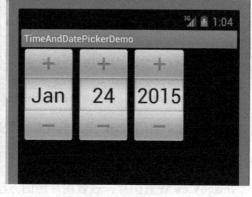

图 4-13　时间选择界面　　　　　　　　图 4-14　日期选择界面

4.5　图片控件简介

4.5.1　图片控件 ImageView 及应用

1. 图片控件 ImageView 简介

图片控件 ImageView 是最常用的组件之一，继承自 android.view.View，它的已知直接子类有：ImageButton, QuickContactBadge；已知间接子类有 ZoomButton，类图继承结

构如图 4-15 所示。

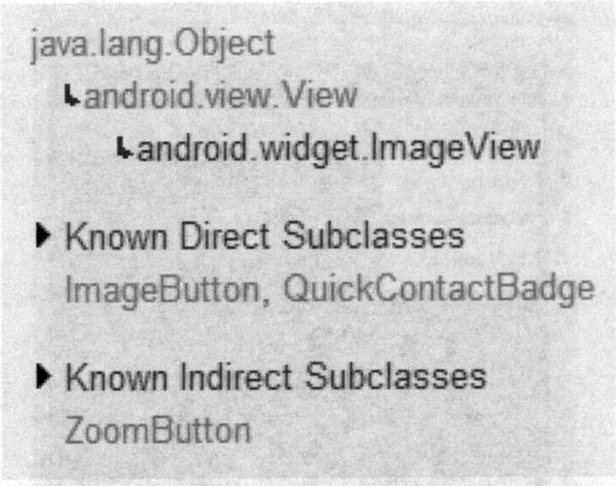

```
java.lang.Object
    ↳android.view.View
        ↳android.widget.ImageView

▶ Known Direct Subclasses
  ImageButton, QuickContactBadge

▶ Known Indirect Subclasses
  ZoomButton
```

图 4-15 ImageView 类继承图

ImageView 控件可显示任意图像，例如图标。ImageView 类可以加载各种来源的图片（如资源或图片库），其图片的来源可以是在资源文件中的 id，也可以是 Drawable 对象或者位图对象。还可以是 Content Provider 的 URI。需要计算图像的尺寸，以便它可以在其他布局中使用，并提供例如缩放和着色（渲染）等各种显示选项。

ImageView 通常的用法：

1）用 XML 来描述 ImageView

```xml
<ImageView
android:id="@+id/imagebutton"
android:src="@drawable/wjj "
android:layout_width="wrap_content"
android:layout_height="wrap_content"/>
```

2）在程序中引用 XML 描述的控件并处理

```java
ImageView  image1 = (ImageView) findViewById(R.id.img1);
```

ImageView 设置图片、设置图片源的方法，主要有以下三种。

1）设定图片相对路径

```java
ImageView iv;
String fileName = "/data/com.test/aa.png";
Bitmap bm = BitmapFactory.decodeFile(fileName);
iv.setImageBitmap(bm);
```

2）通过传递 context 访问特定的图片源

```java
ImageView iv = new ImageView(context);
iv.setImageResource(iv[position]);
iv.setScaleType(ImageView.ScaleType.FIT_XY);
iv.setLayoutParams(new Gallery.LayoutParams(136,88));
```

3）通过获取 XML 描述中设定的图片或图片源

```
mImageView = (ImageView)this.findViewById(R.id.myImageView1);
mImageView.setImageDrawable(getResources().getDrawable(R.drawable.right));
```

ImageView 常用的属性和方法如表 4-7 和表 4-8 所示。

表 4-7 ImageView 常用的属性及描述

属 性 名 称	描 述
android：adjustViewBounds	是否保持宽高比。需要与 maxWidthMaxHeight 一起使用，否则单独使用没有效果
android:cropToPadding	是否截取指定区域用空白代替。单独设置无效果，需要与 scrollY 一起使用
android:tint	将图片渲染成指定的颜色
android：maxHeight	最大高度
android：maxWidth	最大宽度
android：src	图片路径
android：scaleType	调整或移动图片

表 4-8 ImageView 常用方法及对应 XML 属性和描述说明

方 法	对应 XML 属性	描 述
setAlpha（int）		设置 ImageView 透明度
setImageBitmap(Bitmap)		设置位图作为该 ImageView 的内容
setImageDrawable(Drawable)		设置 ImageView 所显示内容为 Drawable
setImageURI（Uri）		设置 ImageView 所显示内容为 Uri
setSelected(boolean)		设置 ImageView 的选择状态
setImageResource(int)	android:src	通过资源 ID 设置可绘制对象为该 ImageView 显示的内容
setBaselineAlignBottom (boolean aligned)	android:baselineAlignBottom	设置是否设置视图底部的视图基线。设置这个值覆盖 setBaseline() 的所有调用
setAdjustViewBounds （boolean adjustViewBounds）	android:adjustViewBounds	当需要在 ImageView 调整边框时保持可绘制对象的比例时，将该值设为真
setScaleType (ImageView.ScaleType scaleType)	android:scaleType	控制图像应该如何缩放和移动，以使图像与 ImageView 一致
getScaleType ()	android:scaleType	返回当前 ImageView 使用的缩放类型

2. ImageView 应用

前面章节介绍了 ImageView 的基本用法、常用属性等知识，下面通过一个案例应用，熟悉和掌握图片控件 ImageView 的应用。具体步骤如下。

(1) 创建一个新的 Android 工程，工程名为 ImageViewDemo，目标 API 选择 17（即 Android 4.2 版本），应用程序名为 ImageViewDemo，包名为 com.bcpl.activity，创建的 Activity 的名字为 ImageViewActivity，最小 SDK 版本根据选择的目标 API 会自动添加为 8。

(2) 打开项目工程中 res→layout 目录下的 activity_main.xml 文件，设置线性布局，添加一个 Button，并设置相关属性，代码如下所示。

```
1.  <?xml version="1.0" encoding="utf-8"?>
2.  <LinearLayout xmlns:android="http://schemas.android.com/apk/res/android"
3.      android:orientation="vertical"
4.      android:layout_width="fill_parent"
5.      android:layout_height="fill_parent"
6.      >
7.      <Button android:text="显示图片"
8.          android:id="@+id/button1"
9.          android:layout_width="wrap_content"
10.         android:layout_height="wrap_content">
11.     </Button>
12. </LinearLayout>
```

(3) 在项目工程中 res→layout 目录下，创建 imageview.xml 文件，设置线性布局，添加一个 ImageView 控件描述，并设置相关属性，代码如下所示。

```
1.  <?xml version="1.0" encoding="utf-8"?>
2.  <LinearLayout
3.      xmlns:android="http://schemas.android.com/apk/res/android"
4.      android:orientation="vertical"
5.      android:layout_width="match_parent"
6.      android:layout_height="match_parent">
7.      <ImageView android:layout_height="wrap_content"
8.          android:id="@+id/imageView1"
9.          android:layout_width="wrap_content"
10.         android:src="@drawable/android_logo">
11.     </ImageView>
12. </LinearLayout>
```

(4) 修改 src 目录中 com.bcpl.activity 包下的 ImageViewActivity.java 文件，代码如下。

```
1.  package com.bcpl.activity;

2.  import android.app.Activity;
```

```
3.   import android.os.Bundle;

4.   public class ImageViewActivity extends Activity
5.   {
6.       @Override
7.       public void onCreate(Bundle savedInstanceState)
8.       {
9.           super.onCreate(savedInstanceState);
10.          setContentView(R.layout.imageview);
11.      }
12.  }
```

（5）在 src 目录中 com.bcpl.activity 包下创建 MainActivity.java 文件，代码如下。

```
1.   package com.bcpl.activity;

2.   import android.app.Activity;
3.   import android.content.Intent;
4.   import android.os.Bundle;
5.   import android.view.View;
6.   import android.view.View.OnClickListener;
7.   import android.widget.Button;
8.
9.   public class MainActivity extends Activity {

10.      private Button button1;

11.      @Override
12.      protected void onCreate(Bundle savedInstanceState) {
13.          super.onCreate(savedInstanceState);
14.          setContentView(R.layout.main);
15.
16.          button1 = (Button) this.findViewById(R.id.button1);
17.
18.          //给 button1 设置监听
19.          button1.setOnClickListener(new OnClickListener() {
20.
21.              public void onClick(View v) {
22.                  //通过 Intent 跳转到 ImageViewActivity
23.                  Intent intent = new Intent();
24.                  intent.setClass(MainActivity.this, ImageViewActivity.class);
25.                  startActivity(intent);
26.              }
27.          });
```

28.   }
29. }

（6）把 android_logo.jpg 图片文件复制到资源 res 目录下 drawable-hdpi、drawable-ldpi、drawable-mdpi、drawable-xhdpi、drawable-xxhdpi 文件夹中。

（7）部署运行 ImageViewDemo 项目工程，项目运行效果如图 4-16 所示。

单击"显示图片"按钮，显示结果如图 4-17 所示。

图 4-16 ImageViewDemo 运行结果

图 4-17 图片按钮显示

4.5.2 切换图片控件 ImageSwitcher、Gallery 应用

1．切换图片控件简介

ImageSwitcher 是 Android 中控制图片展示效果的一个控件，如幻灯片效果，继承自 android.widget.ViewSwitcher，控件继承结构如图 4-18 所示。

图 4-18 ImageSwitcher 类继承图

Gallery 控件是一个锁定中心条目并且拥有水平滚动列表的视图，它继承自 android.widget，其控件继承结构如图 4-19 所示，这个控件目前已经被 Android 系统弃用，不再被长期系统支持。系统库支持的水平滚动部件有 HorizontalScrollView 和 ViewPager。

```
java.lang.Object
  ↳android.view.View
    ↳android.view.ViewGroup
      ↳android.widget.AdapterView<T extends android.widget.Adapter>
        ↳   android.widget.AbsSpinner
          ↳    android.widget.Gallery
```

图 4-19 Gallery 类继承图

Gallery 使用 Theme_galleryItemBackground 作为 Gallery 适配器中的各视图的默认参数。如果没有设置，就需要调整一些 Gallery（画廊）的属性，比如间距等。其常用的属性及对应方法如表 4-9 所示。

Gallery 中的视图应该使用 Gallery.LayoutParams 作为它们的布局参数类型。

表 4-9 Gallery 常用的属性及对应方法、描述

属 性 名 称	对 应 方 法	描 述
android:animationDuration	setAnimationDuration(int)	设置当子视图改变位置时动画转换时间。仅限于动画开始时生效
android:gravity	setGravity(int)	描述子视图的对齐方式
android:spacing	setSpacing(int)	设置 Gallery 中项的间距
android:unselectedAlpha	setUnselectedAlpha(float)	设置 Gallery 中未选中项的透明度 (alpha) 值

注意：通常情况下，ImageSwitcher 组件和 Gallery 组件配合使用。

2．ImageSwitcher、Gallery 综合应用

前面章节中介绍了 ImageSwitcher、Gallery 的常用属性和功能，下面通过一个案例应用，熟悉和掌握 ImageSwitcher、Gallery 控件组合应用效果。具体步骤如下。

（1）创建一个新的 Android 工程，工程名为 ImageSwitcherAndGalleryDemo，目标 API 选择 17（即 Android 4.2 版本），应用程序名为 ImageSwitcherAndGalleryDemo，包名为 com.hisoft.activity，创建的 Activity 的名字为 ImageSwitcherAndGalleryActivity。

（2）打开项目工程中 res→layout 目录下的 activity_main.xml 文件，设置线性布局，添加一个 Button，并设置相关属性，代码如下所示。

```
1.   <?xml version="1.0" encoding="utf-8"?>
2.   <LinearLayout
xmlns:android="http://schemas.android.com/apk/res/android"
3.       android:orientation="vertical"
4.       android:layout_width="fill_parent"
5.       android:layout_height="fill_parent"
6.       >
7.       <Button android:text="浏览图片"
```

```
8.          android:id="@+id/button1"
9.          android:layout_width="wrap_content"
10.         android:layout_height="wrap_content">
11.     </Button>
12. </LinearLayout>
```

(3) 在项目工程中 res→layout 目录下,创建 imageswitchergallery.xml 文件,设置相对布局,添加一个 Gallery 控件描述,并设置相关属性,代码如下所示。

```
1.  <?xml version="1.0" encoding="utf-8"?>
2.  <RelativeLayout
3.      xmlns:android="http://schemas.android.com/apk/res/android"
4.      android:layout_width="match_parent"
5.      android:layout_height="match_parent">
6.      <ImageSwitcher
7.          android:id="@+id/switcher"
8.              android:layout_width="fill_parent"
9.              android:layout_height="fill_parent"
10.             android:layout_alignParentTop="true"
11.             android:layout_alignParentLeft="true" />
12.
13.     <Gallery android:id="@+id/gallery"
14.             android:background="#55000000"
15.             android:layout_width="fill_parent"
16.             android:layout_height="60dp"
17.             android:layout_alignParentBottom="true"
18.             android:layout_alignParentLeft="true"
19.             android:gravity="center_vertical"
20.             android:spacing="16dp" />
21. </RelativeLayout>
```

(4) 修改 src 目录中 com.bcpl.activity 包下的 ImageSwitcherAndGalleryActivity.java 文件,代码如下。

```
1.  package com.bcpl.activity;
2.
3.  import android.app.Activity;
4.  import android.content.Context;
5.  import android.os.Bundle;
6.  import android.view.View;
7.  import android.view.ViewGroup;
8.  import android.view.Window;
9.  import android.view.animation.AnimationUtils;
```

```
10. import android.widget.AdapterView;
11. import android.widget.BaseAdapter;
12. import android.widget.Button;
13. import android.widget.CheckBox;
14. import android.widget.EditText;
15. import android.widget.Gallery;
16. import android.widget.ImageSwitcher;
17. import android.widget.ImageView;
18. import android.widget.TextView;
19. import android.widget.ViewSwitcher;
20. import android.widget.Gallery.LayoutParams;
21.
22.
23. public class ImageSwitcherAndGalleryActivity   extends Activity implements AdapterView.OnItemSelectedListener, ViewSwitcher.ViewFactory {
24.
25.     @Override
26.     public void onCreate(Bundle savedInstanceState) {
27.         super.onCreate(savedInstanceState);
28.         requestWindowFeature(Window.FEATURE_NO_TITLE);
29.
30.         setContentView(R.layout.imageswitchergallery);
31.         setTitle("ImageShowActivity");
32.
33.         mSwitcher = (ImageSwitcher) findViewById(R.id.switcher);
34.         mSwitcher.setFactory(this);
35.         mSwitcher.setInAnimation(AnimationUtils.loadAnimation(this,
36.                 android.R.anim.fade_in));
37.         mSwitcher.setOutAnimation(AnimationUtils.loadAnimation (this,
38.                 android.R.anim.fade_out));
39.
40.
41.         Gallery g = (Gallery) findViewById(R.id.gallery);
42.         g.setAdapter(new ImageAdapter(this));
43.         g.setOnItemSelectedListener(this);
44.     }
45.
46.     public void onItemSelected(AdapterView parent, View v, int position, long id) {
47.         mSwitcher.setImageResource(mImageIds[position]);
48.     }
49.
50.     public void onNothingSelected(AdapterView parent) {
51.     }
```

```
52.
53.    public View makeView() {
54.        ImageView i = new ImageView(this);
55.        i.setBackgroundColor(0xFF000000);
56.        i.setScaleType(ImageView.ScaleType.FIT_CENTER);
57.        i.setLayoutParams(new
ImageSwitcher.LayoutParams(LayoutParams.FILL_PARENT,
58.            LayoutParams.FILL_PARENT));
59.        return i;
60.    }
61.
62.    private ImageSwitcher mSwitcher;
63.
64.    public class ImageAdapter extends BaseAdapter {
65.        public ImageAdapter(Context c) {
66.            mContext = c;
67.        }
68.
69.        public int getCount() {
70.            return mThumbIds.length;
71.        }
72.
73.        public Object getItem(int position) {
74.            return position;
75.        }
76.
77.        public long getItemId(int position) {
78.            return position;
79.        }
80.
81.        public View getView(int position, View convertView, ViewGroup
           parent) {
82.            ImageView i = new ImageView(mContext);
83.
84.            i.setImageResource(mThumbIds[position]);
85.            i.setAdjustViewBounds(true);
86.            i.setLayoutParams(new Gallery.LayoutParams(
87.                LayoutParams.WRAP_CONTENT,
88.                LayoutParams.WRAP_CONTENT));
89.            i.setBackgroundResource(R.drawable.picture_frame);
90.            return i;
91.        }
92.
93.        private Context mContext;
```

```
94.
95.        }
96.
97.    private Integer[] mThumbIds = {
98.            R.drawable.sample_thumb_0,
99.            R.drawable.sample_thumb_1,
100.           R.drawable.sample_thumb_2,
101.           R.drawable.sample_thumb_3,
102.           R.drawable.sample_thumb_4,
103.           R.drawable.sample_thumb_5,
104.           R.drawable.sample_thumb_6,
105.           R.drawable.sample_thumb_7};
106.
107.   private Integer[] mImageIds = {
108.           R.drawable.sample_0, R.drawable.sample_1,
109.           R.drawable.sample_2,
110.           R.drawable.sample_3, R.drawable.sample_4,
111.           R.drawable.sample_5,
112.           R.drawable.sample_6, R.drawable.sample_7};
113.   }
```

（5）在 src 目录中 com.bcpl.activity 包下创建 MainActivity.java 文件，代码如下。

```
1.  package com.bcpl.activity;
2.  import android.app.Activity;
3.  import android.content.Intent;
4.  import android.os.Bundle;
5.  import android.view.View;
6.  import android.view.View.OnClickListener;
7.  import android.widget.Button;

8.  public class MainActivity extends Activity {

9.      private Button button1;

10.     @Override
11.     protected void onCreate(Bundle savedInstanceState) {
12.         super.onCreate(savedInstanceState);
13.         setContentView(R.layout.main);
14.
15.         button1 = (Button) this.findViewById(R.id.button1);
16.
17.         //给 button1 设置监听
18.         button1.setOnClickListener(new OnClickListener() {
```

```
19.
20.            public void onClick(View v) {
21.                //通过Intent跳转到ImageViewActivity
22.                Intent intent = new Intent();
23.                intent.setClass(MainActivity.this,    ImageSwitcher
                       AndGalleryActivity.class);
24.                startActivity(intent);
25.            }
26.        });
27.    }
28. }
```

（6）把图片资源文件复制到资源 res 目录下 drawable-hdpi、drawable-ldpi、drawable-mdpi 文件夹中。

（7）部署运行 ImageSwitcherAndGalleryDemo 项目工程，项目运行效果如图 4-20 所示。

单击"浏览图片"按钮，显示结果如图 4-21 所示。

图 4-20　ImageSwitcherAndGalleryDemo 运行结果

图 4-21　图片相册效果

4.6　时钟控件简介

4.6.1　模拟时钟 AnalogClock 与数字时钟 DigitalClock

时钟控件包括 AnalogClock 和 DigitalClock，它们都负责显示时钟，所不同的是 AnalogClock 控件显示模拟时钟，且只显示时针和分针，而 DigitalClock 显示数字时钟，可精确到秒。

AnalogClock 和 DigitalClock 控件的类继承结构不同，AnalogClock 控件继承自 android.view.View，AnalogClock 的类结构如图 4-22 所示。

DigitalClock 控件继承自 android.widget.TextView，DigitalClock 的类结构如图 4-23 所示。

```
java.lang.Object
  ↳android.view.View
      ↳android.widget.AnalogClock
```

```
java.lang.Object
  ↳android.view.View
      ↳android.widget.TextView
          ↳android.widget.DigitalClock
```

图 4-22　AnalogClock 类继承图　　　　图 4-23　DigitalClock 类继承图

AnalogClock 和 DigitalClock 两个时钟都不需要用户编写 Java 代码，只要在 res→layout 目录下的 XML 里插入以下代码即可自动调用显示时间。

（1）AnalogClock 控件在 XML 中添加的代码如下：

```
1.         <!-- 模拟时钟控件 -->
2.     <AnalogClock android:id="@+id/analogClock"
3.         android:layout_width="wrap_content"
4.         android:layout_height="wrap_content"
5.         android:layout_gravity="center_horizontal"/>
```

（2）DigitalClock 控件在 XML 中添加的代码如下：

```
1.         <!-- 数字时钟控件 -->
2.     <DigitalClock android:id="@+id/digitalClock"
3.         android:layout_width="wrap_content"
4.         android:layout_height="wrap_content"
5.         android:layout_gravity="center_horizontal"/>
```

4.6.2　AnalogClock 和 DigitalClock 应用

前面章节介绍了 AnalogClock 和 DigitalClock 的常用方法，下面通过一个案例时钟控制应用，熟悉和掌握 AnalogClock 和 DigitalClock 控件应用效果。具体步骤如下。

（1）创建一个新的 Android 工程，工程名为 AnalogAndDigitalClockDemo，目标 API 选择 17（即 Android 4.2 版本），应用程序名为 AnalogAndDigitalClockDemo，包名为 com.bcpl.activity，创建的 Activity 的名字为 AnalogActivity，最小 SDK 版本根据选择的目标 API 会自动添加为 8。

（2）打开项目工程中 res→layout 目录下的 activity_main.xml 文件，设置线性布局，添加两个 Button，并设置相关属性，代码如下所示。

```
1.     <?xml version="1.0" encoding="utf-8"?>
```

```
2.   <LinearLayout
    xmlns:android="http://schemas.android.com/apk/res/android"
3.       android:orientation="vertical"
4.       android:layout_width="fill_parent"
5.       android:layout_height="fill_parent"
6.       >
7.       <Button android:text="to AnalogClock"
8.           android:id="@+id/bn_analog"
9.           android:layout_width="wrap_content"
10.          android:layout_height="wrap_content">
11.      </Button>
12.      <Button android:text="to DigitalClock"
13.          android:id="@+id/bn_digital"
14.          android:layout_width="wrap_content"
15.          android:layout_height="wrap_content">
16.      </Button>
17.  </LinearLayout>
```

(3) 在项目工程中 res→layout 目录下，创建 analog.xml 文件，设置相对布局，添加一个 AnalogClock 控件描述，并设置相关属性，代码如下所示。

```
1.   <?xml version="1.0" encoding="utf-8"?>
2.   <LinearLayout
3.    xmlns:android="http://schemas.android.com/apk/res/android"
4.    android:orientation="vertical"
5.    android:layout_width="match_parent"
6.    android:layout_height="match_parent">
7.     <AnalogClock android:id="@+id/analogClock1"
8.         android:layout_width="wrap_content"
9.         android:layout_height="wrap_content">
10.    </AnalogClock>
11.  </LinearLayout>
```

(4) 在项目工程中 res→layout 目录下，创建 digital.xml 文件，设置相对布局，添加一个 DigitalClock 控件描述，并设置相关属性，代码如下所示。

```
1.   <?xml version="1.0" encoding="utf-8"?>
2.   <LinearLayout
3.    xmlns:android="http://schemas.android.com/apk/res/android"
4.    android:orientation="vertical"
5.    android:layout_width="match_parent"
6.    android:layout_height="match_parent">
7.     <DigitalClock android:text="DigitalClock"
8.         android:id="@+id/digitalClock1"
```

```
9.             android:layout_width="wrap_content"
10.            android:layout_height="wrap_content">
11.    </DigitalClock>
12. </LinearLayout>
```

（5）修改 src 目录中 com.bcpl.activity 包下的 AnalogActivity.java 文件，代码如下。

```
1.  package com.bcpl.activity;
2.
3.  import android.app.Activity;
4.  import android.os.Bundle;
5.  public class AnalogActivity extends Activity {
6.      @Override
7.      protected void onCreate(Bundle savedInstanceState) {
8.          super.onCreate(savedInstanceState);
9.          this.setContentView(R.layout.analog);
10.     }
11. }
```

（6）在 src 目录中 com.bcpl.activity 包下创建 MainActivity.java 文件，代码如下。

```
1.  package com.bcpl.activity;
2.
3.  import android.app.Activity;
4.  import android.content.Intent;
5.  import android.os.Bundle;
6.  import android.view.View;
7.  import android.view.View.OnClickListener;
8.  import android.widget.Button;

9.  public class MainActivity extends Activity {

10.     private Button bn_analog, bn_digital;

11.     @Override
12.     protected void onCreate(Bundle savedInstanceState) {
13.         super.onCreate(savedInstanceState);
14.         setContentView(R.layout.main);

15.         this.bn_analog            =          (Button) this.findViewById(R.id.bn_analog);
16.         this.bn_digital           =          (Button) this.findViewById(R.id.bn_digital);
```

```
17.         MyListener ml = new MyListener();
18.         this.bn_analog.setOnClickListener(ml);
19.         this.bn_digital.setOnClickListener(ml);
20.     }
21.     class MyListener implements OnClickListener{
22.
23.         private Intent intent = new Intent();
24.
25.         public void onClick(View v) {
26.             if(v == bn_analog){
27.                 intent.setClass(MainActivity.this,
AnalogActivity.class);
28.             }
29.             if(v == bn_digital){
30.                 intent.setClass(MainActivity.this,
DigitalActivity.class);
31.             }
32.             startActivity(intent);
33.         }
34.     }
35. }
```

（7）在 src 目录中 com.bcpl.activity 包下创建 DigitalActivity.java 文件，代码如下。

```
1.  package com.bcpl.activity;
2.
3.  import android.app.Activity;
4.  import android.os.Bundle;
5.
6.  public class DigitalActivity extends Activity {
7.
8.      @Override
9.      protected void onCreate(Bundle savedInstanceState) {
10.         super.onCreate(savedInstanceState);
11.         this.setContentView(R.layout.digital);
12.     }
13. }
```

（8）部署运行 AnalogAndDigitalClockDemo 项目工程，项目运行效果如图 4-24 所示。单击 to AnalogClock 按钮，如图 4-25 所示。

图 4-24 AnalogAndDigitalClockDemo 运行结果

图 4-25 模拟时钟

单击 to DigitalClock 按钮，如图 4-26 所示。

图 4-26 数字时钟

4.7 项目案例

学习目标：学习 Android 界面系统基础控件的基本方法、属性的设置等应用。

案例描述：使用 RelativeLayout 相对布局、TextView 控件、EditText 控件、Button 按钮，并设置相对父控件的位置、控件之间相对位置的属性，实现用户登录界面。

案例要点： RelativeLayout 相对布局、控件的属性设置，以及控件之间位置关系的属性设置。

案例步骤：

（1）创建工程 Project_Chapter_4，选择 Android 4.2 作为目标平台。

（2）创建 login.xml 文件，将文件存放在 res/layout 下，代码如下。

```
1.  <?xml version="1.0" encoding="utf-8"?>
2.  <RelativeLayout xmlns:android="http://schemas.android.com/apk/res/android"
3.      android:orientation="vertical"
4.      android:layout_width="fill_parent"
5.      android:layout_height="fill_parent"
```

```
6.      >
7.  <TextView
8.      android:id="@+id/TextView"
9.      android:layout_width="fill_parent"
10.     android:layout_height="wrap_content"
11.     android:text=""
12.
13.     />
14.
15. <TextView
16.     android:id="@+id/TextView01"
17.     android:layout_below="@id/TextView"
18.     android:layout_width="wrap_content"
19.     android:layout_height="wrap_content"
20.     android:text="@string/login"
21.     android:layout_marginLeft="100dip"
22.     android:textSize="15px">
23. </TextView>
24.
25. <TextView
26.     android:id="@+id/TextView02"
27.     android:layout_marginTop="12px"
28.     android:layout_marginLeft="5dip"
29.     android:layout_below="@id/TextView01"
30.     android:layout_width="wrap_content"
31.     android:layout_height="wrap_content"
32.     android:text="@string/input_username"
33.     android:textSize="10px">
34. </TextView>
35.
36. <EditText
37.     android:id="@+id/username"
38.     android:layout_alignTop="@id/TextView02"
39.     android:layout_toRightOf="@id/TextView02"
40.     android:layout_width="fill_parent"
41.     android:layout_height="wrap_content"
42.     android:singleLine="true"
43.     android:background="@android:drawable/editbox_background"
44.     />
45.
46.
```

```xml
47. <TextView
48.     android:id="@+id/TextView03"
49.     android:layout_below="@id/username"
50.     android:layout_alignLeft="@id/TextView02"
51.     android:layout_width="wrap_content"
52.     android:layout_height="wrap_content"
53.     android:text="@string/input_userpwd"
54.     android:textSize="13px">
55. </TextView>
56.
57. <EditText
58.     android:id="@+id/password"
59.     android:layout_alignTop="@id/TextView03"
60.     android:layout_toRightOf="@id/TextView03"
61.     android:layout_alignLeft="@id/username"
62.     android:layout_width="fill_parent"
63.     android:layout_height="wrap_content"
64.     android:password="true"
65.     android:singleLine="true"
66.     android:background="@android:drawable/editbox_background"
67.     />
68.
69.
70. <Button
71.     android:id="@+id/login"
72.     android:layout_marginTop="10px"
73.     android:layout_width="50px"
74.     android:layout_height="wrap_content"
75.     android:layout_below="@id/password"
76.     android:layout_alignLeft="@id/password"
77.     android:text="@string/bt_login"
78.     android:textSize="13px"
79.     >
80. </Button>
81.
82. <Button
83.     android:id="@+id/exit"
84.     android:layout_marginLeft="10px"
85.     android:layout_width="50px"
86.     android:layout_height="wrap_content"
87.     android:text="@string/bt_exit"
```

```
88.        android:layout_toRightOf="@id/login"
89.        android:layout_alignTop="@id/login"
90.        android:textSize="10px"
91.        >
92. </Button>
93. </RelativeLayout>
```

(3) 在 src 目录下 com.bcpl.activity 包下，创建 login.java，代码如下。

```
1.  package com.bcpl.activity;
2.  import android.app.Activity;
3.  import android.os.Bundle;
4.  /*************************************************************************
5.   * 程序名称：LoginActivity.java
6.   * 功能：显示用户登录窗口，可登录和退出系统                                *
7.   *************************************************************************/
8.  public class LoginActivity extends Activity {
9.
10.     /**
11.      * 显示登录框页面
12.      */
13.     @Override
14.     public void onCreate(Bundle savedInstanceState) {
15.         super.onCreate(savedInstanceState);
16.         setContentView(R.layout.login);
17.
18.     }
19.
20.
21. }
```

(4) 修改 res 目录下 values 文件夹中的 strings.xml 文件，代码如下。

```
1.  <?xml version="1.0" encoding="utf-8"?>
2.  <resources>
3.      <string name="app_name">用户登录系统</string>
4.      <string name="login">用户登录界面 </string>
5.      <string name="input_username">用户名：</string>
6.      <string name="input_userpwd">密码：</string>
7.      <string name="bt_login">登录</string>
8.      <string name="bt_exit">退出</string>
9.  </resources>
```

(5) 部署项目工程，项目运行效果如图 4-27 所示。

注意：选择的显示设备尺寸为 3.4WQVGA(240×432)，其他尺寸设备显示有可能变形，想一想使用的是相对布局，为什么还有可能变形无法自适应设备的情况发生呢？布局设计中存在什么问题？

图 4-27　用户登录界面

习　题

一、简答题

1. 简述 TextView 和 EditText 控件的功能及用途，以及使用方法。
2. Button 按钮和 ImageButton 按钮分别什么时候使用？它们之间的区别是什么？
3. 单选按键 RadioButton 和复选按钮 CheckBox 的常用用法是什么？使用步骤包含哪些？
4. ImageView 控件如何设置图片或图片源？
5. 模拟时钟和数字时钟在用法上有什么不同？它们通用的方式是什么？

二、实训

要求：

在本章项目案例的基础上，完成用户界面注册功能，以及"用户登录系统"由右到左的滚动。

第 5 章

Android 界面系统高级控件

学习目标

本章介绍了 Android 界面系统高级控件、Android 事件处理监听器、事件处理的机制、消息传递机制、Android 音视频播录等基本应用。读者通过本章的学习,能够深入掌握以下知识要点。
(1)系统高级控件的常用属性及属性设置及描述、常用的方法。
(2)Android 事件类型、事件传递及处理原则、事件处理机制及事件处理常用方法。
(3)Android 消息传递机制及应用。
(4)Android 音视频播放、录制常用的类控件、属性设置及方法应用。

5.1 列表控件简介

5.1.1 列表控件 ListView 及应用

1. ListView 简介

ListView 是一种用于垂直显示的列表控件,它以列表的形式展示具体内容,如果 ListView 控件显示内容过多,则会出现垂直滚动条,并且能够根据数据的长度自适应显示。列表的显示需要以下三个元素。
(1)ListVeiw 用来展示列表的 View。
(2)适配器用来把数据映射到 ListView 上的中介。
(3)数据,指被映射的字符串、图片或者基本组件。

根据列表的适配器类型,列表分为三种:ArrayAdapter,SimpleAdapter 和 SimpleCursorAdapter。

ListView 能够通过适配器将数据和自身绑定,在有限的屏幕上提供大量内容供用户选择,所以是经常使用的用户界面控件。

其中以 ArrayAdapter 最为简单,只能展示一行字。SimpleAdapter 有最好的扩充性,可以自定义出各种效果。SimpleCursorAdapter 可以认为是 SimpleAdapter 对数据库的简单结合,可以方便地把数据库中的内容以列表的形式展示出来。

ListView 支持点击事件处理,用户可以用少量的代码实现复杂的选择功能。ListView 常用的 XML 属性及描述如表 5-1 所示。

表 5-1 ListView 常用的 XML 属性及描述

属 性 名 称	描　　　述
android:dividerHeight	分隔符的高度。若没有指明高度,则用此分隔符固有的高度。必须为带单位的浮点数,如"14.5sp"。可用的单位如 px, dp, sp, in, mm
android:entries	引用一个将使用在此 ListView 里的数组。若数组是固定的,使用此属性将比在程序中写入更为简单
android:footerDividersEnabled	设成 false 时,此 ListView 将不会在页脚视图前画分隔符。此属性默认值为 true。属性值必须设置为 true 或 false
android:headerDividersEnabled	设成 false 时,此 ListView 将不会在页眉视图后画分隔符。此属性默认值为 true。属性值必须设置为 true 或 false
android:choiceMode	规定此 ListView 所使用的选择模式。默认状态下,没有选择模式。属性值必须设置为下列常量之一:none,值为 0,表示无选择模式;singleChoice,值为 1,表示最多可以有一项被选中;multipleChoice,值为 2,表示可以有多项被选中

在布局文件中,用 XML 描述的 ListView 控件,代码如下。

```
1.<ListView
2.        android:id="@+id/myListView01"
3.        android:layout_width="fill_parent"
4.        android:layout_height="287dip"
5.        android:fadingEdge="none"
6.        android:divider="@drawable/list_driver"
7.        android:scrollingCache="false"
8.         android:background="@drawable/list">
9.    </ListView>
```

第 5 行是消除 ListView 的上边和下边黑色的阴影。

第 7 行是消除 ListView 在拖动的时候背景图片消失变成黑色背景。

第 8 行是 ListView 的每一项之间设置一个图片作为间隔。其中,@drawable/list 是一个图片资源。

2. ListView 应用

上面介绍了 ListView 的常用方法和属性,下面通过一个 ListView 案例应用,熟悉和掌握 ListView 控件应用效果。

ListView 控件编写程序的通常步骤如下。
(1) 在布局文件中声明 ListView 控件。
(2) 使用一维或多维动态数组保存 ListView 要显示的数据。
(3) 构建适配器 Adapter，将数据与显示数据的布局页面绑定。
(4) 通过 setAdapter()方法把适配器设置给 ListView。

案例具体步骤如下。

(1) 创建一个新的 Android 工程，工程名为 ListDemo，目标 API 选择 17（即 Android 4.2 版本），应用程序名为 ListViewDemo，包名为 com.bcpl.cn，创建的 Activity 的名字为 MainActivity，最小 SDK 版本根据选择的目标 API 会自动添加为 8。

(2) 修改布局文件 activity_main.xml，添加三个 TextView 和 ListView 实现整体布局。具体代码如下。

```xml
1.  <?xml version="1.0" encoding="utf-8"?>
2.  <LinearLayout xmlns:android="http://schemas.android.com/apk/res/android"
3.      android:orientation="vertical"
4.      android:layout_width="fill_parent"
5.      android:layout_height="fill_parent"
6.      >
7.      <TextView android:layout_width="fill_parent"
8.          android:layout_height="wrap_content"
9.          android:text="@string/tv1"
10.         android:background="#FFF"
11.         android:textColor="#888"
12.         android:gravity="center"/>
13.     <ListView android:id="@+id/lvCheckedTextView"
14.         android:layout_width="fill_parent"
15.         android:layout_height="wrap_content"/>
16.
17.     <TextView android:layout_width="fill_parent"
18.         android:layout_height="wrap_content"
19.         android:text="@string/tv2"
20.         android:background="#FFF"
21.         android:textColor="#888"
22.         android:gravity="center"/>
23.     <ListView android:id="@+id/lvRadioButton"
24.         android:layout_width="fill_parent"
25.         android:layout_height="wrap_content"/>
26.
27.     <TextView android:layout_width="fill_parent"
28.         android:layout_height="wrap_content"
29.         android:text="@string/tv3"
```

```
30.            android:background="#FFF"
31.            android:textColor="#888"
32.            android:gravity="center"/>
33.     <ListView android:id="@+id/lvCheckedButton"
34.            android:layout_width="fill_parent"
35.            android:layout_height="wrap_content"/>
36. </LinearLayout>
```

（3）修改 strings.xml 文件，具体代码如下。

```
1.  <?xml version="1.0" encoding="utf-8"?>
2.  <resources>
3.      <string name="hello">Hello World, Listview02Activity!</string>
4.      <string name="app_name">ListDemo</string>
5.      <string name="tv1">单项选择应用</string>
6.      <string name="tv2">RadioButton 应用</string>
7.      <string name="tv3">CheckBox 应用</string>
8.  </resources>
```

（4）修改 src 目录中 com.bcpl.cn 包下的 MainActivity.java 文件，代码如下。

```
1.  package com.bcpl.cn;
2.  import android.app.Activity;
3.  import android.os.Bundle;
4.  import android.widget.ArrayAdapter;
5.  import android.widget.ListView;
6.  public class MainActivity extends Activity {
7.
8.      @Override
9.      protected void onCreate(Bundle savedInstanceState) {
10.         super.onCreate(savedInstanceState);
11.         setContentView(R.layout.activity_main);
12.         ListView view1=(ListView)findViewById(R.id.iv);
13.         ListView view2=(ListView)findViewById(R.id.ivRadioButton);
14.         ListView view3=(ListView)findViewById(R.id.ivCheckButton);
15.         String[] data=new String[]{"北京","政法集团"};
16.         ArrayAdapter<String> lvAdapter=new ArrayAdapter<String>
                (this,android.R.layout.simple_list_item_checked,data);
17.         view1.setAdapter(lvAdapter);
18.         view1.setChoiceMode(ListView.CHOICE_MODE_SINGLE);
19.         String[] data1=new String[]{"women","man"};
20.         ArrayAdapter<String> lvAdapter1=new ArrayAdapter<String>
                (this,android.R.layout.simple_list_item_single_choice,
                data1);
21.         view2.setAdapter(lvAdapter1);
```

```
22.            view2.setChoiceMode(ListView.CHOICE_MODE_SINGLE);
23.
24.            ArrayAdapter<String> lvAdapter2=new ArrayAdapter<String>
        (this,android.R.layout.simple_list_item_multiple_choice,
            data);
25.            view3.setAdapter(lvAdapter2);
26.            view3.setChoiceMode(ListView.CHOICE_MODE_MULTIPLE);
27.
28.
29.        }
30.
31.        @Override
32.        public boolean onCreateOptionsMenu(Menu menu) {
33.            // Inflate the menu; this adds items to the action bar if it
                is present.
34.            getMenuInflater().inflate(R.menu.main, menu);
35.            return true;
36.        }
37.
38. }
```

（5）部署运行 ListViewDemo 项目工程，项目运行效果如图 5-1 所示。

图 5-1　ListViewDemo 工程运行结果

5.1.2　下拉列表控件 Spinner 及应用

1. 下拉列表控件 Spinner 简介

下拉列表（Spinner）是 AdapterView 的子类，是一个每次只能选择所有项中一项的部件。它的项来自于与之相关联的适配器中。类似于桌面程序的组合框（ComboBox），但没有组合框的下拉菜单，而是使用浮动菜单为用户提供选择。Spinner 数据由 Adapter 提供，通过 Spinner.getItemAtPosition(Spinner.getSelectedItemPosition());获取下拉列表框的值。

调用 setOnItemSelectedListener()方法，处理下拉列表框被选择事件，把 AdapterView.OnItemSelectedListener 实例作为参数传入。

Spinner 的类继承结构和常用的 XML 属性及对应方法描述如表 5-2 所示。

表 5-2 Spinner 常用的 XML 属性及对应方法、描述

属 性 名 称	对应的方法	描 述
android:prompt		该提示在下拉列表对话框或菜单显示时显示，如对话框的标题：setPrompt("选择颜色");
android:dropDownHorizontalOffset	setDropDownHorizontalOffset(int)	在 spinnerMode 为下拉菜单（dropdown）时，设置下拉列表的水平偏移
android:dropDownVerticalOffset	setDropDownVerticalOffset(int)	设置下拉列表和文本框的垂直偏移
android:dropDownWidth	setDropDownWidth(int)	设置下拉列表的宽度
android:gravity	setGravity(int)	设置 ListView 中当前选择的 item 位置
android:popupBackground	setPopupBackgroundResource(int)	在 spinnerMode 为下拉菜单（dropdown）时，设置下拉列表的背景

Spinner 控件应用的通常用法，具体如下。

（1）用 XML 描述的一个 Spinner 控件。

```
<?xml version="1.0" encoding="utf-8"?>
<LinearLayout
      xmlns:android="http://schemas.android.com/apk/res/android"
android:layout_width="fill_parent"
      android:layout_height="wrap_content">
      <Spinner android:id="@+id/spinner"
          android:layout_height="wrap_content"
          android:layout_width="fill_parent"/>
</LinearLayout>
```

（2）引用 XML 描述的 Spinner 控件。

```
public class SpinnerActivity extends Activity {
    private static final String TAG = "SpinnerActivity";
    @Override
    public void onCreate(Bundle savedInstanceState) {
        super.onCreate(savedInstanceState);
```

```
setContentView(R.layout.spinner);
//第二个参数为下拉列表框每一项的界面样式,该界面样式由 Android 系统提供,当
//然也可以自定义
ArrayAdapter<String> adapter = new ArrayAdapter<String>(this,
android.R.layout.simple_spinner_item);
adapter.setDropDownViewResource(android.R.layout.simple_spinner_
dropdown_item);
adapter.add("java");
adapter.add("dotNet");
adapter.add("php");
Spinner spinner = (Spinner) findViewById(R.id.spinner);
spinner.setAdapter(adapter);
spinner.setOnItemSelectedListener(newAdapterView.OnItemSelected
Listener() {
@Override
public void onItemSelected(AdapterView<?> adapterView, View view, int
position, long id) {
    Spinner spinner = (Spinner)adapterView;
    String                       itemContent            =
(String)adapterView.getItemAtPosition(position);
}
@Override
public void onNothingSelected(AdapterView<?> view) {
    Log.i(TAG, view.getClass().getName());
}
});
}
```

2. Spinner 应用

前面介绍了 Spinner 的常用方法和属性,下面通过一个 ListView 应用,熟悉和掌握 Spinner 控件应用效果。

创建一个 Spinner 控件应用的步骤如下。

(1) 在布局文件当中声明:<Spinner>。

(2) 在 string.xml 当中声明一个数组:<string-array>。

(3) 创建一个 ArrayAdapter(默认显示样式与下拉菜单中每个 item 的样式)。

(4) 得到 Spinner 对象,并设置数据:建立连接与提示。

应用具体步骤如下。

(1) 创建一个新的 Android 工程,工程名为 SpinnerDemo,目标 API 选择 17(即 Android 4.2 版本),应用程序名为 SpinnerDemo,包名为 com.bcpl.cn,创建的 Activity 的名字为 MainActivity,最小 SDK 版本根据选择的目标 API 会自动添加为 10。

（2）打开项目工程中 res→layout 目录下的 activity_main.xml 文件，设置线性布局，添加一个 TextView 和一个 Spinner 控件，并设置相关属性，代码如下所示。

```xml
1.  <?xml version="1.0" encoding="utf-8"?>
2.  <LinearLayout xmlns:android="http://schemas.android.com/apk/res/android"
3.      android:layout_width="fill_parent"
4.      android:layout_height="fill_parent"
5.      android:orientation="horizontal">
6.      <TextView android:text="爱好："
7.          android:id="@+id/textView1"
8.          android:layout_width="wrap_content"
9.          android:layout_height="wrap_content">
10.     </TextView>
11.     <Spinner android:id="@+id/favorite"
12.         android:layout_height="wrap_content"
13.         android:layout_width="wrap_content">
14.     </Spinner>
15. </LinearLayout>
```

（3）修改 strings.xml 文件。添加<string-arry>数组，具体代码如下。

```xml
1.  <?xml version="1.0" encoding="utf-8"?>
2.  <resources>
3.      <string name="hello">Hello World, MainActivity!</string>
4.      <string name="app_name">SpinnerDemo</string>
5.      <string-array name="favorite">
6.          <item>music</item>
7.          <item>sport</item>
8.          <item>programming</item>
9.          <item>watch TV</item>
10.         <item>shopping</item>
11.     </string-array>
12. </resources>
```

（4）修改 src 目录中 com.bcpl.cn 包下的 MainActivity.java 文件，代码如下。

```java
1.  package com.bcpl.cn;

2.  import android.app.Activity;
3.  import android.os.Bundle;
4.  import android.widget.ArrayAdapter;
5.  import android.widget.Spinner;

6.  public class MainActivity extends Activity {
```

```
7.
8.       private Spinner favorite;
9.
10.      /** Called when the activity is first created. */
11.      @Override
12.      public void onCreate(Bundle savedInstanceState) {
13.          super.onCreate(savedInstanceState);
14.          setContentView(R.layout.main);

15.          this.favorite = (Spinner) this.findViewById(R.id.favorite);
16.          ArrayAdapter<CharSequence> adapter = ArrayAdapter.CreateFromResource(
17.                  this, R.array.favorite,
18.                  android.R.layout.simple_spinner_item);
19.  adapter.setDropDownViewResource(android.R.layout.select_dialog_multichoice);
20.          this.favorite.setAdapter(adapter);

21.      }
22. }
```

（5）部署运行 SpinnerDemo 项目工程，项目运行效果如图 5-2 所示。
单击下拉按钮，弹出下拉列表菜单，如图 5-3 所示。

图 5-2　SpinnerDemo 项目运行结果　　　　图 5-3　下拉列表

5.2 进度条与滑块控件简介

5.2.1 进度条 ProgressBar 及应用

1. 进度条 ProgressBar 简介

该类型进度条表示运转的过程,例如发送短信、连接网络等,表示一个过程正在执行中。其位于 android.widget 包下。

ProgressBar 控件通用的方法,具体如下。

(1) 在布局 XML 文件中添加进度条代码。

```
<ProgressBar android:layout_width="fill_parent"
android:layout_height="20px"
    style="?android:attr/progressBarStyleHorizontal"
    android:id="@+id/downloadbar"/>
```

(2) 引用 XML 文件控件,在代码中操作进度条。

```
ProgressBar.setMax(100);//设置总长度为100
ProgressBar.setProgress(0);//设置已经开启长度为0,假设设置为50,进度条将进行到一半
```

ProgressBar 常用的 XML 属性及描述,如表 5-3 所示。

表 5-3 ProgressBar 常用的 XML 属性及描述

属性名称	描述
android:maxWidth	设置进度条的最大宽度
android:maxHeight	设置进度条的最大高度
android:max	设置进度条的最大值
android:progress	设置默认的进度值,取值为 0 到最大之间
android:progressDrawable	绘制进度模式
android:secondaryProgress	设置第二进度值,取值为 0 到最大之间

2. ProgressBar 应用

本节先对 ProgressBar 的应用和属性进行讲解,下面通过一个案例,熟悉和掌握 ProgressBar 控件应用效果。具体步骤如下。

(1) 创建一个新的 Android 工程,工程名为 ProgressBarDemo,目标 API 选择 17(即 Android 4.2 版本),应用程序名为 ProgressBarDemo,包名为 com.bcpl.cn,创建的 Activity 的名字为 MainActivity,最小 SDK 版本根据选择的目标 API 会自动添加为 8,创建项目工程。

(2) 打开项目工程中 res→layout 目录下的 main.xml 文件，设置相对布局，添加一个 ProgressBar 和一个 Button 控件，并设置相关属性，代码如下所示。

```
1.  <?xml version="1.0" encoding="utf-8"?>
2.  <LinearLayout
3.      xmlns:android="http://schemas.android.com/apk/res/android"
4.      android:orientation="vertical"
5.      android:layout_width="match_parent"
6.      android:layout_height="match_parent">
7.      <ProgressBar android:layout_width="fill_parent"
8.              android:layout_height="wrap_content"
9.              style="?android:attr/progressBarStyleHorizontal"
10.             android:id="@+id/progressBar1"
11.             android:max="100"
12.             android:progress="50"
13.             android:secondaryProgress="70"
14. 
15.     </ProgressBar>
16. 
17.     <Button android:text="Click Button"
18.             android:id="@+id/button1"
19.             android:layout_width="wrap_content"
20.             android:layout_height="wrap_content">
21.     </Button>
22. 
23. </LinearLayout>
```

(3) 修改 src 目录中 com.bcpl.cn 包下的 MainActivity.java 文件，代码如下。

```
1.   package com.bcpl.cn;
2.   import android.app.Activity;
3.   import android.os.Bundle;
4.   import android.view.View;
5.   import android.view.View.OnClickListener;
6.   import android.widget.Button;
7.   import android.widget.ProgressBar;
8.   
9.   public class MainActivity extends Activity {
10.  
11.      private Button b1;
12.      private ProgressBar pb;
13.      private int currentValue = 0;
14.  
15.      @Override
```

```
16.    protected void onCreate(Bundle savedInstanceState) {
17.        // TODO Auto-generated method stub
18.        super.onCreate(savedInstanceState);
19.        this.setContentView(R.layout.main);
20.
21.        this.b1 = (Button) this.findViewById(R.id.button1);
22.        this.pb = (ProgressBar) this.findViewById(R.id. progress
           Bar1);
23.        this.b1.setOnClickListener(new OnClickListener() {
24.
25.            @Override
26.            public void onClick(View v) {
27.
28.                new Thread(new ProgressThread()).start();
29.
30.            }
31.        });
32.    }
33.
34.    class ProgressThread implements Runnable{
35.
36.        @Override
37.        public void run() {
38.            while(currentValue <= 100){
39.                pb.setProgress(currentValue);
40.                try {
41.                    Thread.sleep(100);
42.                } catch (InterruptedException e) {
43.                    // TODO Auto-generated catch block
44.                    e.printStackTrace();
45.                }
46.                currentValue += 10;
47.            }
48.            currentValue = 0;
49.
50.        }
51.
52.    }
53. }
```

（4）部署运行 ProgressBarDemo 项目工程，项目运行效果如图 5-4 所示。
单击 Click Button 按钮，执行 ProgressBar，效果如图 5-5 所示。

图 5-4　ProgressBarDemo 运行结果

图 5-5　ProgressBar 效果

5.2.2　滑块 SeekBar 及应用

1. 滑块 SeekBar 简介

SeekBar 是 ProgressBar 的扩展，位于 android.widget 包中，在其基础上增加了一个可拖动的 thumb（就是那个可拖动的图标）。用户可以触摸 thumb 并向左或向右拖动，或者可以使用方向键设置当前的进度等级。不建议把可以获取焦点的 widget 放在 SeekBar 的左边或右边。

SeekBar 可以附加一个 SeekBar.OnSeekBarChangeListener 以获得用户操作的通知。

通过 SeekBar.getProgress()方法获取拖动条当前值。

通过调用 setOnSeekBarChangeListener()方法，处理拖动条值变化事件，把 SeekBar.OnSeekBarChangeListener 实例作为参数传入。

SeekBar 通用的方法及步骤如下。

（1）用 XML 描述 SeekBar 控件。

```xml
<?xml version="1.0" encoding="utf-8"?>
<LinearLayout
  xmlns:android="http://schemas.android.com/apk/res/android"
  android:layout_width="fill_parent"
  android:layout_height="fill_parent"
  android:orientation="vertical">
  <SeekBar
    android:id="@+id/seekBar"
    android:layout_height="wrap_content"
    android:layout_width="fill_parent"/>

    <Button android:id="@+id/seekBarButton"
    android:layout_height="wrap_content"
    android:layout_width="wrap_content"
    android:text="获取值"
    />
```

```
</LinearLayout>
```

(2) 引用 XML 描述的控件。

```java
public class SeekBarActivity extends Activity {
    private SeekBar seekBar;
    @Override
    public void onCreate(Bundle savedInstanceState) {
        super.onCreate(savedInstanceState);
        setContentView(R.layout.seekbar);
        seekBar = (SeekBar) findViewById(R.id.seekBar);
        seekBar.setMax(100);//设置最大刻度
        seekBar.setProgress(30);//设置当前刻度
        seekBar.setOnSeekBarChangeListener(new   SeekBar.OnSeekBarChange
        Listener() {
           @Override
           public void onProgressChanged(SeekBar seekBar, int progress,
           boolean fromTouch) {
               Log.v("onProgressChanged()", String.valueOf(progress) + ",
               " + String.valueOf(fromTouch));
           }
           @Override
           public void onStartTrackingTouch(SeekBar seekBar) {//开始拖动
               Log.v("onStartTrackingTouch()",  String.valueOf(seekBar.
                   GetProgress()));
           }
           @Override
           public void onStopTrackingTouch(SeekBar seekBar) {//结束拖动
               Log.v("onStopTrackingTouch()", String.valueOf(seekBar.get
                   Progress()));
           }
        });
        Button button = (Button)this.findViewById(R.id.seekBarButton);
        button.setOnClickListener(new View.OnClickListener() {

    @Override
    public void onClick(View v) {
    Toast.makeText(SeekBarActivity.this, String.valueOf(seekBar.getPro
    gress()), 1).show();
    }
        });
    }
}
```

2. SeekBar 应用

在上面介绍 SeekBar 的常用方法和应用基础上，下面通过一个应用，熟悉和掌握

SeekBar 控件应用效果。

（1）创建一个新的 Android 工程，工程名为 SeekBarDemo，目标 API 选择 10（即 Android 2.3.3 版本），应用程序名为 SeekBarDemo，包名为 com.hisoft.activity，创建的 Activity 的名字为 MainActivity，最小 SDK 版本根据选择的目标 API 会自动添加为 10。

（2）打开项目工程中 res→layout 目录下的 main.xml 文件，设置相对布局，添加一个 SeekBar 和一个 EditText 控件，并设置相关属性，代码如下所示。

```
1.  <?xml version="1.0" encoding="utf-8"?>
2.  <LinearLayout
    xmlns:android="http://schemas.android.com/apk/res/android"
3.      android:orientation="vertical"
4.      android:layout_width="fill_parent"
5.      android:layout_height="fill_parent"
6.      >
7.      <EditText android:layout_height="wrap_content"
8.          android:layout_width="match_parent"
9.          android:id="@+id/editText1"
10.         android:text="当前值: ">
11.         <requestFocus></requestFocus>
12.     </EditText>
13.     <SeekBar android:id="@+id/seekBar1"
14.         android:layout_height="wrap_content"
15.         android:layout_width="match_parent">
16.     </SeekBar>
17. </LinearLayout>
```

（3）修改 src 目录中 com.bcpl.activity 包下的 MainActivity.java 文件，代码如下。

```
1.  package com.bcpl.activity;
2.
3.  import android.app.Activity;
4.  import android.os.Bundle;
5.  import android.widget.EditText;
6.  import android.widget.SeekBar;
7.  import android.widget.SeekBar.OnSeekBarChangeListener;
8.
9.  public class MainActivity extends Activity {
10.
11.     private EditText et;
12.     private SeekBar sb;
13.
14.     /** Called when the activity is first created. */
15.     @Override
16.     public void onCreate(Bundle savedInstanceState) {
17.         super.onCreate(savedInstanceState);
18.         setContentView(R.layout.main);
19.
```

```
20.        this.et = (EditText) this.findViewById(R.id.editText1);
21.        this.sb = (SeekBar) this.findViewById(R.id.seekBar1);
22.
23.        this.sb.setOnSeekBarChangeListener(new
OnSeekBarChangeListener() {
24.
25.            @Override
26.            public void onStopTrackingTouch(SeekBar seekBar) {
27.
28.
29.            }
30.
31.            @Override
32.            public void onStartTrackingTouch(SeekBar seekBar) {
33.
34.
35.            }
36.
37.            @Override
38.            public void onProgressChanged(SeekBar seekBar, int progress,
39.                    boolean fromUser) {
40.
41.                et.setText("当前值: " + progress);
42.            }
43.        });
44.    }
45. }
```

（4）部署运行 SeekBarDemo 项目工程，项目运行效果如图 5-6 所示。

图 5-6　SeekBarDemo 运行效果

5.3　评分控件及应用

RatingBar 位于 android.widget 包中，是基于 SeekBar 和 ProgressBar 的扩展，用星型来显示等级评定。使用 RatingBar 的默认大小时，用户可以触摸/拖动或使用键来设置评分,它有两种样式(小风格用 ratingBarStyleSmall,大风格用 ratingBarStyleIndicator),其中大的只适合指示,不适合与用户交互。

当使用可以支持用户交互的 RatingBar 时,无论将控件(widgets)放在它的左边还是右边都是不合适的。

只有当布局的宽被设置为wrapcontent时,设置的星星数量(通过函数setNumStars(int)或者在 XML 的布局文件中定义)将显示出来(如果设置为另一种布局宽,后果无法预知)。

次级进度一般不应该被修改,因为它只是被当作星型部分内部的填充背景。

RatingBar 控件的 XML 属性如表 5-4 所示。

表 5-4 RatingBar 控件的 XML 属性

属 性 名 称	描 述
android:isIndicator	指示 RatingBar 是否是一个指示器(用户无法进行更改)
android:numStars	显示的星型数量,必须是一个整型值,如"100"
android:rating	默认的评分,必须是浮点类型,如"1.2"
android:stepSize	评分的步长,必须是浮点类型,如"1.2"

本节介绍了 RatingBar 的基础知识和常用方法,关于其案例应用,此处不再赘述。

5.4 自动完成文本控件及应用

1. AutoCompleteTextView 控件简介

AutoCompleteTextView 继承于编辑框 EditText,位于 android.widget 包下,能够完成自动提示功能。

AutoCompleteTextView 是一个用户输入时,能够通过显示一个下拉菜单自动提示一些与用户输入相关的文字提示信息,并可编辑的文本框。用户可以在下拉菜单列表中选择一项,简化输入。下拉菜单列表显示的数据,一般是从一个数据适配器进行获取。AutoCompleteTextView 的 XML 属性及对应方法如表 5-5 所示。

表 5-5 AutoCompleteTextView 的 XML 属性及对应方法、描述

属 性 名 称	对 应 方 法	描 述
android:completionHint	setCompletionHint(CharSequence)	提示信息可以直接显示在提示下拉框中
android:completionThreshold	setThreshold(int)	定义用户在下拉提示菜单出来之前,需要输入的字符数,默认最多提示 20 条
android:dropDownAnchor	setDropDownAnchor(int)	设定 View 下弹出下拉菜单提示,它的值是 View 的 ID
android:dropDownHeight	setDropDownHeight(int)	设置下拉菜单的高度
android:dropDownWidth	setDropDownWidth(int)	设置下拉菜单的宽度

AutoCompleteTextView 的通常用法及步骤如下。

（1）用 XML 描述一个 AutoCompleteTextView。

```
<AutoCompleteTextView android:id="@+id/auto_complete"
android:layout_width="fill_parent"
android:layout_height="wrap_content"/>
```

（2）使用时需要设置一个 ArrayAdapter。

```
AutoCompleteTextView textView = (AutoCompleteTextView)findView ById
(R.id.auto_ complete);
ArrayAdapter adapter = new ArrayAdapter(this,android.R.layout.simple_
dropdown_ item_1line, COUNTRIES);
//定义匹配源的 adapter，COUNTRIES 是数据数组
textView.setAdapter(adapter);
```

2．AutoCompleteTextView 应用

在上述 AutoCompleteTextView 控件讲解的基础之上，下面通过案例熟悉 AutoCompleteTextView 控件的属性和用法，具体步骤如下。

（1）创建一个新的 Android 工程，工程名为 AutoCompleteTextViewDemo，目标 API 选择 17（即 Android 4.2 版本），应用程序名为 AutoCompleteTextViewDemo，包名为 com.bcpl.cn，创建的 Activity 的名字为 MainActivity，最小 SDK 版本根据选择的目标 API 会自动添加为 8。

（2）打开项目工程中 res→layout 目录下的 activity_main.xml 文件，设置相对布局，添加一个 AutoCompleteTextView 和一个 TextView 控件，并设置相关属性，代码如下所示。

```
1.   <?xml version="1.0" encoding="utf-8"?>
2.   <LinearLayout
xmlns:android="http://schemas.android.com/apk/res/android"
3.       android:orientation="vertical"
4.       android:layout_width="fill_parent"
5.       android:layout_height="fill_parent"
6.       >
7.       <TextView android:text="请输入查询信息："
8.           android:id="@+id/textView1"
9.           android:layout_width="wrap_content"
10.          android:layout_height="wrap_content">
11.      </TextView>
12.
13.      <AutoCompleteTextView android:layout_height="wrap_content"
14.              android:layout_width="match_parent"
15.              android:id="@+id/autoCompleteTextView1"
16.              android:text="">
17.          <requestFocus></requestFocus>
18.      </AutoCompleteTextView>
```

19. </LinearLayout>

（3）修改 src 目录中 com.bcpl.cn 包下的 MainActivity.java 文件，代码如下。

```
1.  package com.bcpl.cn;

2.  import android.app.Activity;
3.  import android.os.Bundle;
4.  import android.widget.ArrayAdapter;
5.  import android.widget.AutoCompleteTextView;

6.  public class MainActivity extends Activity {

7.      private AutoCompleteTextView actv;

8.      /** Called when the activity is first created. */
9.      @Override
10.     public void onCreate(Bundle savedInstanceState) {
11.         super.onCreate(savedInstanceState);
12.         setContentView(R.layout.main);

13.         this.actv = (AutoCompleteTextView) this.findViewById(R.id.autoCompleteTextView1);
14.
15.         String[] items = {"android handy", "android pad", "android computer", "android tv"};
16.         ArrayAdapter<String> aa = new ArrayAdapter<String>(this, android.R.layout.simple_dropdown_item_1line,items);
17.         this.actv.setAdapter(aa);
18.     }
19. }
```

（4）部署运行 AutoCompleteTextViewDemo 项目工程，项目运行效果如图 5-7 所示。

图 5-7　AutoCompleteTextViewDemo 运行效果

5.5 Tabhost 控件及应用

1. Tabhost 控件简介

Tabhost 是提供选项卡（Tab 页）的窗口视图容器。此控件对象包含两个子对象：一组是用户可以选择指定 Tab 页的标签；另一组是 FrameLayout 用来显示该 Tab 页的内容。个别元素通常控制使用这个容器对象，而不是设置子元素本身的值。Tabhost 是界面设计时经常使用的界面控件，可以实现多个分页之间的快速切换，每个分页可以显示不同的内容。

Tabhost 控件通常的使用用法及步骤如下。
（1）首先要设计所有的分页的界面布局。
（2）在分页设计完成后，使用代码建立 Tab 标签页，并给每个分页添加标识和标题。
（3）最后确定每个分页所显示的界面布局。

每个分页建立一个 XML 文件，用以编辑和保存分页的界面布局，使用的方法与设计普通用户界面没有什么区别。

Tabhost 的实现方式有以下两种。
（1）继承 TabActivity，从 TabActivity 中用 getTabHost()方法获取 TabHost。只要定义具体 Tab 内容布局即可。
（2）不用继承 TabActivity，在布局文件中定义 TabHost 即可，但是 TabWidget 的 id 必须是@android:id/tabs，FrameLayout 的 id 必须是@android:id/tabcontent。TabHost 的 id 可以自定义。

注意：在使用 Tab 标签页时，可以将不同分页的界面布局保存在不同的 XML 文件中，也可以将所有分页的布局保存在同一个 XML 文件中。

第一种方法有利于在 Eclipse 开发环境中进行可视化设计，并且不同分页的界面布局在不同的文件中更加易于管理。

第二种方法则可以产生较少的 XML 文件，同时编码时的代码也会更加简洁。

2. Tabhost 应用

前面介绍了 Tabhost 的基础知识和常用方法，下面通过一个案例应用，熟悉和掌握 Tabhost 控件应用效果。

（1）创建一个新的 Android 工程，工程名为 TabhostDemo，目标 API 选择 17（即 Android 4.2 版本），应用程序名为 TabhostDemo，包名为 com.bcpl.cn，创建的 Activity 的名字为 MainActivity，最小 SDK 版本根据选择的目标 API 会自动添加为 8。

（2）打开项目工程中 res→layout 目录下的 activity_main.xml 文件，设置 Tabhost 控件，并在其中设置三个帧布局，每一个帧布局中添加一个 ImageView 控件，并设置相关属性，代码如下所示。

```xml
1.  <?xml version="1.0" encoding="utf-8"?>
2.  <TabHost xmlns:android="http://schemas.android.com/apk/res/android"
3.      android:layout_width="fill_parent"
4.      android:layout_height="fill_parent"
5.      >
6.      <!-- 定义第一个标签页的内容 -->
7.      <FrameLayout
8.          android:id="@+id/frameLayout1"
9.          android:layout_width="fill_parent"
10.         android:layout_height="fill_parent">
11.         <ImageView android:src="@drawable/sample_0"
12.             android:id="@+id/imageView1"
13.             android:layout_width="wrap_content"
14.             android:layout_height="wrap_content">
15.         </ImageView>
16.     </FrameLayout>
17.     <!-- 定义第二个标签页的内容 -->
18.     <FrameLayout
19.         android:id="@+id/frameLayout2"
20.         android:layout_width="fill_parent"
21.         android:layout_height="fill_parent">
22.         <ImageView android:src="@drawable/sample_1"
23.             android:id="@+id/imageView2"
24.             android:layout_width="wrap_content"
25.             android:layout_height="wrap_content">
26.         </ImageView>
27.     </FrameLayout>
28.     <!-- 定义第三个标签页的内容 -->
29.     <FrameLayout
30.         android:id="@+id/frameLayout3"
31.         android:layout_width="fill_parent"
32.         android:layout_height="fill_parent">
33.         <ImageView android:src="@drawable/sample_2"
34.             android:id="@+id/imageView3"
35.             android:layout_width="wrap_content"
36.             android:layout_height="wrap_content">
37.         </ImageView>
38.     </FrameLayout>
39. </TabHost>
```

（3）在 res 目录下 drawable-hdpi、drawable-ldpi、drawable-mdpi 三个文件中，添加

sample_0、sample_1、sample_2 三张图片。

（4）修改 src 目录中 com.bcpl.cn 包下的 MainActivity.java 文件，代码如下。

```
1.   package com.hisoft.activity;
2.   import android.app.TabActivity;
3.   import android.os.Bundle;
4.   import android.view.LayoutInflater;
5.   import android.widget.TabHost;

6.   public class MainActivity extends TabActivity {
7.       /** Called when the activity is first created. */
8.       @Override
9.       public void onCreate(Bundle savedInstanceState) {
10.          super.onCreate(savedInstanceState);
11.
12.          TabHost tabHost = getTabHost();
13.          //设置使用TabHost布局
14.          LayoutInflater.from(this).inflate(R.layout.main,
15.              tabHost.getTabContentView(), true);
16.          //添加第一个标签页
17.          tabHost.addTab(tabHost.newTabSpec("tab1")
18.
     .setIndicator("image1",this.getResources().getDrawable(R.drawable
     .sample_0))
19.              .setContent(R.id.frameLayout1));
20.          //添加第二个标签页
21.          tabHost.addTab(tabHost.newTabSpec("tab2")
22.
     .setIndicator("image2",this.getResources().getDrawable(R.drawable
     .sample_1))
23.              .setContent(R.id.frameLayout2));
24.          //添加第三个标签页
25.          tabHost.addTab(tabHost.newTabSpec("tab3")
26.
     .setIndicator("image3",this.getResources().getDrawable(R.drawable
     .sample_2))
27.              .setContent(R.id.frameLayout3));
28.      }
29.  }
```

（5）部署 TabhostDemo 项目工程，然后运行其 App。

5.6 视图控件应用

5.6.1 滚动视图控件 ScrollView 及应用

1．ScrollView 控件简介

ScrollView 控件是一种可供用户滚动的层次结构布局容器，允许显示比实际多的内容。ScrollView 是一种 FrameLayout，意味需要在其上放置有自己滚动内容的子元素。子元素可以是一个复杂的对象的布局管理器。通常用的子元素是垂直方向的 LinearLayout，显示在最上层的垂直方向是可以让用户滚动的箭头。

TextView 类也有自己的滚动功能，所以不需要使用 ScrollView，但是只有两者结合使用，才能保证显示较多内容时候的效率。

ScrollView 只支持垂直方向的滚动。

此类继承自 FrameLayout 类，因此内部要加入相应的布局控件才好用，否则就都堆在一起了，如在 ScrollView 内部加入一个 LinearLayout 或是 RelativeLayout 等。

2．ScrollView 应用

前面介绍了 ScrollView 常用方法和基础，下面通过一个案例应用，熟悉和掌握 ScrollView 控件应用效果。

（1）创建一个新的 Android 工程，工程名为 ScrollViewDemo，目标 API 选择 17（即 Android 4.2 版本），应用程序名为 ScrollViewDemo，包名为 com.bcpl.cn，创建的 Activity 的名字为 MainActivity，最小 SDK 版本根据选择的目标 API 会自动添加为 8。

（2）打开项目工程中 res→layout 目录下的 activity_main.xml 文件，设置 ScrollView 控件，并在其中设置 HorizontalScrollView，然后在 HorizontalScrollView 控件中设置线性布局，接着在线性布局中添加多个 TextView 控件，并设置相关属性，代码如下所示。

```
1.   <?xml version="1.0" encoding="utf-8"?>
2.   <!-- 定义 ScrollView,为里面的组件添加垂直滚动条 -->
3.   <ScrollView xmlns:android="http://schemas.android.com/apk/res/ android"
4.       android:layout_width="fill_parent"
5.       android:layout_height="fill_parent"
6.       >
7.   <!-- 定义 HorizontalScrollView,为里面的组件添加水平滚动条 -->
8.   <HorizontalScrollView
9.       android:layout_width="fill_parent"
10.      android:layout_height="wrap_content">
```

```
11. <LinearLayout android:orientation="vertical"
12.     android:layout_width="fill_parent"
13.     android:layout_height="fill_parent">
14. <TextView android:layout_width="wrap_content"
15.     android:layout_height="wrap_content"
16.     android:text="aaaaaaaaaaa"
17.     android:textSize="30dp" />
18. <TextView android:layout_width="wrap_content"
19.     android:layout_height="wrap_content"
20.     android:text="bbbbbbbbbbb"
21.     android:textSize="30dp" />
22. <TextView android:layout_width="wrap_content"
23.     android:layout_height="wrap_content"
24.     android:text="ccccccccccc"
25.     android:textSize="30dp" />
26. <TextView android:layout_width="wrap_content"
27.     android:layout_height="wrap_content"
28.     android:text="ddddddddddd"
29.     android:textSize="30dp" />
30. <TextView android:layout_width="wrap_content"
31.     android:layout_height="wrap_content"
32.     android:text="eeeeeeeeeee"
33.     android:textSize="30dp" />
34. <TextView android:layout_width="wrap_content"
35.     android:layout_height="wrap_content"
36.     android:text="fffffffffff"
37.     android:textSize="30dp" />
38. <TextView android:layout_width="wrap_content"
39.     android:layout_height="wrap_content"
40.     android:text="ggggggggggg"
41.     android:textSize="30dp" />
42. <TextView android:layout_width="wrap_content"
43.     android:layout_height="wrap_content"
44.     android:text="hhhhhhhhhhh"
45.     android:textSize="30dp" />
46. <TextView android:layout_width="wrap_content"
47.     android:layout_height="wrap_content"
48.     android:text="iiiiiiiiiii"
49.     android:textSize="30dp" />
50. <TextView android:layout_width="wrap_content"
51.     android:layout_height="wrap_content"
52.     android:text="jjjjjjjjjjj"
53.     android:textSize="30dp" />
54. <TextView android:layout_width="wrap_content"
```

```
55.        android:layout_height="wrap_content"
56.        android:text="kkkkkkkkkkkk"
57.        android:textSize="30dp" />
58. <TextView android:layout_width="wrap_content"
59.        android:layout_height="wrap_content"
60.        android:text="llllllllllll"
61.        android:textSize="30dp" />
62. <TextView android:layout_width="wrap_content"
63.        android:layout_height="wrap_content"
64.        android:text="mmmmmmmmmmmm"
65.        android:textSize="30dp" />
66. </LinearLayout>
67. </HorizontalScrollView>
68. </ScrollView>
```

(3) src 目录中 com.bcpl.cn 包下的 MainActivity.java 文件不做修改。

(4) 部署运行 ScrollViewDemo 项目工程, 项目运行效果如图 5-8 所示。

图 5-8　ScrollViewDemo 运行效果

5.6.2　网格视图控件 GridView 及应用

1. GridView 控件简介

GridView 控件视图以二维滚动网格的格式显示其包含的子项控件, 这些子项控件全部来自于与视图相关的 ListAdapter 适配器。它位于 android.widget 包下，其 XML 属性及对应方法，如表 5-6 所示。

表 5-6 GridView 控件 XML 属性及对应方法、描述

属性名称	对应方法	描述
android:columnWidth	setColumnWidth(int)	设置列的宽度
android:gravity	setGravity(int)	设置元素的对齐方式
android:horizontalSpacing	setHorizontalSpacing(int)	设置列之间的水平间距
android:numColumns	setNumColumns(int)	设置列数
android:stretchMode	setStretchMode(int)	设置可自动填充空间的列数
android:verticalSpacing	setVerticalSpacing(int)	设置行之间默认的垂直间距

2．GridView 应用

前面介绍了 GridView 的基础知识和常用方法，下面通过一个案例应用，熟悉和掌握 GridView 控件应用效果。

（1）创建一个新的 Android 工程，工程名为 GridViewDemo，目标 API 选择 17（即 Android 4.2 版本），应用程序名为 GridViewDemo，包名为 com.bcpl.cn，创建的 Activity 的名字为 MainActivity，最小 SDK 版本根据选择的目标 API 会自动添加为 8。

（2）打开项目工程中 res→layout 目录下的 activity_main.xml 文件，设置相对布局，添加一个 GridView 控件，并设置相关属性，代码如下所示。

```
1.    <?xml version="1.0" encoding="utf-8"?>
2.    <GridView
      xmlns:android="http://schemas.android.com/apk/res/android"
3.        android:id="@+id/gridview"
4.        android:layout_width="fill_parent"
5.        android:layout_height="fill_parent"
6.        android:columnWidth="90dp"
7.        android:numColumns="auto_fit"
8.        android:verticalSpacing="10dp"
9.        android:horizontalSpacing="10dp"
10.       android:stretchMode="columnWidth"
11.       android:gravity="center" />
```

（3）修改 src 目录中 com.bcpl 包下的 MainActivity.java 文件，代码如下。

```
1.    package com.bcpl.cn;

2.    import android.app.Activity;
3.    import android.os.Bundle;
4.    import android.view.View;
5.    import android.view.View.OnClickListener;
6.    import android.view.ViewGroup;
```

```
7.  import android.widget.BaseAdapter;
8.  import android.widget.GridView;
9.  import android.widget.ImageView;
10. import android.widget.Toast;
11.
12. public class MainActivity extends Activity {
13.
14.     private GridView gridview = null;
15.
16.     /** Called when the activity is first created. */
17.     @Override
18.     public void onCreate(Bundle savedInstanceState) {
19.         super.onCreate(savedInstanceState);
20.         setContentView(R.layout.main);
21.
22.         this.gridview = (GridView) findViewById(R.id.gridview);
23.         this.gridview.setAdapter(new MyAdapter());
24.     }
25.     class MyAdapter extends BaseAdapter {
26.         int[] images = { R.drawable.photo1, R.drawable.photo2,
27. R.drawable.photo3, R.drawable.photo4, R.drawable.photo5,
28. R.drawable.photo6, R.drawable.sample_0, R.drawable.sample_1,
29. R.drawable.sample_2,R.drawable.sample_3, R.drawable.sample_4,
30. R.drawable.sample_5,R.drawable.sample_6,R.drawable.sample_7, };

31.         @Override
32.         public int getCount() {
33.             // TODO Auto-generated method stub
34.             return this.images.length;
35.         }

36.         @Override
37.         public Object getItem(int arg0) {
38.             // TODO Auto-generated method stub
39.             return null;
40.         }

41.         @Override
42.         public long getItemId(int arg0) {
43.             // TODO Auto-generated method stub
44.             return 0;
45.         }
```

```
46.        @Override
47. public View getView(final int arg0, View arg1, ViewGroup arg2) {
48.        ImageView iv = new ImageView(MainActivity.this);
49.        iv.setImageResource(this.images[arg0]);
50.        iv.setLayoutParams(new GridView.LayoutParams(85, 85));
51.        iv.setScaleType(ImageView.ScaleType.CENTER_CROP);
52.        iv.setPadding(8, 8, 8, 8);
53.        iv.setOnClickListener(new OnClickListener() {
54.
55.            @Override
56.            public void onClick(View v) {
57.                Toast.makeText(MainActivity.this, arg0+ " ",
                        Toast.LENGTH_SHORT).show();
58.
59.            }
60.        });
61.        return iv;
62.    }
63.
64.  }
65. }
```

（4）部署运行 GridViewDemo 项目工程，项目运行效果如图 5-9 所示。

图 5-9　GridViewDemo 运行效果

5.7 Android 事件处理

在 Android 系统中，存在多种界面事件，如点击事件、触摸事件、焦点事件和菜单事件等。在这些界面事件发生时，Android 界面框架调用界面控件的事件处理函数对事件进行处理。

5.7.1 Android 事件和监听器

Android 中的事件按类型可以分为：按键事件和屏幕触摸事件。在 MVC 模型中，控制器根据界面事件（UI Event）类型不同，将事件传递给界面控件不同的事件处理函数。

（1）按键事件（KeyEvent）：将传递给 onKey()函数进行处理。

（2）触摸事件（TouchEvent）：将传递给 onTouch()函数进行处理。

Android 系统界面事件的传递和处理遵循如下规则。

（1）如果界面控件设置了事件监听器，则事件将先传递给事件监听器。

（2）如果界面控件没有设置事件监听器，界面事件则会直接传递给界面控件的其他事件处理函数。

（3）即使界面控件设置了事件监听器，界面事件也可以再次传递给其他事件处理函数。

（4）是否继续传递事件给其他处理函数是由事件监听器处理函数的返回值决定的。

（5）如果监听器处理函数的返回值为 true，表示该事件已经完成处理过程，不需要其他处理函数参与处理过程，这样事件就不会再继续进行传递。

（6）如果监听器处理函数的返回值为 false，则表示该事件没有完成处理过程，或需要其他处理函数捕获到该事件，事件会被传递给其他的事件处理函数。

下面以 EditText 控件中的按键事件为例，说明 Android 系统界面事件传递和处理过程，假设 EditText 控件已经设置了按键事件监听器。

（1）当用户按下键盘上的某个按键时，控制器将产生 KeyEvent 按键事件。

（2）Android 系统会首先判断 EditText 控件是否设置了按键事件监听器，因为 EditText 控件已经设置按键事件监听器 OnKeyListener，所以按键事件先传递到监听器的事件处理函数 onKey()中。

（3）事件能够继续传递给 EditText 控件的其他事件处理函数，完全根据 onKey()函数的返回值来确定。

（4）如果 onKey()函数返回 false，事件将继续传递，这样 EditText 控件就可以捕获到该事件，将按键的内容显示在 EditText 控件中。

（5）如果 onKey()函数返回 true，将阻止按键事件的继续传递，这样 EditText 控件就不能够捕获到按键事件，也就不能够将按键内容显示在 EditText 控件中。

5.7.2 Android 事件处理机制

前面讲述了 Android 事件的类型和事件处理的原则，按 Android 事件类别的处理可分为：Android 按键事件处理和屏幕触摸处理两类。不论是按键事件还是屏幕触摸处理，它们的 Android 事件处理模型都分为：基于监听接口的事件处理和基于回调机制的事件处理两类。此外，还有 Handler 消息传递机制，用于解决 Android 系统平台不允许新启动的线程访问该 Activity 里面的界面组件 Widget 问题，用户使用 Handler 可以完成 Activity 的界面组件与应用程序中线程之间的交互。

下面对 Android 的按键事件处理和屏幕触摸事件处理机制进行介绍，然后介绍 Android 事件处理机制中常用的方法。

1. Android 按键事件处理

Android 按键事件处理主要着重于 View 和 Activity 两个级别。

按键事件的处理如下。

（1）默认情况下，如果没有 View 获得焦点，事件将传递给 Activity 处理。

（2）如果 View 获得焦点，事件首先传递到 View 的回调方法中。View 回调方法返回 false，事件继续传递到 Activity 处理。反之，事件不会继续传递。

使用 View.SetFocusable(true)设置可以获得焦点。

public boolean onKeyDown(int keyCode, KeyEvent msg) 用来处理键按下事件。

public boolean onKeyUp(int keyCode, KeyEvent msg) 用来处理键抬起事件。

注意：

（1）要使按键可以被响应，需要在构造函数中调用 this.setFocusable(true);。

（2）按键的 onKeyDown 和 onKeyUp 是相互独立的，不会相互影响。

（3）无论是在 View 还是 Activity 中，建议重写事件回调方法时，只对处理过的按键返回 true，没有处理的事件应该调用其父类方法。否则，其他未处理事件不会被传递到合适的目标组件中，例如，Back 按键失效问题。

下面就 Android 按键事件的监听及信息传递给处理函数举例如下。

为了处理 Android 控件的按键事件，先需要设置按键事件的监听器，并重载 onKey()函数。

```
1.  entryText.setOnKeyListener(new OnKeyListener(){
2.      @Override
3.  public boolean onKey(View view, int keyCode, KeyEvent keyEvent) {
4.          //过程代码…
5.          return true/false;
6.      }
```

第 1 行代码是设置控件的按键事件监听器。

第 3 行代码的 onKey ()函数中的参数如下。

（1）第1个参数 view 表示产生按键事件的界面控件；

（2）第2个参数 keyCode 表示按键代码；

（3）第3个参数 keyEvent 则包含事件的详细信息，如按键的重复次数、硬件编码和按键标志等。

第5行代码是 onKey()函数的返回值。

（1）返回 true，阻止事件传递。

（2）返回 false，允许继续传递按键事件。

2. Android 屏幕触摸事件处理

在 Android 系统中，touch 事件是屏幕触摸事件的基础事件。对于多层用户界面（UI）嵌套情况，如果用户点击的 UI 部分没有重叠，只是属于单独的某个 UI（如点击父 View 没有重叠的部分），那么只有这个单独的 UI 能够捕获到 touch 事件。如果用户点击了 UI 重叠的部分，首先捕获到 touch 事件的是父类 View，然后再根据特定方法的返回值，决定 touch 事件的处理者。

在 Android 系统中，每个 View 的子类都有三个和 TouchEvent 处理密切相关的方法，分别如下。

（1）public boolean dispatchTouchEvent(MotionEvent ev); //这个方法用来分发 TouchEvent。

（2）public boolean onInterceptTouchEvent(MotionEvent ev); //这个方法用来拦截 TouchEvent。

（3）public boolean onTouchEvent(MotionEvent ev); //这个方法用来处理 TouchEvent。

其中，onTouchEvent 方法定义在 View 类中，当 touch 事件发生，首先传递到该 View，由该 View 处理时，该方法将会被执行。

dispatchTouchEvent、onInterceptTouchEvnet 这两个方法定义在 ViewGroup 中，因为只有 ViewGroup 才会包含子 View 和子 ViewGroup，才需要在 UI 多层嵌套时，通过上述的两个方法去决定是否监听处理连续 touch 动作和 touch 动作由谁去截获处理。

（1）dispatchTouchEvent 方法，默认返回值为 false。如果返回值为 false，表示捕获到一个 touch 事件，View 便会调用 onInterceptTouchEvnet 方法进行处理，而忽略掉后面的事件。如果返回值为 true，View 将监听和处理一连串的事件。如用户点击 UI 会产生几次的 touch 事件，如果该方法返回值为 false，View 将会处理第一次 touch 事件，而忽略后续的 touch 事件。如果返回值为 true，View 将处理所有的 touch 事件。如果在第一个 touch 事件的处理中，某个 View 的 onInterceptTouchEvnet 方法返回值为 true，把事件截获并处理，那么后续的 touch 事件处理将不会调用该 View 的 onInterceptTouchEvnet 方法，而是直接调用该 View 的 onTouchEvent 方法。

（2）onInterceptTouchEvnet 方法，默认返回值为 false。如果返回值为 false，该 View 将不处理传递过来的 touch 事件，而把事件传递给子 View。如果返回值为 true，该 View 将把事件截获并进行处理，不会把事件传递给子 View。因为 onInterceptTouchEvnet 方法的默认返回值为 false，所以在默认情况下，touch 事件将由处于最里层的 View 的 onTouchEvent 方法去处理。如果有相邻 View 重叠,将由处于底下的 View 的 onTouchEvent 方法处理。父类 View 把事件传递给子类 View 后，子类 View 与父类 View 一样需要完成 dispatchTouchEvent、onInterceptTouchEvnet 的流程。如果 View 的 onInterceptTouchEvnet 方法返回 true，把 touch 事件截获，将会调用自身的 onTouchEvent 事件进行处理。

（3）onTouchEvent 方法，默认返回值为 true。如果返回值为 true，表示事件处理完毕，将等待下一次事件。如果返回值为 false，则会返回调用重叠的处于上层的相邻 View 的 onTouchEvent 方法。如果没有重叠相邻 View，将返回调用父 View 的 onTouchEvent 方法。如果到了最外层的父 View 的 onTouchEvent 方法还是返回 false，则 touch 事件消失。

如果为 View 设置了 OnTouchListener，而且 touch 事件由该 View 进行处理时，监听器里面的 onTouch 方法将先于 View 自身的 onTouchEvent 方法执行。如果 onTouch 方法返回 true, onTouchEvent 方法将不会执行。

当 TouchEvent 发生时，首先 Activity 将 TouchEvent 传递给最顶层的 View，TouchEvent 最先到达最顶层 View 的 dispatchTouchEvent,然后由 dispatchTouchEvent 方法进行分发，如果 dispatchTouchEvent 返回 true，则交给这 View 的 onTouchEvent 处理，如果 dispatchTouchEvent 返回 false，则交给这个 View 的 interceptTouchEvent 方法来决定是否要拦截这个事件，如果 interceptTouchEvent 返回 true，也就是拦截掉了，则交给它的 onTouchEvent 来处理，如果 interceptTouchEvent 返回 false，那么就传递给子 View，由子 View 的 dispatchTouchEvent 再来重新开始这个事件的分发。如果事件传递到某一层的子 View 的 onTouchEvent 上了，这个方法返回了 false，那么这个事件会从这个 View 往上传递，都是 onTouchEvent 来接收。而如果传递到最上面的 onTouchEvent 也返回 false，则这个事件就会"消失"，系统认为事件处于阻塞状态，不再传递下一次事件。处理流程如图 5-10 所示。

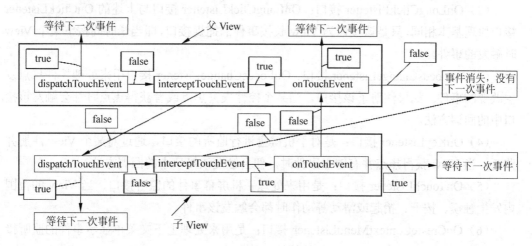

图 5-10　事件处理流程

前面已经介绍了 Android 系统的事件处理机制有基于回调机制的和基于监听接口的两种，它们在事件处理中常用的方法如下所述。

1. 基于回调机制的事件处理

Android 提供了 onKeyDown、onKeyUp、onTouchEvent、onTrackBallEvent、onFocusChanged 等回调方法供用户使用。

（1）onKeyDown：该方法是接口 KeyEvent.Callback 中的抽象方法，所有的 View 全部实现了该接口并重写了该方法，该方法用来捕捉手机键盘被按下的事件。

（2）onKeyUp：该方法也是接口 KeyEvent.Callback 中的一个抽象方法，并且所有的 View 同样全部实现了该接口并重写了该方法，onKeyUp 方法用来捕捉手机键盘按键抬起的事件。

（3）onTouchEvent：该方法在 View 类中的定义，并且所有的 View 子类全部重写了该方法，应用程序可以通过该方法处理手机屏幕的触摸事件。

（4）onTrackBallEvent：是手机中轨迹球的处理方法 onTrackBallEvent。所有的 View 同样全部实现了该方法。

（5）onFocusChanged：该方法是焦点改变的回调方法，当某个控件重写了该方法后，当焦点发生变化时，会自动调用该方法来处理焦点改变的事件。

2. 基于监听接口的事件处理

Android 提供的基于事件的监听接口有 OnClickListener、OnLongClickListener、OnFocusChangeListener、OnKeyListener、OnTouchListener、OnCreateContextMenuListener 等。

（1）OnClickListener 接口：该接口处理的是点击事件。在触摸模式下，是在某个 View 上按下并抬起的组合动作，而在键盘模式下，是某个 View 获得焦点后按"确定"按钮或者按下轨迹球事件。

（2）OnLongClickListener 接口：OnLongClickListener 接口与上述的 OnClickListener 接口原理基本相同，只是该接口为 View 长按事件的捕捉接口，即当长时间按下某个 View 时触发的事件。

（3）OnFocusChangeListener 接口：OnFocusChangeListener 接口用来处理控件焦点发生改变的事件。如果注册了该接口，当某个控件失去焦点或者获得焦点时都会触发该接口中的回调方法。

（4）OnKeyListener 接口：是对手机键盘进行监听的接口，通过对某个 View 注册并监听，当 View 获得焦点并有键盘事件时，便会触发该接口中的回调方法。

（5）OnTouchListener 接口：是用来处理手机屏幕事件的监听接口，当 View 的范围内发生触摸、按下、抬起或滑动等动作时都会触发该事件。

（6）OnCreateContextMenuListener 接口：是用来处理上下文菜单显示事件的监听接口。该方法是定义和注册上下文菜单的另一种方式。

5.7.3 Android 事件处理机制应用

前面讲述了 Android 事件处理机制的基础及详细原理，本节将通过一个案例，帮助读者理解和掌握 Android 的事件处理机制，具体步骤如下。

（1）创建一个新的 Android 工程，工程名为 TestTouchEventApp，目标 API 选择 17（即 Android 4.2 版本），应用程序名为 TestTouchEventApp，包名为 com.hisoft.activity，创建的 Activity 的名字为 TestTouchEventAppActivity，最小 SDK 版本根据选择的目标 API 会自动添加为 8。

（2）打开项目工程中 res→目录下的 activity_main.xml 文件，设置自定义线性布局和自定义 TextView，并设置相关属性，代码如下所示。

```
1.  <?xml version="1.0" encoding="utf-8"?>
2.  <com.bcpl.activity.MyLinearLayout xmlns:android="http://schemas.
    android.com/apk/res/android"
3.      android:orientation="vertical"
4.      android:layout_width="fill_parent"
5.      android:layout_height="fill_parent"
6.      android:gravity="center"
7.      >
8.  <com.bcpl.activity.MyTextView
9.          android:layout_width="200px"
10.         android:layout_height="200px"
11.         android:id="@+id/tv"
12.         android:text="bjzf"
13.         android:textSize="40sp"
14.         android:textStyle="bold"
15.         android:background="#FFFFFF"
16.         android:textColor="#0000FF"
17.     />
18. </com.bcpl.activity.MyLinearLayout>
```

（3）在 src 目录下的包 com.bcpl.activity 下，创建 MyLinearLayout.java 文件，代码如下所示。

```
1.  package com.bcpl.activity;

2.  import android.content.Context;
3.  import android.util.AttributeSet;
4.  import android.util.Log;
5.  import android.view.MotionEvent;
6.  import android.widget.LinearLayout;
7.
```

```
8.   public class MyLinearLayout extends LinearLayout {
9.       private final String TAG = "MyLinearLayout";
10.      public MyLinearLayout(Context context, AttributeSet attrs) {
11.          super(context, attrs);
12.          Log.d(TAG, TAG);
13.      }
14.      @Override
15.      public boolean dispatchTouchEvent(MotionEvent ev) {
16.          int action = ev.getAction();
17.          switch (action) {
18.          case MotionEvent.ACTION_DOWN:
19.              Log.d(TAG, "dispatchTouchEvent action:ACTION_DOWN");
20.              break;
21.          case MotionEvent.ACTION_MOVE:
22.              Log.d(TAG, "dispatchTouchEvent action:ACTION_MOVE");
23.              break;
24.          case MotionEvent.ACTION_UP:
25.              Log.d(TAG, "dispatchTouchEvent action:ACTION_UP");
26.              break;
27.          case MotionEvent.ACTION_CANCEL:
28.              Log.d(TAG, "dispatchTouchEvent action:ACTION_CANCEL");
29.              break;
30.          }
31.          return super.dispatchTouchEvent(ev);
32.      }
33.      @Override
34.      public boolean onInterceptTouchEvent(MotionEvent ev) {
35.          int action = ev.getAction();
36.          switch (action) {
37.          case MotionEvent.ACTION_DOWN:
38.              Log.d(TAG, "onInterceptTouchEvent action:ACTION_DOWN");
39.              break;
40.          case MotionEvent.ACTION_MOVE:
41.              Log.d(TAG, "onInterceptTouchEvent action:ACTION_MOVE");
42.              break;
43.          case MotionEvent.ACTION_UP:
44.              Log.d(TAG, "onInterceptTouchEvent action:ACTION_UP");
45.              break;
46.          case MotionEvent.ACTION_CANCEL:
47.              Log.d(TAG, "onInterceptTouchEvent action:ACTION_CANCEL");
48.              break;
49.          }
```

```
50.        return false;
51.    }
52.    @Override
53.    public boolean onTouchEvent(MotionEvent ev) {
54.        int action = ev.getAction();
55.        switch (action) {
56.        case MotionEvent.ACTION_DOWN:
57.            Log.d(TAG, "---onTouchEvent action:ACTION_DOWN");
58.            break;
59.        case MotionEvent.ACTION_MOVE:
60.            Log.d(TAG, "---onTouchEvent action:ACTION_MOVE");
61.            break;
62.        case MotionEvent.ACTION_UP:
63.            Log.d(TAG, "---onTouchEvent action:ACTION_UP");
64.            break;
65.        case MotionEvent.ACTION_CANCEL:
66.            Log.d(TAG, "---onTouchEvent action:ACTION_CANCEL");
67.            break;
68.        }
69.        return true;
70.    }
71. }
```

（4）在 src 目录下的包 com.bcpl.activity 下，创建 MyTextView.java 文件，代码如下所示。

```
1.  package com.bcpl.activity;

2.  import android.content.Context;
3.  import android.util.AttributeSet;
4.  import android.util.Log;
5.  import android.view.MotionEvent;
6.  import android.widget.TextView;

7.  public class MyTextView extends TextView {

8.      private final String TAG = "MyTextView";
9.      public MyTextView(Context context, AttributeSet attrs) {
10.         super(context, attrs);
11.     }
12.     @Override
13.     public boolean dispatchTouchEvent(MotionEvent ev) {
14.         int action = ev.getAction();
```

```
15.     switch (action) {
16.     case MotionEvent.ACTION_DOWN:
17.         Log.d(TAG, "dispatchTouchEvent action:ACTION_DOWN");
18.         break;
19.     case MotionEvent.ACTION_MOVE:
20.         Log.d(TAG, "dispatchTouchEvent action:ACTION_MOVE");
21.         break;
22.     case MotionEvent.ACTION_UP:
23.         Log.d(TAG, "dispatchTouchEvent action:ACTION_UP");
24.         break;
25.     case MotionEvent.ACTION_CANCEL:
26.         Log.d(TAG, "onTouchEvent action:ACTION_CANCEL");
27.         break;
28.     }
29.     return super.dispatchTouchEvent(ev);
30. }
31. @Override
32. public boolean onTouchEvent(MotionEvent ev) {
33.     int action = ev.getAction();
34.     switch (action) {
35.     case MotionEvent.ACTION_DOWN:
36.         Log.d(TAG, "---onTouchEvent action:ACTION_DOWN");
37.         break;
38.     case MotionEvent.ACTION_MOVE:
39.         Log.d(TAG, "---onTouchEvent action:ACTION_MOVE");
40.         break;
41.     case MotionEvent.ACTION_UP:
42.         Log.d(TAG, "---onTouchEvent action:ACTION_UP");
43.         break;
44.     case MotionEvent.ACTION_CANCEL:
45.         Log.d(TAG, "---onTouchEvent action:ACTION_CANCEL");
46.         break;
47.     }
48.     return true;
49. }
50. }
```

（5）部署运行 TestTouchEventApp 项目工程，项目运行效果如图 5-11 所示。

（6）在下面给定的条件下，通过程序运行时输出的 Log 来说明在不同条件下调用的时间顺序。

① 在 MyLinearLayout.onInterceptTouchEvent=false、MyLinearLayout.onTouchEvent=true、MyTextView.onTouchEvent=true 的条件下，输出的 Log 如图 5-12 所示，表明 TouchEvent 完全由 TextView 处理。

图 5-11　TestTouchEventApp 运行效果

图 5-12　TouchEvent 完全由 TextView 处理

② 在 MyLinearLayout.onInterceptTouchEvent=false、MyLinearLayout.onTouchEvent=true、MyTextView.onTouchEvent=false 的条件下，输出的 Log 如图 5-13 所示，表明 TextView 只处理了 ACTION_DOWN 事件，LinearLayout 处理了所有的 TouchEvent。

图 5-13　TextView 只处理了 ACTION_DOWN 事件

③ 在 MyLinearLayout.onInterceptTouchEvent=true、MyLinearLayout.onTouchEvent=true 的条件下，输出的 Log 如图 5-14 所示，表明 LinearLayout 处理了所有的 TouchEvent。

```
Time                    pid    Message
08-17 13:00:36.454   D  507    dispatchTouchEvent action:ACTION_DOWN
08-17 13:00:36.464   D  507    onInterceptTouchEvent action:ACTION_DOWN
08-17 13:00:36.487   D  507    ---onTouchEvent action:ACTION_DOWN
08-17 13:00:36.524   D  507    dispatchTouchEvent action:ACTION_MOVE
08-17 13:00:36.524   D  507    ---onTouchEvent action:ACTION_MOVE
08-17 13:00:36.554   D  507    ---onTouchEvent action:ACTION_MOVE
08-17 13:00:36.577   D  507    ---onTouchEvent action:ACTION_MOVE
08-17 13:00:36.586   D  507    dispatchTouchEvent action:ACTION_MOVE
08-17 13:00:36.586   D  507    ---onTouchEvent action:ACTION_MOVE
08-17 13:00:36.774   D  507    dispatchTouchEvent action:ACTION_UP
08-17 13:00:36.774   D  507    ---onTouchEvent action:ACTION_UP
```

图 5-14 LinearLayout 处理了所有的 TouchEvent

④ 在 MyLinearLayout.onInterceptTouchEvent=true、MyLinearLayout.onTouchEvent=false 的条件下，输出的 Log 如图 5-15 所示，LinearLayout 只处理了 ACTION_DOWN 事件，其他的 TouchEvent 被 LinearLayout 最外层的 Activity 处理了。

```
Time                    pid    Message
08-17 13:06:56.485   D  507    dispatchTouchEvent action:ACTION_DOWN
08-17 13:06:56.494   D  507    onInterceptTouchEvent action:ACTION_DOWN
08-17 13:06:56.516   D  507    ---onTouchEvent action:ACTION_DOWN
```

图 5-15 其他的 TouchEvent 被 LinearLayout 最外层的 Activity 处理

5.7.4 按键事件应用

前面介绍了 Android 事件按键及处理机制的基础知识和常用方法，下面通过一个按钮移动红色小球案例应用，熟悉和掌握 Android 事件处理流程和方法。

（1）创建一个新的 Android 工程，工程名为 KeyEventDemo，目标 API 选择 17（即 Android 4.2 版本），应用程序名为 KeyEventDemo，包名为 com.bcpl.activity，创建的 Activity 的名字为 MainActivity。

（2）打开项目工程中 res→layout 目录下的 activity_main.xml 文件，设置线性布局，并设置相关属性，代码如下所示。

```
1.  <?xml version="1.0" encoding="utf-8"?>
2.  <LinearLayout xmlns:android="http://schemas.android.com/apk/res/
    android"
3.      android:orientation="vertical"
4.      android:layout_width="fill_parent"
5.      android:layout_height="fill_parent"
6.      >
7.
8.  </LinearLayout>
```

（3）修改 src 目录中 com.bcpl.activity 包下的 MainActivity.java 文件，代码如下。

```
1.   package com.bcpl.activity;
2.
3.   import android.app.Activity;
4.   import android.content.Context;
5.   import android.graphics.Canvas;
6.   import android.graphics.Color;
7.   import android.graphics.Paint;
8.   import android.os.Bundle;
9.   import android.view.Display;
10.  import android.view.KeyEvent;
11.  import android.view.View;
12.  import android.view.View.OnKeyListener;
13.  import android.view.WindowManager;
14.  import android.widget.Toast;
15.
16.  public class MainActivity extends Activity {
17.      /** Called when the activity is first created. */
18.      @Override
19.      public void onCreate(Bundle savedInstanceState) {
20.          super.onCreate(savedInstanceState);
21.
22.          // 获取窗口管理器
23.          WindowManager windowManager = getWindowManager();
24.          Display display = windowManager.getDefaultDisplay();
25.          // 获得屏幕宽和高
26.          int screenWidth = display.getWidth();
27.          int screenHeight = display.getHeight();
28.          // 设置小球的初始位置
29.          int radius = 20;
30.          int x = screenWidth / 2;
31.          int y = screenHeight / 2 - radius;
32.
33.          final BallView bv = new BallView(this, x, y, radius);
34.          this.setContentView(bv);
35.
36.          //监听上下左右键
37.          bv.setOnKeyListener(new OnKeyListener() {
38.
39.              @Override
40.              public boolean onKey(View v, int keyCode, KeyEvent event) {
41.                  switch (keyCode) {
```

```java
42.             case KeyEvent.KEYCODE_DPAD_DOWN:
43.                 bv.y += 10;
44.                 break;
45.             case KeyEvent.KEYCODE_DPAD_UP:
46.                 bv.y -= 10;
47.                 break;
48.             case KeyEvent.KEYCODE_DPAD_LEFT:
49.                 bv.x -= 10;
50.                 break;
51.             case KeyEvent.KEYCODE_DPAD_RIGHT:
52.                 bv.x += 10;
53.                 break;
54.
55.             }
56.
57.             bv.invalidate(); // 重画
58.
59.             return true;
60.         }
61.     });
62. }
63. class BallView extends View {
64.
65.     private int x, y; // 代表圆心
66.     private int radius; // 半径
67.
68.     public BallView(Context context, int x, int y, int radius) {
69.         super(context);
70.         this.x = x;
71.         this.y = y;
72.         this.radius = radius;
73.         this.setFocusable(true);
74.     }
75.     @Override
76.     protected void onDraw(Canvas canvas) {
77.         // TODO Auto-generated method stub
78.         super.onDraw(canvas);
79.         canvas.drawColor(Color.WHITE);
80.         Paint p = new Paint();
81.         p.setStyle(Paint.Style.FILL);
82.         p.setColor(Color.RED);
83.         canvas.drawCircle(x, y, radius, p);
```

84.   }

85.   }
86. }

（4）部署运行 KeyEventDemo 项目工程，项目运行效果如图 5-16 所示，通过→、↓、←、↑按键，可以移动小球到指定位置。

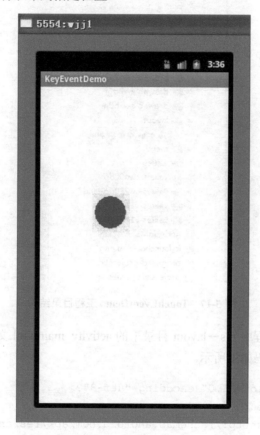

图 5-16 小球运行效果

5.7.5 触摸事件应用

前面章节中已经介绍了 Android 触摸事件监听、处理机制及常用方法，本节通过屏幕触摸并获取触摸位置坐标案例，介绍触摸事件及其监听处理的通用方法，具体步骤如下。

（1）创建一个新的 Android 工程，工程名为 TouchEventDemo，目标 API 选择 17（即 Android 4.2 版本），应用程序名为 TouchEventDemo，包名为 com.bcpl.activity，创建的 Activity 的名字为 MainActivity，最小 SDK 版本根据选择的目标 API 会自动添加为 8，创建项目工程，如图 5-17 所示。

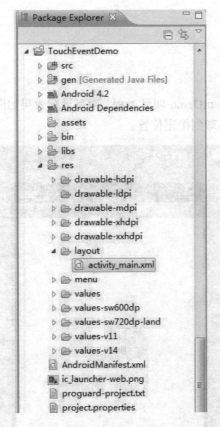

图 5-17 TouchEventDemo 工程目录结构

（2）打开项目工程中 res→layout 目录下的 activity_main.xml 文件，设置线性布局，并设置相关属性，代码如下所示。

```
1.  <?xml version="1.0" encoding="utf-8"?>
2.  <LinearLayout xmlns:android="http://schemas.android.com/apk/res/android"
3.      android:orientation="vertical"
4.      android:layout_width="fill_parent"
5.      android:layout_height="fill_parent"
6.      >
7.  </LinearLayout>
```

（3）修改 src 目录中 com.bcpl.activity 包下的 MainActivity.java 文件，代码如下。

```
1.  package com.bcpl.activity;
2.
3.  import android.app.Activity;
4.  import android.content.Context;
5.  import android.graphics.Canvas;
6.  import android.graphics.Color;
```

```
7.  import android.graphics.Paint;
8.  import android.os.Bundle;
9.  import android.view.Display;
10. import android.view.MotionEvent;
11. import android.view.View;
12. import android.view.WindowManager;
13.
14. public class MainActivity extends Activity {
15.     /** Called when the activity is first created. */
16.     @Override
17.     public void onCreate(Bundle savedInstanceState) {
18.         super.onCreate(savedInstanceState);
19.
20.         // 获取窗口管理器
21.         WindowManager windowManager = getWindowManager();
22.         Display display = windowManager.getDefaultDisplay();
23.         // 获得屏幕宽和高
24.         int screenWidth = display.getWidth();
25.         int screenHeight = display.getHeight();
26.         // 设置初始位置
27.         int x = screenWidth / 2;
28.         int y = screenHeight / 2;
29.
30.         setContentView(new TouchView(this, x, y, "( )"));
31.     }
32.
33.     class TouchView extends View {
34.
35.         private float x, y; // 初始位置坐标
36.         private String str;
37.
38.         public TouchView(Context context, float x, float y, String str) {
39.             super(context);
40.             this.x = x;
41.             this.y = y;
42.             this.str = str;
43.         }
44.
45.         @Override
46.         protected void onDraw(Canvas canvas) {
47.             // TODO Auto-generated method stub
48.             super.onDraw(canvas);
49.             canvas.drawColor(Color.WHITE);
50.             Paint p = new Paint();
```

```
51.            p.setStyle(Paint.Style.FILL);
52.            p.setColor(Color.BLACK);
53.            canvas.drawText(str, x, y, p);
54.
55.        }
56.
57.        @Override
58.        public boolean onTouchEvent(MotionEvent event) {
59.            this.x = event.getX();
60.            this.y = event.getY();
61.            this.str = "(" + x + "," + y + ")";
62.            this.invalidate();
63.            return true;
64.        }
65.    }
66. }
```

（4）部署运行 TouchEventDemo 项目工程，移动鼠标点击屏幕，屏幕显示当前位置的坐标，项目运行效果如图 5-18 所示。

图 5-18　鼠标坐标显示效果

5.8 Android 消息传递机制

在 Android 3.0 之后的版本中，为了使 Android UI 能够更加流畅，不允许用户通过 UI 的主线程访问网络资源，而是强制用户开辟一个子线程，在子线程中完成耗时的下载操作，子线程中下载完成后，再将结果推送给 UI，此时 UI 的主线程与下载的子线程通过异步的方式完成操作的整个过程。

在 Android 中主线程与子线程之间除了异步的通信机制之外，还有 Handler、Message、MessageQueue、Looper 等涉及线程操作的消息处理类及模块，它们主要是为了解决 Android UI 组件导致的线程操作安全问题，即 UI 线程只能修改 Activity 中的 UI 组件的问题。如直接在 UI 线程中开启子线程来更新 TextView 显示的内容，会显示错误。下面就对它们分别进行介绍。

5.8.1 异步任务

异步任务 AsyncTask 是线程操作的框架，它允许用户在接口上执行异步工作，在自己的工作线程上执行可能阻塞的操作，然后再将结果推送到 UI 主线程。AsyncTask 类是一个抽象类，它是由三个泛型类型和 4 个回调方法来组成，自己创建的类继承 AsyncTask 类，需要进行实现。实现 doInBackground() 回调接口，可实现运行在后台的线程池中，如果想更新自己的 UI，需要实现 onPostExecute()方法，通过它可以将 doInBackground()方法中的结果传递到 UI 主线程中，通过 UI 主线程中的 execute()方法实现异步任务的执行。

1．AsyncTask 类的泛型类型

（1）Params：启动任务执行的输入参数，比如 HTTP 请求的 URL。
（2）Progress：后台任务执行的进度值（百分比）会传递到 UI 主线程。
（3）Result：后台执行任务最终返回的结果，比如 String、Integer 等。
如果上述的三个泛型没有类型或不进行指定，也可以用 Void 进行代替，如：

```
private class MyTask extends AsyncTask<Void, Void, Void> {}
```

2．AsyncTask 类的回调方法

（1）onPreExecute()：在任务执行之前在 UI 主线程中进行调用。通常是用来做初始化任务的准备，如获得一个显示进度条的实例等。
（2）doInBackground(Params…)：在 onPreExecute()执行完成后被后台的进程进行调用，用来处理耗时的操作，异步任务的输入参数也会传递到这里。计算得到结果会通过后面的执行方法(onPostExecute()方法)推送到 UI 主线程中。此外，还可以应用 publishProgress(Progress…) 方法来显示进度刻度。这些刻度会在 UI 主线程中通过

onProgressUpdate(Progress…)方法实时显示。

（3）onProgressUpdate(Progress…)：在 publishProgress(Progress…)方法执行之后会被 UI 主线程调用，用来在 UI 主线程中实时显示计算刻度。

（4）onPostExecute(Result)：在后台计算完成之后被 UI 主线程调用。doInBackground()方法返回的结果会作为它的一个参数来推送到 UI 主线程中。

AsyncTask 的取消，可以通过调用 cancel(boolean)来取消，调用之后 isCancelled()方法都是返回值为 true，取消之后执行完 doInBackground(Object[])后，onCancelled(Object)方法会代替 onPostExecute(Object)方法被执行。一般为了使任务快速取消，可以通过周期性检查 doInBackground(Object[])方法中 isCancelled()的返回值来进行实现。

注意：
（1）AsyncTask 类必须在 UI 主线程中被加载（Android 4.1 之后会自动添加）。
（2）AsyncTask 类的实例必须在 UI 主线程中创建。
（3）execute(Params…) 必须在 UI 主线程中被调用。
（4）调用 onPreExecute()，onPostExecute(Result)，doInBackground(Params…)，onProgressUpdate(Progress…)这些方法尽量由系统负责。
（5）每个任务只能被执行一次(如果第二次执行会抛出一个异常)。

3．AsyncTask 应用

（1）创建一个新的 Android 工程，工程名为 AsyncTaskDemo，目标 API 选择 17（即 Android 4.2 版本），应用程序名为 AsyncTaskDemo，包名为 com.bcpl.activity，创建的 Activity 的名字为 MainActivity，最小 SDK 版本根据选择的目标 API 会自动添加为 8，MainActivity 的代码如下。

```
1.package com.bcpl.activity;
2.import java.io.IOException;
3.import org.apache.http.HttpEntity;
4.import org.apache.http.HttpResponse;
5.import org.apache.http.client.ClientProtocolException;
6.import org.apache.http.client.HttpClient;
7.import org.apache.http.client.methods.HttpGet;
8.import org.apache.http.impl.client.DefaultHttpClient;
9.import org.apache.http.util.EntityUtils;
10. import android.os.AsyncTask;
11. import android.os.Bundle;
12. import android.app.Activity;
13. import android.app.ProgressDialog;
14. import android.graphics.Bitmap;
15. import android.graphics.BitmapFactory;
16. import android.view.Menu;
17. import android.view.View;
18. import android.widget.Button;
```

```java
19.    import android.widget.ImageView;
20.    public class MainActivity extends Activity {
21.
22.        private Button btn;
23.        private ImageView img;
24.        private String htmlPath = " http://a.hiphotos.baidu. com/zhidao
       /pic/item/5366d0160924ab1807f29aea34fae6cd7a890b02.jpg";
25.        private ProgressDialog dialog;
26.
27.        @Override
28.        protected void onCreate(Bundle savedInstanceState) {
29.            super.onCreate(savedInstanceState);
30.            setContentView(R.layout.activity_main);
31.            initComponent();
32.            dialog = new ProgressDialog(this);
33.            dialog.setTitle("消息信息");
34.            dialog.setMessage("正在下载,请等待...");
35.            btn.setOnClickListener(new View.OnClickListener() {
36.
37.                @Override
38.                public void onClick(View v) {
39.                    new MyTask().execute(htmlPath);
40.                }
41.            });
42.        }
43.        public class MyTask extends AsyncTask<String, Void, Bitmap>{
44.
45.            // 任务执行之前的准备工作
46.            @Override
47.            protected void onPreExecute() {
48.                // TODO Auto-generated method stub
49.                super.onPreExecute();
50.                dialog.show();
51.            }
52.
53.            // 完成耗时操作,将结果推送到onPostExecute()方法中
54.            // String... params : 表示可以传递多个String类型的参数,我们只取
               一个所以用params[0]
55.            @Override
56.            protected Bitmap doInBackground(String... params) {
57.                // TODO Auto-generated method stub
58.                // 使用网络链接类 HttpClient 类完成对网络数据的提取
59.                HttpClient httpClient = new DefaultHttpClient();
60.                HttpGet httpGet = new HttpGet(params[0]);
```

```
61.        Bitmap bitmap = null;
62.        try {
63.            HttpResponse httpResponse = httpClient.execute(httpGet);
64.            if(httpResponse.getStatusLine().getStatusCode() == 200){
65.                HttpEntity httpEntity = httpResponse.getEntity();
                   // 取出 HTTP 实体
66.                byte[] data = EntityUtils.toByteArray(httpEntity);
                   //转换成字节数组
67.                // 字节数组转换成 Bitmap 对象
68.                bitmap = BitmapFactory.decodeByteArray(data, 0, data.length);
69.            }
70.        } catch (ClientProtocolException e) {
71.            // TODO Auto-generated catch block
72.            e.printStackTrace();
73.        } catch (IOException e) {
74.            // TODO Auto-generated catch block
75.            e.printStackTrace();
76.        }
77.        // 返回 bitmap 对象,最终会作为参数到 onPostExecute()方法//中,用这个方法将其推送到 UI 主线程中
78.        return bitmap;
79.    }
80.
81.    @Override
82.    protected void onProgressUpdate(Void... values) {
83.        // TODO Auto-generated method stub
84.        super.onProgressUpdate(values);
85.    }
86.
87.    // 更新 UI 线程
88.    @Override
89.    protected void onPostExecute(Bitmap result) {
90.        // TODO Auto-generated method stub
91.        super.onPostExecute(result);
92.        img.setImageBitmap(result);
93.        dialog.dismiss();
94.    }
95. }
96.
97. @Override
98. public boolean onCreateOptionsMenu(Menu menu) {
```

```
99.         // Inflate the menu; this adds items to the action bar //if it
            is present.
100.        getMenuInflater().inflate(R.menu.main, menu);
101.        return true;
102.    }
103.
104.    private void initComponent(){
105.        btn = (Button)findViewById(R.id.button1);
106.        img = (ImageView)findViewById(R.id.imageView1);
107.    }
108. }
```

（2）在 AndroidManifest.xml 文件中添加网络访问权限：

`<uses-permission android:name="android.permission.INTERNET"/>`

（3）打开项目工程中 res→layout 目录下的 activity_main.xml 文件，设置线性布局，并添加设置相关属性，代码如下所示。

```
1.  <?xml version="1.0" encoding="utf-8"?>
2.  <LinearLayout xmlns:android="http://schemas.
3.  android.com/apk/res/android"
4.  android:orientation="vertical"
5.  android:layout_width="fill_parent"
6.  android:layout_height="fill_parent"
7.  android:gravity="center"     >
8.   <!-- 添加 ImageView 控件-->
9.   <ImageView
10.  android:id="@+id/myImageView"
11.  android:layout_width="fill_parent"
12.  android:layout_height="wrap_content"
13.  android:src="@drawable/wjj"
14.  android:gravity="center" />
15.  <Button
16.  android:id="@+id/but"
17.  android:layout_width="wrap_content"
18.  android:layout_height="wrap_content"
19.  android:text="@string/but"
20.  </LinearLayout>
```

如果设置进度条对话框，可以通过修改 onCreate()方法，添加如下代码：
dialog.setProgressStyle(ProgressDialog.STYLE_HORIZONTAL); // 设置进度条对话框的样式

```
1.  public class MyTask extends AsyncTask<String, Integer, Bitmap>{
```

```
2.
3.       // 任务执行之前的准备工作
4.       @Override
5.       protected void onPreExecute() {
6.           // TODO Auto-generated method stub
7.           super.onPreExecute();
8.           dialog.show();
9.       }
10.
11.      // 完成耗时操作，将结果推送到onPostExecute()方法中
12.      // String... params : 表示可以传递多个String类型的参数，我们只
         // 取一个所以用params[0]
13.      @Override
14.      protected Bitmap doInBackground(String... params) {
15.          // 完成图片的下载功能
16.          Bitmap bitmap = null;
17.          ByteArrayOutputStream outputStream = new ByteArrayOutput-
             Stream();
18.          InputStream inputStream = null;
19.          try {
20.              HttpClient httpClient = new DefaultHttpClient();
21.              HttpGet httpGet = new HttpGet(params[0]);
22.              HttpResponse httpResponse = httpClient.execute
                 (httpGet);
23.              if(200 == httpResponse.getStatusLine().
                 getStatusCode()){
24.                  inputStream = httpResponse.getEntity().
                     getContent();
25.                  // 先要获得文件的总长度
26.                  long file_length = httpResponse.getEntity().
                     getContentLength();
27.                  // 每次读取字节的长度
28.                  int len = 0;
29.                  // 读取字节长度的总和
30.                  int total_length = 0;
31.                  byte[] data = new byte[1024];
32.                  while(-1 != (len = inputStream.read(data))){
33.                      total_length += len; //每次下载的长度进行叠加
34.                      /*
35.                       * 计算机每次下载完的部分占全部文件长度的百分比
36.                       * 计算公式如下： (int)((i/(float)count) * 100)
37.                       * 得到的结果就是它的刻度值了
38.                       */
39.                      int value = (int)((total_length / (float)file_
```

```
40.                    // 使用 publishProgress(value)方法把刻度发布出去，
                       它会发布到 onProgressUpdate()方法中
41.                    publishProgress(value);
42.                    outputStream.write(data, 0, len);
43.                }
44.                // outputStream 有一个特性，它可以将流里面的数据转换成一
                   个字节数组,通过字节数据可以转换成Bitmap 图像
45.                byte[] result = outputStream.toByteArray();
46.                // 将字节数组流转换成Bitmap 的图片格式
47.                bitmap = BitmapFactory.decodeByteArray(result, 0,
                   result.length);
48.            }
49.        } catch (Exception e) {
50.            // TODO: handle exception
51.        } finally{
52.            // outputStream 是特殊的流，它作用在内存中可以不用关闭
53.            if(inputStream != null){
54.                try {
55.                    inputStream.close();
56.                } catch (IOException e) {
57.                    // TODO Auto-generated catch block
58.                    e.printStackTrace();
59.                }
60.            }
61.        }
62.        return bitmap;
63.    }
64.
65.    @Override
66.    protected void onProgressUpdate(Integer... values) {
67.        // TODO Auto-generated method stub
68.        super.onProgressUpdate(values);
69.        dialog.setProgress(values[0]);
70.    }
71.
72.    // 更新UI 线程
73.    @Override
74.    protected void onPostExecute(Bitmap result) {
75.        // TODO Auto-generated method stub
76.        super.onPostExecute(result);
77.        img.setImageBitmap(result);
78.        dialog.dismiss();
79.    }
```

80. }

5.8.2 Handler 类应用

Handler 是线程通信工具类，主要用于传递消息，它有两个队列，分别为消息队列和线程队列。

消息队列使用 sendMessage 和 HandleMessage 方法的组合来发送和处理消息。

线程队列类似一段代码，或者说一个方法的委托（用户传递方法），使用 post、postDelayed 添加委托，使用 removeCallbacks 移除委托。

handler 类似一个容器对象，它携带了消息的集合和委托的集合。此外，handler 在另外的线程和主线程之间传递消息和可执行的代码，它不仅携带了数据，而且封装了一些操作行为，比如在适当的时机来执行线程队列里的"委托"的代码等。

一个线程的 Handle 指向线程核心对象，只要 Handler 对象以主线程的 Looper 创建，通过调用 Handler 的 sendMessage 等接口，将会把消息放入队列都将是放入主线程的消息队列，并且将会在 Handler 主线程中调用该 handler 的 handleMessage 接口来处理消息。

在主线程队列中，如果处理一个消息超过 5 秒，Android 就会抛出一个 ANP（无响应）的消息，因而需要把一些比较长的消息放在一个单独的线程里面处理，再把处理完的结果返回给主线程运行，那么主线程和单独线程之间的通信就需要 Handler 来进行消息的传递和通信。Message、Handler、MessageQueue 和 Looper 的功能如下。

Message：消息，其中包含消息 ID、消息处理对象以及处理的数据等，由 MessageQueue 统一列队，由 Handler 处理。

Handler：处理者，负责 Message 的发送及处理。使用 Handler 时，需要实现 handleMessage(Message msg)方法来对特定的 Message 进行处理，如更新 UI 等。

MessageQueue：消息队列，用来存放 Handler 发送过来的消息，并按照 FIFO 规则执行。当然，存放 Message 并非实际意义上的保存，而是将 Message 以链表的方式串联起来，等待 Looper 的抽取。

Looper：不断地从 MessageQueue 中读取 Message 交给相应的 Handler 执行。因此，一个 MessageQueue 需要一个 Looper，每个线程只能拥有一个 Looper。

Thread：线程，负责调度整个消息循环。

Handler、MessageQueue 和 Looper 之间的关系和工作机制，如图 5-19 和图 5-20 所示。

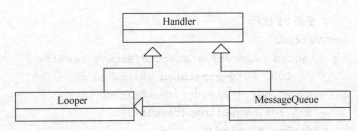

图 5-19 Handler、MessageQueue 和 Looper 的关系

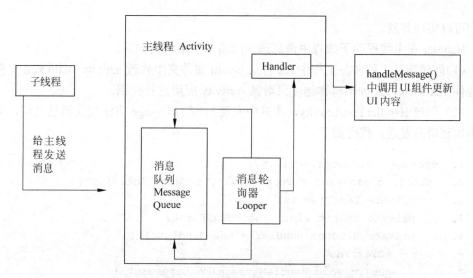

图 5-20 Handler、MessageQueue 和 Looper 工作机制

Looper 创建时，MessageQueue 也会创建，Looper 和 MessageQueue 之间一一对应，Handler 与 Looper、MessageQueue 之间只是聚集关系，即在 Handler 里会引用当前线程里的特定 Looper 和 MessageQueue。多个 Handler 可以共享同一 Looper 和 MessageQueue。但这些 Handler 在同一个线程中运行。

子线程通过 handler 的 sendMessage()方法发送消息给主线程 handler 接收进行消息处理，然后消息被加入到消息队列，主线程中消息轮询器 Looper 不断查询消息队列，如发现新的消息加入，就调用 Handler 的 handleMessage()方法处理消息，这样就可以在 handleMessage()方法中调用 UI 组件，更新 UI 内容。

一个 Activity 中可以创建多个工作线程或者其他的组件，如果这些线程或者组件把它们的消息放入 Activity 的主线程消息队列，那么该消息就会在主线程中进行处理。因为主线程一般负责界面的更新操作，并且 Android 系统中 widget 不是线程安全的，所以这种方式可以很好地实现 Android 界面更新。Android 会自动替主线程（UI 线程）建立 MessageQueue，但在子线程里并没有建立 Message Queue，所以调用 Looper.getMainLooper()得到的主线程的 Looper 不为 NULL，但调用 Looper.myLooper()得到当前线程的 Looper 就有可能为 NULL。

Handler 与 Thread 的主要区别为：Handler 与调用者处于同一线程，如果在 Handler 中做耗时的动作，调用者线程会阻塞。Android UI 操作不是线程安全的，并且这些操作必须在 UI 线程中执行。

Android 提供了几种基本的可以在其他线程中处理 UI 操作的方案，包括 Activity 的 runOnUiThread(Runnable)、View 的 post 以及工具类 AsyncTask 等方式，但都采用了 Handler。Handler 中的 post 对线程的处理并不是真正启动一个新的线程，而是直接调用了线程的 run 方法。

Handler 对于 Message 的处理方式是线性的而非并发的，即一个 Looper 只有处理完一条 Message 才会读取下一条，所以消息的处理是阻塞形式的。但如采用不同的 Looper

轮询可以实现并发。

Handler 在主线程和子线程中消息通信应用的简单示例如下。

（1）创建项目工程和在工程中 res 目录 layout 文件夹中修改 activity_main.xml 文件添加控件 TextView，此处不再详述，只对其 Activity 应用进行介绍。

（2）创建 HandlerTestActivity，在其中实现 handleMessage 方法接受消息处理，多线程中实现消息发送，代码如下。

```
1.   package   com.bcpl.activity;
2.   public class HandlerTestActivity extends Activity {
3.      private TextView tv;
4.      private static final int UPDATE = 0;
5.      private Handler handler = new Handler() {
6.         @Override
7.         public void handleMessage(Message msg) {
8.            // 接收消息并且去更新 UI 线程上的控件内容
9.            if (msg.what == UPDATE) {
10.              // Bundle b = msg.getData();
11.              // tv.setText(b.getString("num"));
12.              tv.setText(String.valueOf(msg.obj));
13.           }
14.           super.handleMessage(msg);
15.        }
16.     };
17.
18.     @Override
19.     public void onCreate(Bundle savedInstanceState) {
20.        super.onCreate(savedInstanceState);
21.        setContentView(R.layout.main);
22.        tv = (TextView) findViewById(R.id.tv);
23.
24.        new Thread() {
25.           @Override
26.           public void run() {
27. //子线程中通过 handler 发送消息给 handler 接收，由 handler 去更新 TextView 的值
28.              try {
29.                 for (int i = 0; i < 50; i++) {
30.                    Thread.sleep(500);
31.                    Message msg = new Message();
32.                    msg.what = UPDATE;
33.                    // Bundle b = new Bundle();
34.                    // b.putString("num", "新数值: " + i);
35.                    // msg.setData(b);
36.                    msg.obj = "新数值: " + i;
```

```
37.                     handler.sendMessage(msg);
38.                 }
39.             } catch (InterruptedException e) {
40.                 e.printStackTrace();
41.             }
42.         }
43.     }.start();
44. }
45. }
```

注意：获得 Message 的构造方法最好的方式是调用 obtainMessage () 和 Handler.obtainMessage()方法，以便能够更好地被回收池所回收，而不是直接用 new Message 的方式创建来获得 Message 对象。

5.9 Android 音视频播录应用

Android 系统提供了 API 对音频、视频的常用文件格式进行支持，通常支持的音频格式有 MP3、WAV、3GP 等，视频格式有 MP4、3GP 等。在 Android 中，音频和视频播放实现的方式和引用 API 都不相同，音频播放的方式有两种，分别是使用 MediaPlayer 类、SoundPool 类；音频的录制使用 MediaRecorder 类实现；视频播放的方式有 4 种，分别为使用 VideoView 来播放，使用 MediaPlayer 类和 SurfaceView 来实现，使用其自带的播放器，使用 WebView 来播放。下面就音视频播放的实现方式进行简要介绍。

5.9.1 音频播放应用

1. MediaPlayer 控制播放音频

在 Android 系统中，音频文件的装载方式有两种：一种是通过实例化对象过程中，调用其无参构造方法实现音频文件的引用装载；另一种是通过创建 MediaPlayer 类的对象，调用其静态方法 create()实现音频文件的装载。

（1）MediaPlayer 无参构造方法实现音频文件的引用装载，主要步骤如下。

```
MediaPlayer mp=new MediaPlayer();    //创建无参对象
mp.setDataSource("/mnt/sdcard/test.mp3");   //指定要装载的音频文件
mp.prepare();    //加载音频资源
```

如加载资源文件夹 raw 中的 vi.mp3 文件，其代码如下：

```
MediaPlayer mp=MediaPlayer.create(this,R.raw.vi);
```

或加载网络音频资源，如下：

```
MediaPlayer
```

```
mp=MediaPlayer.create(this,Uri.parse("http://yinyueshiting.baidu.com/
data2/music/134441710/13132725248400128.mp3)
```

使用 MediaPlayer 进行音频播放、设置，常用方法如表 5-7 所示。

表 5-7 MediaPlayer 常用方法

方 法	说 明
static MediaPlayer create(Context context, int resid)	从资源 ID 对应的资源文件中来装载音乐文件，并返回新创建的 MediaPlyaer 对象
static MediaPlayer create(Context context, Uri uri)	从 Uri 加载音频文件，并返回新创建的 MediaPlayer 对象
static MediaPlayer create(Context context, Uri uri, SurfaceHolder holder)	从资源 ID 对应的资源文件中来加载音乐文件，同时指定 SurfaceHolder 对象并返回 MediaPlyaer 对象
int getCurrentPosition()	获取当前播放的位置
int getDuration()	获取音频的时长
void pause()	暂停播放
void start()	开始播放
void stop()	停止播放
int setOnCompletionListener(MediaPlayer.OnCompletionListener listener)	为 MediaPlayer 的播放完成事件绑定事件监听器
void setDataSource(String path)	装载 path 路径所代表的文件
void setDataSource(FileDescriptor fd)	装载 fd 所代表的文件

注：MediaPlayer 根据资源文件存储位置的不同，使用不同的方法进行播放控制，上述介绍了本地和网络两种方法，此外，还可以使用表 5-7 中的 setDataSource 调用资源文件或路径的方式进行实现。

（2）MediaPlayer 播放音频应用。

创建工程，在 res 目录中 layout 文件下的 activity_main.xml 文件中，添加两个 Button，分别为"开始"和"停止"，在 raw 目录下添加需要播放的音频资源文件。由于其应用比较简单，下面只介绍 MainActivity 类代码。

```
1.  package com.bcpl.mediaplayer;
2.  import java.io.IOException;
3.  import android.app.Activity;
4.  import android.media.MediaPlayer;
5.  import android.os.Bundle;
6.  import android.view.View;
7.  import android.view.View.OnClickListener;
8.  import android.widget.Button;
9.
10. public class MainActivity extends Activity {
11.
12.     private MediaPlayer player=null;
```

```
13.     private Button start=null,stop=null;
14.     /** Called when the activity is first created. */
15.     @Override
16.     public void onCreate(Bundle savedInstanceState) {
17.         super.onCreate(savedInstanceState);
18.         setContentView(R.layout.main);
19.         player=MediaPlayer.create(this, R.raw.tru);
20.         start=(Button) this.findViewById(R.id.button1);
21.         stop=(Button) this.findViewById(R.id.button2);
22.         start.setOnClickListener(new OnClickListener() {
23.
24.             @Override
25.             public void onClick(View v) {
26.                 // TODO Auto-generated method stub
27.                 try {
28.                     player.prepare();
29.                 } catch (IllegalStateException e) {
30.                     // TODO Auto-generated catch block
31.                     e.printStackTrace();
32.                 } catch (IOException e) {
33.                     // TODO Auto-generated catch block
34.                     e.printStackTrace();
35.                 }
36.                 player.start();
37.             }
38.         });
39.
40.         stop.setOnClickListener(new OnClickListener() {
41.             @Override
42.             public void onClick(View v) {
43.                 // TODO Auto-generated method stub
44.                 if(player.isPlaying()){
45.                     player.stop();
46.                     player.release();
47.                     player=null;
48.                 }
49.
50.             }
51.         });
52.     }
53. }
```

后续拓展中可以实现"暂停/继续"、"继续"功能，单击"暂停"按钮时"开始"按钮可用，"暂停"按钮上的文字变为"继续"；单击"继续"按钮，其按钮文字变为"暂停"；单击"停止"按钮，"暂停"、"停止"按钮不可用，"开始"按钮由灰色变为可用状态等功能。

2. 使用 SoundPool 播放音频

SoundPool（音频池）主要用于播放一些较小的音频，与 MediaPlayer 相比，具有占用资源少、反应延迟小、可以同时播放多个音频等优势，它通常被用来管理短小音频的播放，如应用程序中的消息提示音、游戏中的连续爆炸音等。

SoundPool 类用来创建对象的构造方法如下：

```
public SoundPool (int maxStreams, int streamType, int srcQuality)
```

其中，

maxStreams：表示支持音频的个数，SoundPool 对象中允许同时存在的最大流的数量。

streamType：声音类型，在 AudioManager 中定义，分为 STREAM_VOICE_CALL、STREAM_SYSTEM、STREAM_RING、STREAM_MUSIC 和 STREAM_ALARM 5 种类型。

srcQuality：指定声音品质（采样率变换质量），可以设为0。

SoundPool 提供了 4 个 load 方法载入声音资源，通过 load 方法返回声音的 ID，下面就常用的两个 load 方法进行简要说明：

public int load(Context context, int resId, int priority)：从 resId 所对应的资源加载声音。

public int load(String path, int priority)：从 path 对应的文件中加载声音。

上述方法中的 priority 参数，Android 建议将其设置为 1，以保持和未来的兼容性。

通过上述 load 方法载入资源后，接着调用 play()方法实现声音文件的播放。play()方法及其参数说明如下：

```
int play (int soundID, float leftVolume, float rightVolume, int priority, int loop, float rate)
```

其中，

soundID：load()函数返回的声音 ID 号。

leftVolume：左声道音量设置。

rightVolume：右声道音量设置。

priority：播放声音的优先级，数值越高，优先级越大。

loop：是否循环，–1 表示无限循环，0 表示不循环，其他值表示要重复播放的次数。

rate：播放速率，1.0 的播放率可以使声音按照其原始频率播放；而 2.0 的播放速率，可以使声音按照其原始频率的两倍播放；如果为 0.5 的播放率，则播放速率是原始频率的一半。播放速率的取值范围是 0.5~2.0。

其应用步骤如下：

```
SoundPool mSoundPool = null;
```

```
HashMap<Integer, Integer> soundMap = new HashMap<Integer, Integer>();
mSoundPool = new SoundPool(20, AudioManager.STREAM_SYSTEM, 5);
soundMap.put(1 , mSoundPool.load(this, R.raw.bomb , 1));
```

然后在事件监听中实现声音文件的调用播放，如下：

```
mSoundPool.play(soundMap.get(1), 1, 1, 0, 0, 1);
```

5.9.2 视频播放应用

在 Android 中，实现视频播放的方式有 4 种，分别为：使用 Android 自带的播放器，使用 VideoView 来播放，使用 MediaPlayer 类和 SurfaceView 播放，使用 WebView 播放。它们各自的应用场景及侧重点不同，下面对它们的应用分别进行介绍。

（1）Android 自带的播放器应用较为简单，代码如下。

```
Intent intent = new Intent(Intent.ACTION_VIEW);
intent.setDataAndType(Uri, MimeType);
startActivity(intent);
```

指定 Action 为 ACTION_VIEW，Data 为 Uri，Type 为其 MIME 类型，指定所要播放的文件类型。具体实例代码如下。

```
Intent intent = new Intent(Intent.ACTION_VIEW);
//从 SDCard 文件中获取指定文件的 URi
File sdcard = Environment.getExternalStorageDirectory(); //获取 SDCard 路径
File audioFile = new File(sdcard.getPath()+"/music/test.mp3");
//需要获取该文件的 Uri
Uri audioUri = Uri.fromFile(audioFile);
//指定 Uri 和 MIME
intent.setDataAndType(audioUri, "audio/mp3");
startActivity(intent);
```

（2）使用 VideoView 来播放视频。

Android 提供了 VideoView 组件标签在布局文件中使用，然后在 Activity 中获取该组件引用，VideoView 结合 MediaController 来实现对其控制，再应用 setVideoURI()或 setVideoPath()方法加载需要播放的视频，通过调用 VideoView 的 start()方法播放视频。关键代码如下：

```
private void startVideo(String path){
    //指定播放文件路径
    videoView.setVideoURI(Uri.parse(path));
    //设置关联
    videoView.setMediaController(new MediaController(this));
    //开始播放
```

```
            videoView.start();
    }
```

(3) 使用 MediaPlayer 和 SurfaceView 组合实现视频播放。

上述介绍了 MediaPlayer 播放音频的方法，同样 MediaPlayer 也可以播放视频，但其缺陷是没有图像输出界面，因而需要结合 SurfaceView 组件来显示视频图像。SurfaceView 组件可以通过 SurfaceView 组件标签在布局文件中使用，也可以通过 Activity 中创建对象来使用。关键代码如下。

```
String videoPath = "/sdcard/test.3gp"
surfaceHolder = surfaceView.getHolder();
//设置 SurfaceHolder 的监听
    surfaceHolder.addCallback(this);
//设置 SurfaceView 的类型
surfaceHolder.setType(SurfaceHolder.SURFACE_TYPE_PUSH_BUFFERS);
public boolean onOptionsItemSelected(MenuItem item) {
        switch (item.getItemId()) {
        case 0:
            playVideo(videoPath);
            break;
    …
        }
        return super.onOptionsItemSelected(item);
    }
    private void playVideo(String strPath) {
                //加载 Raw 资源中的数据
                mediaPlayer = MediaPlayer.create(this, R.raw.test);
                //设置 Video 影片以 SurfaceHolder 形式播放
                mediaPlayer.setDisplay(surfaceHolder);
                //开始播放
                mediaPlayer.start();
                }
```

(4) 使用 WebView 播放视频。

WebView 通过 Android 系统中 WebKit 渲染引擎加载显示网页，对 HTML5 提供支持，WebView 在实践中有以下两种方法可以实现视频播放。

第一种方法实现步骤如下。

(1) 在 Activity 中实例化 WebView 组件：

```
WebView wv = new WebView(this);
```

(2) 然后通过调用 WebView 的 loadUrl()方法，设置 WevView 要显示的网页：

```
webView.loadUrl("http://www.google.com"); //网络调用
webView.loadUrl("file:///android_asset/XX.html");
```

//本地文件调用，本地文件存放在 assets 文件夹中

（3）在 Activity 中通过 setContentView()方法来显示网页视图、onKeyDown()方法来实现 WebView 回退功能。

（4）在 AndroidManifest.xml 文件中添加访问网络权限，如下：

```
<uses-permission android:name="android.permission.INTERNET" />
```

否则会出现 Web page not available 错误。

Activity 代码如下。

```
1.  import android.app.Activity;
2.  import android.os.Bundle;
3.  import android.view.KeyEvent;
4.  import android.webkit.WebView;
5.   public class MainActivity extends Activity {
6.      private WebView webview;
7.      @Override
8.      public void onCreate(Bundle savedInstanceState) {
9.         super.onCreate(savedInstanceState);
10.        //实例化 WebView 对象
11.        webview = new WebView(this);
12.        //设置 WebView 属性，能够执行 JavaScript 脚本
13.        webview.getSettings().setJavaScriptEnabled(true);
14.        //加载需要显示的网页
15.        webview.loadUrl("http://www.bcpl.cn/");
16.        //设置 Web 视图
17.        setContentView(webview);
18.     }
19.
20.     @Override
21.     //设置回退
22.     //覆盖 Activity 类的 onKeyDown(int keyCoder,KeyEvent event)方法
23.     public boolean onKeyDown(int keyCode, KeyEvent event) {
24.        if ((keyCode==KeyEvent.KEYCODE_BACK) && webview.canGoBack())
{
25.         webview.goBack(); //goBack()表示返回 WebView 的上一个页面
26.            return true;
27.        }
28.        return false;
29. }
```

上述代码中，如果单击链接后由指定的程序处理，而不是 Android 的系统浏览器（Browser）响应，可以通过给 WebView 添加一个事件监听对象（WebViewClient)并重写 shouldOverrideUrlLoading 方法来进行实现。其方法如下：

```
public boolean shouldOverrideUrlLoading (WebView view, String url)
{ view.loadUrl(url);
 return true; }
```

第二种方法实现与第一种的步骤大致相同，不同之处主要有两个方面，一个是在布局文件中声明 WebView 组件标签，并在 Activity 中进行引用；另外一个是通过调用 setWebViewClient( )方法，设置 WebView 视图，让 WebView 能够响应超链接。其关键代码如下。

```
1.   public class MainActivity extends Activity {
2.       private WebView webview;
3.       @Override
4.       public void onCreate(Bundle savedInstanceState) {
5.           super.onCreate(savedInstanceState);
6.           setContentView(R.layout.main);
7.           webview = (WebView) findViewById(R.id.webview);
8.           //设置 WebView 属性，能够执行 JavaScript 脚本
9.           webview.getSettings().setJavaScriptEnabled(true);
10.          //加载需要显示的网页
11.          webview.loadUrl("http://www.bcpl.cn/");
12.          //设置 Web 视图
13.          webview.setWebViewClient(new HelloWebViewClient ());
14.      }
15.
16.      @Override
17.      //设置回退
18.      //覆盖 Activity 类的 onKeyDown(int keyCoder,KeyEvent event)方法
19.      public boolean onKeyDown(int keyCode, KeyEvent event) {
20.          if ((keyCode == KeyEvent.KEYCODE_BACK) && webview.canGoBack())
{
21.              webview.goBack(); //goBack()表示返回 WebView 的上一页面
22.              return true;
23.          }
24.          return false;
25.      }
26.      //定义 Web 视图类
27.      private class HelloWebViewClient extends WebViewClient {
28.          @Override
29.          public boolean shouldOverrideUrlLoading(WebView view, String url) {
30.              view.loadUrl(url);
31.              return true;
32.          }
33.      }
34. }
```

此外，在 WebView 中还可以实现与 JS 的双向交互，对此部分感兴趣的读者，可以阅读相关材料。

5.9.3 音视频录制应用

Android 系统中，通过 MediaRecorder 可以实现音视频的录制，音频录制需要借助硬件麦克风，视频需要硬件摄像头（Camera），下面仅对音频录制的主要步骤进行简要介绍。

（1）实例化 MediaRecorder 类（创建对象）。

（2）调用 MediaRecorder 类的 setAudioSource 方法设置硬件麦克风，调用 setOutputFormat 方法设置输出文件的格式，调用 setAudioEncoder 方法设置音频文件的编码。

（3）设置文件输出位置，准备开始、录音。

```
1.   public class MediaRecorder extends ListActivity
2.   {
3.       private Button mAudioStartBtn;
4.       private Button mAudioStopBtn;
5.       private File mRecAudioFile;            // 录制的音频文件
6.       private File mRecAudioPath;            // 录制的音频文件路径
7.       private MediaRecorder mMediaRecorder;// MediaRecorder 对象
8.       private List<String> mMusicList = new ArrayList<String>();//录音
         文件列表
9.       private String strTempFile = "redio_";// 零时文件的前缀
10.
11.      @Override
12.      protected void onCreate(Bundle savedInstanceState)
13.      {
14.          // TODO Auto-generated method stub
15.          super.onCreate(savedInstanceState);
16.          setContentView(R.layout.mymultimedia_mediarecorder1);
17.
18.          mAudioStartBtn = (Button) findViewById(R.id.mediarecorder1_
             AudioStartBtn);
19.          mAudioStopBtn = (Button) findViewById(R.id.mediarecorder1_
             AudioStopBtn);
20.
21.          /*按钮状态*/
22.          mAudioStartBtn.setEnabled(true);
23.          mAudioStopBtn.setEnabled(false);
24.
25.          /* 检测是否存在 SD 卡 */
26.          if (Environment.getExternalStorageState().equals(android.
```

```
                    os.Environment.MEDIA_MOUNTED))
27.             {
28.                 mRecAudioPath = Environment.getExternalStorageDirectory();
                    // 得到 SD 卡的路径
29.                 musicList();// 更新所有录音文件到 List 中
30.             } else
31.             {
32.                 Toast.makeText(mediarecorder1.this, "没有 SD 卡", Toast.
                    LENGTH_LONG).show();
33.             }
34.
35.             /* "开始"按钮事件监听 */
36.             mAudioStartBtn.setOnClickListener(new Button.OnClickListener()
37.             {
38.                 @Override
39.                 public void onClick(View arg0)
40.                 {
41.                     try
42.                     {
43.                         /* ①Initial：实例化 MediaRecorder 对象 */
44.                         mMediaRecorder = new MediaRecorder();
45.                         /* ②setAudioSource/setVedioSource*/
46.                         mMediaRecorder.setAudioSource(MediaRecorder.
                            AudioSource.MIC);//设置麦克风
47.                         /* ②设置输出文件的格式：THREE_GPP/MPEG-4/RAW_AMR
                            /Default
48.                         * THREE_GPP(3gp 格式，H263 视频/ARM 音频编码)、MPEG-4、
                            RAW_AMR(只支持音频且音频编码要求为 AMR_NB)
49.                         * */
50.                         mMediaRecorder.setOutputFormat(MediaRecorder.
                            OutputFormat.DEFAULT);
51.                         /* ②设置音频文件的编码：AAC/AMR_NB/AMR_MB/Default */
52.                         mMediaRecorder.setAudioEncoder(MediaRecorder.
                            AudioEncoder.DEFAULT);
53.                         /* ②设置输出文件的路径 */
54.                         try
55.                         {
56.                             mRecAudioFile = File.createTempFile(strTempFile,
                                ".amr", mRecAudioPath);
57.
58.                         } catch (Exception e)
59.                         {
60.                             e.printStackTrace();
61.                         }
```

```
62.              mMediaRecorder.setOutputFile(mRecAudioFile.
                     getAbsolutePath());
63.              /* ③准备 */
64.              mMediaRecorder.prepare();
65.              /* ④开始 */
66.              mMediaRecorder.start();
67.              /*按钮状态*/
68.              mAudioStartBtn.setEnabled(false);
69.              mAudioStopBtn.setEnabled(true);
70.          } catch (IOException e)
71.          {
72.              e.printStackTrace();
73.          }
74.      }
75.  });
76.  /* "停止" 按钮事件监听 */
77.  mAudioStopBtn.setOnClickListener(new Button.OnClickListener()
78.  {
79.      @Override
80.      public void onClick(View arg0)
81.      {
82.          // TODO Auto-generated method stub
83.          if (mRecAudioFile != null)
84.          {
85.              /* ⑤停止录音 */
86.              mMediaRecorder.stop();
87.              /* 将录音文件添加到List中 */
88.              mMusicList.add(mRecAudioFile.getName());
89.              ArrayAdapter<String> musicList = new ArrayAdapter
                     <String>(mediarecorder1.this,
90.                  R.layout.list, mMusicList);
91.              setListAdapter(musicList);
92.              /* ⑥释放 MediaRecorder */
93.              mMediaRecorder.release();
94.              mMediaRecorder = null;
95.              /* 按钮状态 */
96.              mAudioStartBtn.setEnabled(true);
97.              mAudioStopBtn.setEnabled(false);
98.          }
99.      }
100. });
101. }
102.
103. /* 播放录音文件 */
```

```java
104.    private void playMusic(File file)
105.    {
106.        Intent intent = new Intent();
107.        intent.addFlags(Intent.FLAG_ACTIVITY_NEW_TASK);
108.        intent.setAction(android.content.Intent.ACTION_VIEW);
109.        /* 设置文件类型 */
110.        intent.setDataAndType(Uri.fromFile(file), "audio");
111.        startActivity(intent);
112.    }
113.
114.    @Override
115.    /* 当单击列表时，播放被单击的音乐 */
116.    protected void onListItemClick(ListView l, View v, int position, long id)
117.    {
118.        /* 得到被单击的文件 */
119.        File playfile = new File(mRecAudioPath.getAbsolutePath() + File.separator
120.                + mMusicList.get(position));
121.        /* 播放 */
122.        playMusic(playfile);
123.    }
124.
125.    /* 播放列表 */
126.    public void musicList()
127.    {
128.        // 取得指定位置的文件设置显示到播放列表
129.        File home = mRecAudioPath;
130.        if (home.listFiles(new MusicFilter()).length > 0)
131.        {
132.            for (File file : home.listFiles(new MusicFilter()))
133.            {
134.                mMusicList.add(file.getName());
135.            }
136.            ArrayAdapter<String> musicList = new ArrayAdapter<String>(mediarecorder1.this,
137.                    R.layout.list, mMusicList);
138.            setListAdapter(musicList);
139.        }
140.    }
141. }
142.
143. /* 过滤文件类型 */
144. class MusicFilter implements FilenameFilter
```

```
145.    {
146.        public boolean accept(File dir, String name)
147.        {
148.            return (name.endsWith(".amr"));
149.        }
150.    }
```

同时在布局文件中添加两个 Button 按钮,分别为"停止录音"、"开始录音"。

5.10 Android 图形应用

Android 系统提供了图形、图像处理的 API 供开发者处理 2D、3D 图形及开发游戏,2D 图形的处理主要使用 Android 内置的 Canavas、Graphics 等组件。除此之外,针对 3D 图形的处理,Android 系统通过内置、调用底层的 OpenGL(Open Graphics Library,开放图形库接口)包,支持 3D 的加速、渲染等。2007 年发布的 OpenGL ES 2.0(Open Graphics Library for Embedded System)是 OpenGL 的子集,是一个 2D/3D 轻量级图形库,专门用于实现嵌入式系统和移动设备的图形可编程开发。其中,以 android.opengl 包下的 GLU、GLUtils、GLSurfaceView 等工具类应用最为广泛。

在 Android 系统中 2D 图形绘制,既可以通过调用 Canavas 组件类的 drawColor()、drawText()、drawLine()等方法进行绘制,也可以通过调用 GLSurfaceView 组件、GL10 组件的相关方法来绘制。下面对它们的主要应用步骤分别进行介绍。

5.10.1 Canavas 组件图形应用

首先自定义继承于 View 的视图类 MyView,在其中定义 Paint(画笔),重写 onDraw()方法,在方法中根据任务要求调用 Canavas 的相关方法。

```
1.  public class MyView extends View {
2.      private Paint paint=null;
3.      public MyView(Context context) {
4.          super(context);
5.          paint=new Paint();
6.      }
7.
8.      @Override
9.      protected void onDraw(Canvas canvas) {
10.         // TODO Auto-generated method stub
11.         super.onDraw(canvas);
12.         //设置画布的颜色为白色
13.         canvas.drawColor(Color.WHITE);
14.         paint.setColor(Color.RED);
15.         canvas.drawText("写点什么呢", 40, 50, paint);
```

```
16.         canvas.drawLine(10, 10, 100, 150, paint);
17.         Bitmap bitmap = BitmapFactory.decodeResource(getResources(),
            R.drawable.ic_launcher);
18.         canvas.drawBitmap(bitmap, 50, 100,null);
19.     }
20. }
```

然后在 Activity 类的 onCreate 方法中创建上述类的对象,并设置视图界面显示,主要代码如下:

```
setContentView(new MyView(this));
```

5.10.2 OpenGL ES 包组件图形应用

使用 OpenGL ES 包组件创建图形应用的主要步骤如下。

(1) 自定义继承于渲染器 GLSurfaceView.Renderer 的类,实现继承的三个接口 onDrawFrame()、onSurfaceChanged()、onSurfaceCreated(),代码如下。

```
1.  public class MySceneRenderer implements GLSurfaceView.Renderer {
2.      @Override
3.      public void onDrawFrame(GL10 gl) {
4.          gl.glEnable(GL10.GL_CULL_FACE);
5.          …
6.      }
7.
8.      @Override
9.      public void onSurfaceChanged(GL10 gl, int width, int height) {
10.         gl.glViewport(0, 0, width, height);
11.         …
12.     }
13.
14.     @Override
15.     public void onSurfaceCreated(GL10 gl, EGLConfig config) {
16.         gl.glDisable(GL10.GL_DITHER);
17.         …
18.     }
19.
20. }
```

(2) 自定义继承于 GLSurfaceView 的视图类,创建对象,设置渲染器,实现屏幕事件的处理方法 onTouchEvent(),主要处理如下。

```
1.  public class MySurfaceView extends GLSurfaceView {
2.      private final float TOUCH_SCALE_FACTOR = 180.0f / 320;
3.      private MySceneRenderer mysRenderer;
```

```
4.      private float myPreviousX;
5.      public MySurfaceView(Context context) {
6.          super(context);
7.          mysRenderer = new MySceneRenderer();
8.          this.setRenderer(mysRenderer);
9.          this.setRenderMode(GLSurfaceView.RENDERMODE_CONTINUOUSLY);
10.     }
11.
12.     @Override
13.     public boolean onTouchEvent(MotionEvent event) {
14.         float x = event.getX();
15.         …
16.         switch (event.getAction()) {
17.         case MotionEvent.ACTION_MOVE:
18.             …
19.             requestRender();
20.             break;
21.         }
22.         myPreviousX = x;
23.         …
24.         return true;
25.     }
26.
27. }
```

上述代码第 9 行 setRenderMode(GLSurfaceView.RENDERMODE_CONTINUOUSLY) 设置 Render 的 mode 为自动循环模式（GL 线程定时自动循环调用 onDrawFrame()方法进行绘制）。

Render 的 mode 可以设置为两种模式，还有一种为手工模式，通过 setRenderMode(GLSurfaceView.RENDERMODE_WHEN_DIRTY)设定；绘制图形时，需要调用 GLSurfaceView.requestRender()方法，此方式可以有效降低 CPU 负载。

（3）调用 GL10 的 glEnableClientState()、glVertexPointer()、glDrawElements()等方法进行图形绘制。

（4）在 Activity 中把上述创建的视图添加到布局中进行显示，代码如下。

```
1. public class MyActivity extends Activity {
2.     private MySurfaceView mySurfaceView;
3.     @Override
4.     public void onCreate(Bundle savedInstanceState) {
5.         super.onCreate(savedInstanceState);
6.         setContentView(R.layout.activity_main);
7.         mySurfaceView = new MySurfaceView(this);
8.         mySurfaceView.requestFocus();
```

```
9.         mySurfaceView.setFocusableInTouchMode(true);
10.        LinearLayout lv = (LinearLayout) this.findViewById
           (R.id.linear);
11.        lv.addView(mySurfaceView);
12.    }
```

5.11 项目案例

学习目标：学习 Android 系统高级控件、事件处理、消息传递等应用。

案例描述：使用 RelativeLayout 相对布局、TextView 控件、ListView 控件，并设置相对父控件的位置、控件之间相对位置的属性，实现"妈咪宝贝"的加载资源界面、欢迎进入界面。

案例要点：ListView 控件、事件响应处理、Handler、ArrayAdapter。

案例步骤：

（1）创建工程 Project_Chapter_5，选择 Android 4.2 作为目标平台。

（2）创建 logo.xml 文件，将文件存放在 res/layout 目录下。

```xml
1.  <?xml version="1.0" encoding="utf-8"?>
2.  <RelativeLayout xmlns:android="http://schemas.android.com/apk/res/android"
3.      android:layout_width="match_parent"
4.      android:layout_height="match_parent"
5.      android:background="@drawable/huanying"
6.      android:orientation="vertical" >
7.
8.      <ProgressBar
9.          android:id="@+id/progressBar"
10.         style="?android:attr/progressBarStyleHorizontal"
11.         android:layout_width="fill_parent"
12.         android:layout_height="wrap_content"
13.         android:layout_alignParentBottom="true"
14.         android:layout_alignParentLeft="true"
15.         android:max="100"
16.         android:progress="30"
17.         android:secondaryProgress="50" />
18. </RelativeLayout>
```

（3）在 src 目录下包 com.bcpl.bady 中，创建 LogoActivity.java 文件，代码如下。

```java
1.  public class LogoActivity extends Activity {
2.      private ProgressBar pr;
3.      private int currentValue=0;
4.      private Handler handler = new Handler() {
```

```
5.
6.          @Override
7.          public void handleMessage(Message msg) {
8.              // TODO Auto-generated method stub
9.              super.handleMessage(msg);
10.             switch (msg.what) {
11.             case 0:
12.                 Toast.makeText(getBaseContext(), "妈咪宝贝", Toast.
                    LENGTH_SHORT).show();
13.                 break;
14.             case 1:
15.                 Toast.makeText(getBaseContext(), "欢迎使用！", Toast.
                    LENGTH_SHORT).show();
16.                 Intent i = new Intent(LogoActivity.this,
                    FenLeiActivity.class);
17.                 startActivity(i);
18.                 LogoActivity.this.finish();
19.                 break;
20.             }
21.         }
22.
23.     };
24.
25.     @Override
26.     protected void onCreate(Bundle savedInstanceState) {
27.         // TODO Auto-generated method stub
28.         super.onCreate(savedInstanceState);
29.         setContentView(R.layout.logo);
30.      pr=(ProgressBar)this.findViewById(R.id.progressBar);
31.
32.         new Thread(new Runnable(){
33.
34.             @Override
35.             public void run() {
36.                 // TODO Auto-generated method stub
37.                 while(currentValue<=100){
38.                     pr.setProgress(currentValue);
39.                     try {
40.                         Thread.sleep(100);
41.                     } catch (InterruptedException e) {
42.                         // TODO Auto-generated catch block
43.                         e.printStackTrace();
44.                     }
45.                     currentValue+=5;
```

```
46.                }
47.                currentValue=0;
48.            }
49.
50.       }).start();
51.
52.       File f = new File(Environment.getExternalStorageDirectory()
          + "/data");
53.
54.       if (!f.exists()) {
55.
56.           new Thread(new Runnable() {
57.
58.               @Override
59.               public void run() {
60.                   // TODO Auto-generated method stub
61.                   try {
62.                       copy();
63.                       Message msg = handler.obtainMessage();
64.                       msg.what = 0;
65.                       handler.sendMessage(msg);
66.                       unzipFile(Environment.
                          getExternalStorageDirectory()
67.                           + "", Environment.
                          getExternalStorageDirectory()
68.                           + "/data.zip");
69.
70.                   } catch (Exception e) {
71.                       // TODO Auto-generated catch block
72.                       e.printStackTrace();
73.                   }
74.               }
75.           }).start();
76.       }else{
77.           new Thread(new Runnable(){
78.
79.               @Override
80.               public void run() {
81.                   // TODO Auto-generated method stub
82.                   try {
83.                       Thread.sleep(2000);
84.                   } catch (InterruptedException e) {
85.                       // TODO Auto-generated catch block
86.                       e.printStackTrace();
```

```
87.              }
88.              Message msg = new Message();
89.              msg.what=1;
90.              handler.sendMessage(msg);
91.          }
92.
93.      }).start();
94.
95.      }
96.  }
97.
98.
99.  public void copy() {
100.     try {
101.         InputStream input;
102.         String outFile = Environment.
                 getExternalStorageDirectory()
103.                 .getPath();
104.         OutputStream output = new FileOutputStream(outFile +
                 "/data.zip");
105.         input = this.getAssets().open("data.zip");
106.         byte[] buf = new byte[1024];
107.         int length;
108.         while ((length = input.read(buf)) > 0) {
109.             output.write(buf, 0, length);
110.         }
111.         input.close();
112.         output.flush();
113.         output.close();
114.     } catch (Exception e) {
115.         e.printStackTrace();
116.     }
117. }
118. //解压缩
119. public void unzipFile(String targetPath,String zipFilePath){
120.
121.     try {
122.         File zipFile = new File(zipFilePath);
123.         InputStream is = new FileInputStream(zipFile);
124.         ZipInputStream zis = new ZipInputStream(is);
125.         ZipEntry entry = null;
126.         System.out.println("开始解压:" + zipFile.getName() +
                 "...");
127.         while ((entry = zis.getNextEntry()) != null) {
```

```
128.                    String zipPath = entry.getName();
129.                    try {
130.
131.                        if (entry.isDirectory()) {
132.                            File zipFolder = new File(targetPath +
                                File.separator
133.                                    + zipPath);
134.                            if (!zipFolder.exists()) {
135.                                zipFolder.mkdirs();
136.                            }
137.                        } else {
138.                            File file = new File(targetPath +
                                File.separator
139.                                    + zipPath);
140.                            if (!file.exists()) {
141.                                File pathDir = file.getParentFile();
142.                                pathDir.mkdirs();
143.                                file.createNewFile();
144.                            }
145.
146.                            FileOutputStream fos = new
                                FileOutputStream(file);
147.                            int bread;
148.                            while ((bread = zis.read()) != -1) {
149.                                fos.write(bread);
150.                            }
151.                            fos.close();
152.
153.                        }
154.                        System.out.println("成功解压:" + zipPath);
155.
156.                    } catch (Exception e) {
157.                        System.out.println("解压"+zipPath+"失败");
158.                        Message msg = new Message();
159.                        msg.what = 1;
160.                        handler.sendMessage(msg);
161.                        continue;
162.                    }
163.                }
164.                zis.close();
165.                is.close();
166.                System.out.println("解压结束");
167.                Message msg = new Message();
168.                msg.what = 1;
```

```
169.            handler.sendMessage(msg);
170.        } catch (Exception e) {
171.            e.printStackTrace();
172.        }
173.
174.    }
175. }
```

（4）创建 fenlei.xml 文件，将文件存放在 res/layout 目录下。

```
1.  <?xml version="1.0" encoding="utf-8"?>
2.  <LinearLayout xmlns:android="http://schemas.android.com/apk/res/android"
3.      android:layout_width="match_parent"
4.      android:layout_height="fill_parent"
5.      android:orientation="vertical" >
6.
7.      <RelativeLayout
8.          android:id="@+id/rlt_top_bar"
9.          android:layout_width="fill_parent"
10.         android:layout_height="83dp" >
11.
12.         <RadioGroup
13.             android:id="@+id/recommend_radiogroup"
14.             android:layout_width="wrap_content"
15.             android:layout_height="wrap_content"
16.             android:layout_centerInParent="true"
17.             android:orientation="horizontal" >
18.         <RadioButton
19.             android:id="@+id/recommend_rbtn_splendrecommend"
20.             android:layout_width="107dp"
21.             android:layout_height="wrap_content"
22.             android:button="@null"
23.             android:checked="true"
24.             android:text="@string/shui"
25.             android:textSize="@dimen/btn_textsize" />
26.
27.         <RadioButton
28.             android:id="@+id/recommend_rbtn_popularserialize"
29.             android:layout_width="107dp"
30.             android:layout_height="wrap_content"
31.             android:button="@null"
32.             android:checked="false"
33.             android:text="寓言故事"
34.             android:textSize="@dimen/btn_textsize" />
```

```
35.
36.        <RadioButton
37.            android:id="@+id/recommend_btn_brandhall"
38.            android:layout_width="107dp"
39.            android:layout_height="wrap_content"
40.            android:button="@null"
41.            android:checked="false"
42.            android:text="童话故事"
43.            android:textSize="@dimen/btn_textsize" />
44.    </RadioGroup>
45.
46. </RelativeLayout>
47.
48. <LinearLayout
49.     android:layout_width="fill_parent"
50.     android:layout_height="fill_parent"
51.     android:layout_weight="2"
52.     android:orientation="horizontal" >
53.
54.     <include layout="@layout/mainleft" />
55.
56.     <ListView
57.         android:id="@+id/list"
58.         android:layout_width="290dp"
59.         android:layout_height="fill_parent"
60.         android:layout_weight="0.50"
61.         android:background="@drawable/bac"
62.         android:cacheColorHint="#FFFFFF"
63.         android:divider="@color/jiangexian"
64.         android:dividerHeight="1dp"
65.         android:fadingEdge="none" >
66.     </ListView>
67.
68.     <include layout="@layout/mainright" />
69. </LinearLayout>
70.
71. </LinearLayout>
```

注意：在上述文件中使用布局包含，包含 mainright.xml 和 mainleft.xml 两个布局文件，这两个文件内容类似，这里只写 mainleft.xml 文件，代码如下。

```
1. <?xml version="1.0" encoding="utf-8"?>
2. <LinearLayout xmlns:android="http://schemas.android.com/apk/res/android"
```

```
3.        android:layout_width="wrap_content"
4.        android:layout_height="fill_parent"
5.        android:id="@+id/mainleft"
6.        android:background="@drawable/main_left"
7.    >
8.    </LinearLayout>
9.
```

（5）在 src 目录下的包 com.bcpl.bady 中，创建 FenLeiActivity.java 文件，代码如下。

```
1.  public class FenleiActivity extends Activity implements
    OnItemClickListener{
2.      private ListView listview;
3.      private Intent intent;
4.      @Override
5.      protected void onCreate(Bundle savedInstanceState) {
6.          // TODO Auto-generated method stub
7.          super.onCreate(savedInstanceState);
8.          setContentView(R.layout.fenlei);
9.          listv();
10.     }
11.
12.     private void listv() {
13.         // TODO Auto-generated method stub
14.
15.          listview = (ListView) findViewById(R.id.list); // 获取列表
                                                             视图
16.         int[] imageId = new int[] { R.drawable.katong, R.drawable.
            lishi,
17.                 R.drawable.mingren, R.drawable.yizhi,R.drawable.
                    minjian ,R.drawable.hui_ben,R.drawable.qi_meng,R.
                    drawable.qing_shang}; // 定义并初始化保存图片 id 的数组
18.         String[] title = new String[] { "童话故事","成语故事","名人
            故事","益智故事",
19.                 "民间传说" ,
20.                 "绘本故事","启蒙故事","情商故事"
21.                 };
22.
23.         // 定义并初始化保存列表项文字的数组
24.         List<Map<String, Object>> listItems = new ArrayList<Map
            <String, Object>>(); // 创建一个 list 集合
25.         // 通过 for 循环将图片 id 和列表项文字放到 Map 中，并添加到 list 集合中
26.         for (int i = 0; i < imageId.length; i++) {
27.             Map<String, Object> map = new HashMap<String, Object>();
                // 实例化 Map 对象
```

```
28.              map.put("image", imageId[i]);
29.              map.put("title", title[i]);
30.
31.              listItems.add(map); // 将 map 对象添加到 List 集合中
32.          }
33.
34.          SimpleAdapter adapter = new SimpleAdapter(this, listItems,
35.              R.layout.itemss, new String[] { "title", "image" },
                 new int[] {
36.                         R.id.title1, R.id.image1 }); // 创建
                            SimpleAdapter
37.
38.          listview.setAdapter(adapter); // 将适配器与 ListView 关联
39.          listview.setOnItemClickListener(this);
40.
41.      }
42.      @Override
43.      public void onItemClick(AdapterView<?> arg0, View v, int arg2,
         long arg3) {
44.          // TODO Auto-generated method stub
45.          intent=new Intent();
46.
47.          HashMap<String,Object> map=(HashMap<String,Object>)
             listview.getItemAtPosition(arg2);
48.          //String title1=map.get("title");
49.          if(arg2==0){
50.              intent.setClass(FenleiActivity.this, RecommendActivity.
                 class);
51. //           Toast.makeText(getApplicationContext(), "你选了 "+arg2+"
                 个", Toast.LENGTH_SHORT).show();
52.
53.          }if(arg2==1){
54.              intent.setClass(FenleiActivity.this, NoneActivity.
                 class);
55.          }if(arg2==2){
56.              intent.setClass(FenleiActivity.this, NoneActivity.
                 class);
57.
58.          } if(arg2==3){
59.              intent.setClass(FenleiActivity.this, NoneActivity.
                 class);
60.          } if(arg2==4){
61.              intent.setClass(FenleiActivity.this, NoneActivity.
                 class);
```

```
62.            }if(arg2==5){
63.                intent.setClass(FenleiActivity.this, NoneActivity.class);
64.            }if(arg2==6){
65.                intent.setClass(FenleiActivity.this, NoneActivity.class);
66.            } if (arg2==7){
67.                intent.setClass(FenleiActivity.this, NoneActivity.class);
68.            }
69.            startActivity(intent);
70.
71.        }
72. }
```

（6）部署运行 Project_Chapter_5 工程，运行效果如图 5-21 和图 5-22 所示。

图 5-21　欢迎界面显示

图 5-22　故事分类界面

习　　题

一、简答题

1．简述 ListView 控件适配器类型及创建步骤。

2．简述 Tabhost 控件与 Fragment 之间的区别。

3．Android 事件处理机制常用的方法有哪些？按键事件和触摸事件处理有什么不同？

4. 通常什么情况下，使用 Android 消息处理类 Handler？

二、实训

要求：

在本章项目案例的基础上，在登录界面和分类界面之间添加欢迎界面，如图 5-23 所示。

图 5-23 欢迎界面

第6章

Android 界面菜单、对话框

学习目标

本章介绍了 Android 界面菜单 Menu、选项菜单、子菜单、快捷菜单、对话框 Dialog、AlertDialog、日期选择对话框 DatePickerDialog、时间选择对话框 TimePickerDialog、ProgressDialog、Toast 控件、Notification 控件等。读者通过本章的学习可掌握以下知识要点。

（1）菜单 Menu 的分类、创建方法。
（2）选项菜单的分类及创建方法。
（3）子菜单的添加方法及应用。
（4）对话框的常用方法及属性设置。
（5）Toast 控件及 Notification 控件的属性设置、通常用法及实现方式。

6.1 菜单控件 Menu

6.1.1 Menu 概述

菜单是应用程序中非常重要的组成部分，能够在不占用界面空间的前提下，为应用程序提供统一的功能和设置界面，并为程序开发人员提供易于使用的编程接口。在 Android 系统中，菜单和前面讲述的控件一样，不仅能够在代码中定义，而且可以像界面布局一样在 XML 文件中进行定义。使用 XML 文件定义界面菜单，将代码与界面设计分类，有助于简化代码的复杂程度，并且更有利于界面的可视化。

Android 系统支持以下三种菜单。
（1）选项菜单（Option Menu）。
（2）子菜单（Submenu）。
（3）上下文菜单（Context Menu）。

在 Activity 中可以通过重写 onCreateOptionsMenu(Menu menu)方法创建选项菜单，然后在用户按下手机的 Menu 按钮时就会显示创建好的菜单，在 onCreateOptionsMenu

(Menu menu)方法内部可以调用 Menu.add()方法实现菜单的添加。

如果处理选择事件，可以通过重写 Activity 的 onMenuItemSelected()方法实现，该方法常用于处理菜单被选择事件。

6.1.2 选项菜单及应用

1. 选项菜单简介

选项菜单是一种经常被使用的 Android 系统菜单。可以通过"菜单"键（Menu）打开浏览或选择。

选项菜单通常分为两类，分别是：图标菜单（Icon Menu）和扩展菜单（Expanded Menu）。

对于 Android 4.0 之后的版本，系统默认的 UI 风格有所变化，如果仍希望采用原有的显示方式，可以通过为 Activity 设置 Theme 指定风格，通过指定 Theme 以及 Theme.Light 均可以使用旧的菜单风格。具体实现方式是通过在 AndroidManifest.xml 中，activity 标签中添加属性 android:theme，显示如图 6-1 所示，代码如下。

```
<activity android:name=".MyMenuTest" android:label="@string/myMenuTest"
    android:theme="@android:style/Theme.Light" />
```

图 6-1 图标菜单（旧风格）

图标菜单在 Android 4.0 之后默认为垂直的列表型菜单，可以同时显示文字和图标的菜单，图标菜单不支持单选框和复选框控件。在创建菜单 Menu 时，如果不采用上述在 XML 中设定显示原有风格的方法，而仅通过 setIcon 方法给菜单添加图标无法显示出来（虽然在 Android 2.3 系统中是可以显示出来的）。其原因在于 4.0 系统中，涉及菜单的源码类 MenuBuilder 做了改变，mOptionalIconsVisible 成员初始值默认为 false（菜单设置图标不进行显示），所以，只要在创建菜单时通过调用 setOptionalIconsVisible 方法设置 mOptionalIconsVisible 为 true 即可显示，如图 6-2 所示。

图 6-2 图标菜单（新风格）

扩展菜单是垂直的列表型菜单，它不支持显示图标，但支持单选框和复选框控件。
1) onCreateOptionMenu()方法

只有在 Activity 中重载 onCreateOptionMenu()方法，才能够在 Android 应用程序中使用选项菜单。第一次使用选项菜单时，会调用 onCreateOptionMenu()方法，用来初始化菜单子项的相关内容（设置菜单子项自身的子项的 ID 和组 ID、菜单子项显示的文字和图片等）。

```
1.   final static int DOWNLOAD = Menu.FIRST;
2.     final static int UPLOAD = Menu.FIRST+1;
3.     @Override
4.     public boolean onCreateOptionsMenu(Menu menu){
5.         menu.add(0,DOWNLOAD,0,"下载");
6.         menu.add(0,UPLOAD,1,"上传");
7.         return true;
8.   }
```

第 1 行和第 2 行代码将菜单子项 ID 定义成静态常量，并使用静态常量 Menu.FIRST（整数类型，值为 1）定义第一个菜单子项，后续的菜单子项仅需在 Menu.FIRST 基础上增加相应的数值即可。

第 4 行 Menu 对象作为一个参数被传递到方法内部，因此在 onCreateOptionsMenu()方法中，用户可以使用 Menu 对象的 add()方法添加设置的菜单子项。

第 7 行代码是 onCreateOptionsMenu()方法返回值，返回 true 将显示在方法中设置的菜单，否则不能够显示菜单。

add()方法的语法：

MenuItem android.view.Menu.add(int groupId, int itemId, int order, CharSequence title)

其中，

groupId 是组 ID，用以批量地对菜单子项进行处理和排序。

itemId 是子项 ID，是每一个菜单子项的唯一标识，通过子项 ID 使应用程序能够定位到用户所选择的菜单子项。

order 是定义菜单子项在选项菜单中的排列顺序。

title 是菜单子项所显示的标题。

添加菜单子项的图标和快捷键：使用 setIcon()方法和 setShortcut()方法。

```
1.  menu.add(0,DOWNLOAD,0,"下载")
2.       .setIcon(R.drawable.download);
3.       .setShortcut(',','d');
```

第 2 行代码中设置新的图像资源，用户将需要使用的图像文件复制到/res/drawable 目录下。

setShortcut()方法中第一个参数是为数字键盘设定的快捷键；第二个参数是为全键盘设定的快捷键，且不区分字母的大小写。

注意：添加前需要在 AndroidManifest.xml 中的 activity 标签中添加属性 android:theme。

2）onPrepareOptionsMenu()方法

重载 Activity 中的 onPrepareOptionsMenu()方法，能够实现动态地添加、删除菜单子项，或修改菜单的标题、图标和可见性等内容。

onPrepareOptionsMenu()方法的返回值含义与 onCreateOptionsMenu()方法相同，即返回 true 显示菜单，返回 false 则不能显示菜单。

2．选项菜单应用

前面介绍了选项菜单 OptionsMenu 所涉及的 Menu、MenuItem 的基础知识，下面通过一个案例应用，熟悉和掌握选项菜单 OptionsMenu 的应用开发过程。

（1）创建一个新的 Android 工程，工程名为 OptionsMenuDemo，目标 API 选择 17（即 Android 4.2 版本），应用程序名为 OptionsMenuDemo，包名为 com.bcpl.activity，创建的 Activity 的名字为 MainActivity，最小 SDK 版本根据选择的目标 API 会自动添加为 8，创建项目工程。

（2）在 res 目录下 menu 文件夹中创建 game_menu.xml 文件，添加 menu 和 item，并设置相关属性，代码如下所示。

```
1.  <?xml version="1.0" encoding="utf-8"?>
2.  <menu xmlns:android="http://schemas.android.com/apk/res/android">
3.     <item android:icon="@drawable/icon" android:title="新游戏"
        android:id="@+id/new_game"/>
4.     <item android:icon="@drawable/icon" android:title="保存进度"
        android:id="@+id/save_game"/>
5.     <item android:icon="@drawable/icon" android:title="载入进度"
        android:id="@+id/load_game"/>
```

```
6.     <item android:icon="@drawable/icon" android:title="退出游戏"
           android:id="@+id/exit_game"/>
7. </menu>
```

（3）在 AndroidManifest.xml 中的 activity 标签中添加属性 android:theme，代码如下：

```
android:theme="@android:style/Theme.Light"
```

（4）修改 src 目录中 com.bcpl.activity 包下的 MainActivity.java 文件，初始化菜单的操作主要通过 onCreateOptionsMenu 方法中使用 MenuInflater 方法自带的 inflate 方法绑定调用 game_menu.xml 文件和 menu 菜单，方法 onOptionsItemSelected 中放置菜单选项被选中时的程序处理代码，本案例使用了 Toast 控件及方法，全部实现代码如下：

```
1.  package com.bcpl.activity;

2.  import android.app.Activity;
3.  import android.os.Bundle;
4.  import android.view.Menu;
5.  import android.view.MenuInflater;
6.  import android.view.MenuItem;
7.  import android.widget.Toast;

8.  public class MainActivity extends Activity {
9.      /** Called when the activity is first created. */
10.     @Override
11.     public void onCreate(Bundle savedInstanceState) {
12.         super.onCreate(savedInstanceState);
13.         setContentView(R.layout.main);
14.     }
15.
16.     @Override
17.     public boolean onCreateOptionsMenu(Menu menu) {
18.         MenuInflater inflater = getMenuInflater();
19.         inflater.inflate(R.menu.game_menu, menu);
20.         return true;
21.     }
22.
23.     @Override
24.     public boolean onOptionsItemSelected(MenuItem item) {
25.         int id = item.getItemId();
26.         switch (id) {
27.         case R.id.new_game:
28.             Toast.makeText(this, item.getTitle(), Toast.LENGTH_
                    LONG).show();
29.             break;
```

```
30.         case R.id.save_game:
31.             Toast.makeText(this, item.getTitle(), Toast.LENGTH_
                LONG).show();
32.             break;
33.         case R.id.load_game:
34.             Toast.makeText(this, item.getTitle(), Toast.LENGTH_
                LONG).show();
35.             break;
36.         case R.id.exit_game:
37.             Toast.makeText(this, item.getTitle(), Toast.LENGTH_
                LONG).show();
38.             break;
39.         default:
40.             break;
41.         }
42.         return super.onOptionsItemSelected(item);
43.     }
44. }
```

（5）部署运行 OptionsMenuDemo 项目工程，单击移动设备上的 menu 按钮，程序运行显示如图 6-3 所示。然后单击"保存进度"菜单，程序运行如图 6-4 所示。

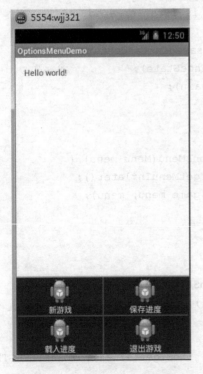

图 6-3 Menu 菜单（旧风格）

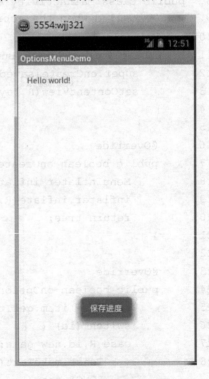

图 6-4 保存进度显示

6.1.3 子菜单及应用

1. 子菜单创建方式简介

子菜单是指能够显示更加详细信息的菜单子项。在子菜单中，菜单子项使用浮动窗体的显示形式，更好地适应了小屏幕的显示。

Android 系统的子菜单使用非常灵活，可以在选项菜单或快捷菜单中使用子菜单，这样有利于将相同或相似的菜单子项组织在一起，便于显示和分类。此外，子菜单不支持嵌套，子菜单的添加是使用 addSubMenu()方法来实现的。

```
1.  SubMenu uploadMenu = (SubMenu) menu.addSubMenu(0, UPLOAD,1,"上传
    ").setIcon(R.drawable.upload);
2.  uploadMenu.setHeaderIcon(R.drawable.upload);
3.  uploadMenu.setHeaderTitle("上传");
4.  uploadMenu.add(0,SUB_UPLOAD_A,0,"上传参数 A");
5.  uploadMenu.add(0,SUB_UPLOAD_B,0,"上传参数 B");
```

第 1 行代码在上述的 onCreateOptionsMenu()方法传递的 menu 对象上调用 addSubMenu()方法，在选项菜单中添加一个菜单子项，用户点击后可以打开子菜单。

addSubMenu()方法与选项菜单中使用过的 add()方法支持相同的参数，同样可以指定菜单子项的 ID、组 ID 和标题等参数，并且能够通过 setIcon()方法设置菜单所显示的图标。

第 2 行代码调用 setHeaderIcon ()方法，定义子菜单的图标。

第 3 行定义子菜单的标题，如果不设定子菜单的标题，子菜单将显示父菜单子项标题，即第 1 行代码中的"上传"。

第 4 行和第 5 行在子菜单中添加了两个菜单子项，菜单子项的更新方法和选择事件处理方法，仍然使用 onPrepareOptionsMenu()方法和 onOptionsItemSelected ()方法。

2. 子菜单应用

前面介绍了子菜单 SubMenu 的常用方法，下面通过一个接受用户菜单选项然后弹出子菜单，选中子菜单项后显示所选项目的案例应用，熟悉和掌握菜单 SubMenu 的应用开发过程。

（1）创建一个新的 Android 工程，工程名为 SubMenuDemo，目标 API 选择 17（即 Android 4.2 版本），应用程序名为 SubMenuDemo，包名为 com.bcpl.activity，创建的 Activity 的名字为 MainActivity，最小 SDK 版本根据选择的目标 API 会自动添加为 8，创建项目工程。

（2）在 res 目录下 menu 文件夹中创建 main.xml 文件，添加 menu 和 item，并设置相关属性，代码如下所示。

```xml
1.  <?xml version="1.0" encoding="utf-8"?>
2.  <menu xmlns:android="http://schemas.android.com/apk/res/android">
3.      <item android:id="@+id/file"
4.          android:icon="@drawable/icon"
5.          android:title="@string/file" >
6.          <!-- "file" submenu -->
7.          <menu>
8.              <item android:id="@+id/create_new"
9.                  android:title="@string/create_new" />
10.             <item android:id="@+id/open"
11.                 android:title="@string/open" />
12.             <item android:id="@+id/save"
13.                 android:title="@string/save" />
14.             <item android:id="@+id/exit"
15.                 android:title="@string/exit" />
16.         </menu>
17.     </item>
18. </menu>
```

（3）修改 res 目录下 values 文件夹中的 strings.xml 文件，代码如下所示。

```xml
1.  <?xml version="1.0" encoding="utf-8"?>
2.  <resources>
3.      <string name="hello">Hello World, MainActivity!</string>
4.      <string name="app_name">SubMenuDemo</string>
5.      <string name="file">文件</string>
6.      <string name="create_new">新建</string>
7.      <string name="open">打开</string>
8.      <string name="save">保存</string>
9.      <string name="exit">退出</string>
10. </resources>
```

（4）修改 src 目录中 com.bcpl.activity 包下的 MainActivity.java 文件，代码如下所示。

```java
1.  package com.bcpl.activity;
2.
3.  import android.app.Activity;
4.  import android.os.Bundle;
5.  import android.view.Menu;
6.  import android.view.MenuInflater;
7.  import android.view.MenuItem;
8.  import android.widget.Toast;
9.
```

```
10. public class MainActivity extends Activity {
11.     /** Called when the activity is first created. */
12.     @Override
13.     public void onCreate(Bundle savedInstanceState) {
14.         super.onCreate(savedInstanceState);
15.         setContentView(R.layout.main);
16.     }
17.
18.     @Override
19.     public boolean onCreateOptionsMenu(Menu menu) {
20.         MenuInflater inflater = getMenuInflater();
21.         inflater.inflate(R.menu.file_submenu, menu);
22.         return true;
23.     }
24.
25.     @Override
26.     public boolean onOptionsItemSelected(MenuItem item) {
27.         int id = item.getItemId();
28.         switch (id) {
29.         case R.id.create_new:
30. Toast.makeText(this,item.getTitle(),Toast.LENGTH_LONG).show();
31.             break;
32.         case R.id.open:
33. Toast.makeText(this,item.getTitle(),Toast.LENGTH_LONG).show();
34.             break;
35.         case R.id.save:
36. Toast.makeText(this,item.getTitle(),Toast.LENGTH_LONG).show();
37.             break;
38.         case R.id.exit:
39. Toast.makeText(this,item.getTitle(),Toast.LENGTH_LONG).show();
40.             break;
41.         default:
42.             break;
43.         }
44.         return super.onOptionsItemSelected(item);
45.     }
46. }
```

（5）部署运行 OptionsMenuDemo 项目工程，单击移动设备上的 menu 按钮，程序运行显示如图 6-5 所示。然后选择"文件"菜单，弹出子菜单，程序运行如图 6-6 所示。

选择"新建"子菜单，程序运行结果如图 6-7 所示。

图 6-5　menu 菜单显示　　　图 6-6　"文件"子菜单　　　图 6-7　"新建"子菜单显示

6.1.4　快捷菜单及应用

1．快捷菜单创建方式简介

快捷菜单同样采用了动窗体的显示方式，与子菜单的实现方式相同，但两种菜单的启动方式不同。快捷菜单类似于普通桌面程序中的"右键菜单"，当用户点击界面元素超过 2s 后，将启动注册到该界面元素的快捷菜单。

快捷菜单与使用选项菜单的方法大致相似，同样需要重载 onCreateContextMenu()方法和 onContextItemSelected()方法。onCreateContextMenu()方法主要用来添加快捷菜单所显示的标题、图标和菜单子项等内容。

选项菜单中的 onCreateOptionsMenu()方法仅在选项菜单第一次启动时被调用一次，而快捷菜单的 onCreateContextMenu()方法在每次启动时都会被调用一次。

```
1.  final static int CONTEXT_MENU_1 = Menu.FIRST;
2.  final static int CONTEXT_MENU_2 = Menu.FIRST+1;
3.  final static int CONTEXT_MENU_3 = Menu.FIRST+2;
4.  @Override
5.  public void onCreateContextMenu(ContextMenu menu, View v,
         ContextMenuInfo menuInfo){
6.    menu.setHeaderTitle("快捷菜单标题");
7.    menu.add(0, CONTEXT_MENU_1, 0,"菜单子项1");
8.    menu.add(0, CONTEXT_MENU_2, 1,"菜单子项2");
```

```
9.         menu.add(0, CONTEXT_MENU_3, 2,"菜单子项3");
10.    }
```

Android 系统中，ContextMenu 类支持 add()方法和 addSubMenu()方法，可以在快捷菜单中添加菜单子项和子菜单。

第 5 行代码的 onCreateContextMenu()方法中：第 1 个参数 menu 是需要显示的快捷菜单；第 2 个参数 v 是用户选择的界面元素；第 3 个参数 menuInfo 是所选择界面元素的额外信息。

菜单选择事件的处理需要重载 onContextItemSelected()方法，该方法在用户选择快捷菜单中的菜单子项后被调用，与 onOptionsItemSelected ()方法的使用方法基本相同。

```
1.  public boolean onContextItemSelected(MenuItem item){
2.       switch(item.getItemId()){
3.           case CONTEXT_MENU_1:
4.                LabelView.setText("子项1");
5.                return true;
6.           case CONTEXT_MENU_2:
7.                LabelView.setText("子项2");
8.                return true;
9.           case CONTEXT_MENU_3:
10.               LabelView.setText("子项3");
11.      return true;
12.      }
13.      return false;
14. }

1.  TextView LabelView = null;
2.     @Override
3.     public void onCreate(Bundle savedInstanceState) {
4.         super.onCreate(savedInstanceState);
5.         setContentView(R.layout.main);
6.         LabelView = (TextView)findViewById(R.id.label);
7.         registerForContextMenu(LabelView);
8.  }
```

第 7 行中使用 registerForContextMenu()方法，将快捷菜单注册到界面控件上。用户在长时间点击该界面控件时，便会启动快捷菜单。

第 6 行中使用 TextView，是为了能够在界面上直接显示用户所选择快捷菜单的菜单子项。

第 1 段代码第 5、8 和 11 行通过更改 TextView 的显示内容，显示用户所选择的菜单子项。

下方代码是/src/layout/main.xml 文件的部分内容，第 1 行声明了 TextView 的 ID 为 label，在上方代码的第 6 行中，通过 R.id.label 将 ID 传递给 findViewById()方法，这样

用户便能够引用该界面元素，并能够修改该界面元素的显示内容。

```
1.  <TextView  android:id="@+id/label"
2.      android:layout_width="fill_parent"
3.      android:layout_height="fill_parent"
4.      android:text="@string/hello"
5.  />
```

需要注意的一点是，上方代码的第 2 行，将 android:layout_width 设置为 fill_parent，这样 TextView 将填充满父节点的所有剩余屏幕空间，用户点击屏幕 TextView 下方任何位置都可以启动快捷菜单。

如果将 android:layout_width 设置为 wrap_content，则用户必须准确点击 TextView 才能启动快捷菜单。

2．快捷菜单应用

前面介绍了子菜单 ContextMenu 的常用方法，下面通过一个接受用户菜单选项然后弹出子菜单，选中子菜单项后显示所选项目的案例应用，熟悉和掌握菜单 ContextMenu 的应用开发过程。

（1）创建一个新的 Android 工程，工程名为 ContextMenuDemo，目标 API 选择 17（即 Android 4.2 版本），应用程序名为 ContextMenuDemo，包名为 com.hisoft.activity，创建的 Activity 的名字为 MainActivity，最小 SDK 版本根据选择的目标 API 会自动添加为 8，创建项目工程。

（2）修改 res 目录下 layout 文件夹中的 main.xml 文件，添加一个 Button 按钮控件描述，并设置相关属性，代码如下所示。

```
1.  <?xml version="1.0" encoding="utf-8"?>
2.  <LinearLayout xmlns:android="http://schemas.android.com/apk/res/
    android"
3.      android:orientation="vertical"
4.      android:layout_width="fill_parent"
5.      android:layout_height="fill_parent"
6.      >
7.      <Button android:text="编辑按钮"
8.          android:id="@+id/btn_edit"
9.          android:layout_width="wrap_content"
10.         android:layout_height="wrap_content">
11.     </Button>
12. </LinearLayout>
```

（3）在 res 目录下，新建 menu 文件夹，然后创建 context_menu.xml 文件，添加 menu、group 和 item 描述，并设置相关属性，代码如下所示。

```
1.  <?xml version="1.0" encoding="utf-8"?>
```

```
2.  <menu xmlns:android="http://schemas.android.com/apk/res/android">
3.    <group android:checkableBehavior="single">
4.      <item android:icon="@drawable/ic_launcher" android:title="剪切" android:id="@+id/cut"/>
5.      <item android:icon="@drawable/ic_launcher" android:title="拷贝" android:id="@+id/copy"/>
6.      <item android:icon="@drawable/ic_launcher" android:title="粘贴" android:id="@+id/paste"/>
7.    </group>
8.  </menu>
```

（4）修改 src 目录中 com.bcpl.activity 包下的 MainActivity.java 文件，代码如下。

```
1.  package com.bcpl.activity;
2.  import android.app.Activity;
3.  import android.os.Bundle;
4.  import android.view.ContextMenu;
5.  import android.view.ContextMenu.ContextMenuInfo;
6.  import android.view.MenuInflater;
7.  import android.view.MenuItem;
8.  import android.view.View;
9.  import android.widget.Button;
10. import android.widget.Toast;
11.
12. public class MainActivity extends Activity {
13.
14.     private Button btn_edit;
15.
16.     /** Called when the activity is first created. */
17.     @Override
18.     public void onCreate(Bundle savedInstanceState) {
19.         super.onCreate(savedInstanceState);
20.         setContentView(R.layout.main);
21.
22.         //获取按钮对象
23.         this.btn_edit = (Button) this.findViewById(R.id.btn_edit);
24.         //为按钮注册上下文菜单
25.         this.registerForContextMenu(btn_edit);
26.     }
27.
28.     //长按按钮时回调此方法
29.     @Override
30.     public void onCreateContextMenu(ContextMenu menu, View v,
31.             ContextMenuInfo menuInfo) {
```

```
32.         super.onCreateContextMenu(menu, v, menuInfo);
33.         MenuInflater inflater = this.getMenuInflater();
34.         inflater.inflate(R.menu.context_menu, menu);
35.
36.     }
37.
38.     //对菜单项添加监听器
39.     @Override
40.     public boolean onContextItemSelected(MenuItem item) {
41.         int id = item.getItemId();
42.         switch (id) {
43.         case R.id.cut:
44.             Toast.makeText(this, item.getTitle(), Toast.LENGTH_
                LONG).show();
45.             break;
46.         case R.id.copy:
47.             Toast.makeText(this, item.getTitle(), Toast.LENGTH_
                LONG).show();
48.             break;
49.         case R.id.paste:
50.             Toast.makeText(this, item.getTitle(), Toast.LENGTH_
                LONG).show();
51.             break;
52.         default:
53.             break;
54.         }
55.         return super.onContextItemSelected(item);
56.     }
57. }
```

（5）部署运行 ContextMenuDemo 项目工程，程序运行结果如图 6-8 所示。长按"编辑按钮"，弹出快捷菜单，如图 6-9 所示。

图 6-8　ContextMenuDemo 运行效果

图 6-9　快捷菜单

6.2　对话框控件 Dialog

6.2.1　对话框 Dialog 简介

对话框 Dialog 是 Android 应用开发中经常用到的用户界面组件，它不属于 View 的子类，它包含的类型有：自定义对话框（继承 Dialog）、提示（或警告）对话框 AlertDialog、进度对话框 ProgressDialog、日期选择对话框 DatePickerDialog、时间选择对话框 TimePickerDialog。

其中，AlertDialog 和 CharacterPickerDialog 是它的直接子类，DatePickerDialog、ProgressDialog、TimePickerDialog 是它的非直接子类。其类的继承关系如图 6-10 所示。

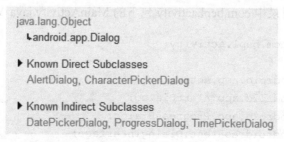

图 6-10　对话框 Dialog 类继承关系

在 Android 实际应用程序开发中，对话框 Dialog 的创建方式有以下两种。
（1）使用 new 操作符创建一个新的 Dialog 对象，然后调用 Dialog 对象的 show 和

dismiss 方法来控制对话框的显示和隐藏。

（2）在 Activity 的 onCreateDialog(int id)方法中创建 Dialog 对象并返回，然后调用 Activty 的 showDialog(int id) 和 dismissDialog(int id)来显示和隐藏对话框，使用 getOwnerActivity()可以返回 Activity 并管理 Dialog。

上述两种方式的区别是，通过第二种方式创建的对话框会继承 Activity 的属性，比如获得 Activity 的 menu 事件等。下面就对它们分别进行介绍。

6.2.2 警告（提示）对话框 AlertDialog 及应用

1. AlertDialog 对话框简介

AlertDialog 对话框是 Dialog 的子类，它有两个或者三个 Button 按钮，用 setMessage() 方法可以在 AlertDialog 对话框上显示一个字符串。它继承自 Dialog，直接子类有：DatePickerDialog，ProgressDialog，TimePickerDialog，其类的继承结构图如图 6-11 所示。

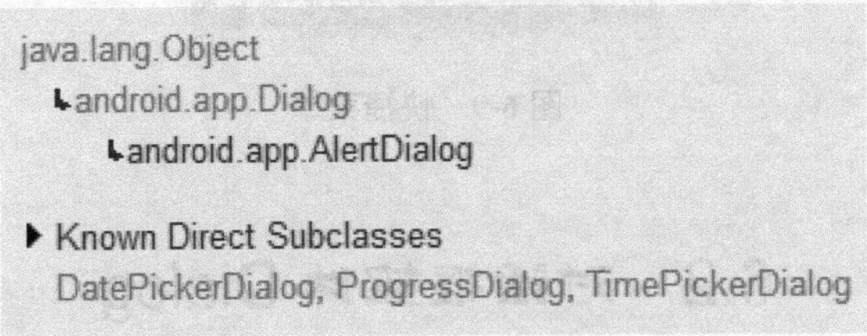

```
java.lang.Object
   ↳android.app.Dialog
       ↳android.app.AlertDialog

▶ Known Direct Subclasses
DatePickerDialog, ProgressDialog, TimePickerDialog
```

图 6-11 AlertDialog 类的继承关系

2. AlertDialog 应用

上述介绍了 Dialog 和 AlertDialog 的基础知识及类之间的关系，下面通过一个提示对话框的案例应用，熟悉和掌握菜单 AlertDialog 的应用开发过程。

（1）创建一个新的 Android 工程，工程名为 AlertDialogDemo，目标 API 选择 17（即 Android 4.2 版本），应用程序名为 AlertDialogDemo，包名为 com.bcpl.activity，创建的 Activity 的名字为 MainActivity。

（2）修改 src 目录中 com.bcpl.activity 包下的 MainActivity.java 文件，代码如下。

```
1.   package com.bcpl.activity;
2.
3.   import android.app.Activity;
4.   import android.app.AlertDialog;
5.   import android.app.ProgressDialog;
6.   import android.content.DialogInterface;
7.   import android.content.DialogInterface.OnClickListener;
8.   import android.os.Bundle;
9.   import android.view.Menu;
10.  import android.view.MenuItem;
```

```
11.
12. public class MainActivity extends Activity {
13.
14.     private static final int EXIT = 1;
15.     private static final int RESTART = 2;
16.
17.     /** Called when the activity is first created. */
18.     @Override
19.     public void onCreate(Bundle savedInstanceState) {
20.         super.onCreate(savedInstanceState);
21.         setContentView(R.layout.main);
22.     }
23.
24.     @Override
25.     public boolean onCreateOptionsMenu(Menu menu) {
26.         menu.add(1, EXIT, 1, "退出程序");
27.         menu.add(1, RESTART, 2, "重启应用");
28.         return true;
29.     }
30.
31.     @Override
32.     public boolean onOptionsItemSelected(MenuItem item) {
33.         if (item.getItemId() == EXIT) {
34.             showAlertDialog();
35.         }
36.
37.         return super.onOptionsItemSelected(item);
38.     }
39.
40.     public void showAlertDialog() {
41.         AlertDialog.Builder builder = new AlertDialog.Builder(this);
42.         builder.setTitle("退出");
43.         builder.setMessage("真的要退出程序吗？");
44.         builder.setPositiveButton("是", new OnClickListener() {
45.
46.             @Override
47.             public void onClick(DialogInterface dialog, int which) {
48.
49.                 MainActivity.this.finish();
50.             }
51.         });
52.
53.         builder.setNegativeButton("否", new OnClickListener() {
54.
```

```
55.            @Override
56.            public void onClick(DialogInterface dialog, int which) {
57.
58.                dialog.cancel();
59.            }
60.        });
61.
62.        AlertDialog alert = builder.create();
63.        alert.show();
64.
65.    }
66. }
```

（3）部署运行 AlertDialogDemo 项目工程，程序运行后，然后单击 menu 按钮，界面下面出现菜单选项，如图 6-12 所示。

图 6-12 菜单选项

单击"重启应用"，程序重置到起始运行状态，菜单消失；单击"退出程序"，弹出对话框，出现"是"或者"否"按钮，如图 6-13 所示。如单击"是"按钮，应用程序退出，单击"否"按钮，应用程序回到上一级状态。

图 6-13 对话框

6.2.3 日期选择对话框 DatePickerDialog 及应用

1. 日期选择对话框简介

在 Android 应用中，日期控件有 DatePicker 和 DatePickerDialog，它们类的继承结构不同，所在的包也不一样，DatePicker 位于 android.widget 包下，继承自 android.widget.FrameLayout，类继承结构如图 6-14 所示；而 DatePickerDialog 位于 android.app 包下，继承自 android.app.AlertDialog，类继承结构如图 6-15 所示。在 DatePickerDialog 类中可以通过 getDatePicker()方法获取包含在这个对话框中的 DatePicker 对象。

DatePickerDialog 的使用要复杂一些，它是以弹出式对话框形式出现的，并需要实现 DialogInterface.OnClickListener 和 DatePicker.OnDateChangedListener 接口。其主要是通过 DatePickerDialog 的 OnDateSetListener 方法实现。DatePicker 类主要是通过 OnDateChangedListener 方法实现用户选择的日期。

图 6-14 DatePicker 类继承关系   图 6-15 DatePickerDialog 类继承关系

2. DatePickerDialog 应用

前面介绍了 DatePicker 和 DatePickerDialog 的区别及联系，下面通过一个设置日期的案例应用，讲解 DatePickerDialog 的应用，具体步骤如下。

(1) 创建一个新的 Android 工程，工程名为 DatePickerDialogDemo，目标 API 选择 17（即 Android 4.2 版本），应用程序名为 DatePickerDialogDemo，包名为 com.bcpl.activity，创建的 Activity 的名字为 MainActivity。

(2) 修改 res 目录下 layout 文件夹中的 activity_main.xml 文件，设置线性布局，添加一个 TextView 控件和 Button 按钮控件描述，并设置相关属性，代码如下所示。

```
1.  <?xml version="1.0" encoding="utf-8"?>
2.  <LinearLayout xmlns:android="http://schemas.android.com/apk/res/android"
3.      android:layout_width="wrap_content"
4.      android:layout_height="wrap_content"
5.      android:orientation="vertical">
6.      <TextView android:id="@+id/dateDisplay"
7.          android:layout_width="wrap_content"
8.          android:layout_height="wrap_content"
9.          android:text=""/>
10.     <Button android:id="@+id/pickDate"
11.         android:layout_width="wrap_content"
12.         android:layout_height="wrap_content"
13.         android:text="Change the date"/>
14. </LinearLayout>
```

(3) 修改 src 目录中 com.bcpl.activity 包下的 MainActivity.java 文件，代码如下。

```
1.  package com.bcpl.activity;

2.  import java.util.Calendar;
3.  import android.app.Activity;
4.  import android.app.DatePickerDialog;
5.  import android.app.Dialog;
6.  import android.os.Bundle;
7.  import android.view.View;
8.  import android.view.View.OnClickListener;
9.  import android.widget.Button;
10. import android.widget.DatePicker;
11. import android.widget.TextView;

12. public class MainActivity extends Activity {
13.     private TextView mDateDisplay;
14.     private Button mPickDate;
15.     private int mYear;
16.     private int mMonth;
17.     private int mDay;
```

```
18.    static final int DATE_DIALOG_ID = 0;
19.    /** Called when the activity is first created. */
20.    @Override
21.    public void onCreate(Bundle savedInstanceState) {
22.        super.onCreate(savedInstanceState);
23.        setContentView(R.layout.main);
24.
25.        mDateDisplay = (TextView) findViewById(R.id.dateDisplay);
26.        mPickDate = (Button) findViewById(R.id.pickDate);
27.
28.        this.mPickDate.setOnClickListener(new OnClickListener() {
29.
30.            @Override
31.            public void onClick(View v) {
32.
33.                MainActivity.this.showDialog(DATE_DIALOG_ID);
34.            }
35.        });
36.
37.        final Calendar c = Calendar.getInstance();
38.        mYear = c.get(Calendar.YEAR);
39.        mMonth = c.get(Calendar.MONTH);
40.        mDay = c.get(Calendar.DAY_OF_MONTH);
41.
42.        updateDisplay();
43.
44.    }
45.
46.    private void updateDisplay() {
47.        mDateDisplay.setText(new StringBuffer()
48.                // Month is 0 based so add 1
49.                .append(mMonth + 1).append("-")
50.                .append(mDay).append("-")
51.                .append(mYear).append(" "));
52.    }
53.
54.    protected Dialog onCreateDialog(int id) {
55.        switch (id) {
56.        case DATE_DIALOG_ID:
57.            return new DatePickerDialog(this,
58.                    mDateSetListener,
59.                    mYear, mMonth, mDay);
60.        }
```

```
61.         return null;
62.     }
63.
64.     private DatePickerDialog.OnDateSetListener mDateSetListener =
65.         new DatePickerDialog.OnDateSetListener() {
66.
67.             public void onDateSet(DatePicker view, int year,
68.                     int monthOfYear, int dayOfMonth) {
69.                 mYear = year;
70.                 mMonth = monthOfYear;
71.                 mDay = dayOfMonth;
72.                 updateDisplay();
73.             }
74.         };
75. }
```

(4) 部署运行 DatePickerDialogDemo 项目工程，程序运行结果如图 6-16 所示。

单击 changer date 按钮，出现如图 6-17 所示界面，可以修改日期，并通过 Done 按钮设定。

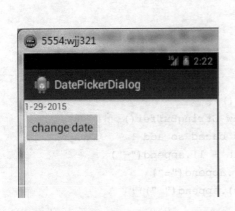

图 6-16　DatePickerDialogDemo 运行效果　　　　图 6-17　日期设置修改界面

6.2.4　时间选择对话框 TimePickerDialog 及应用

同上述的日期选择对话框一样，时间选择对话框也有 TimePickerDialog 和 TimePicker 两个实现。TimePickerDialog 的类继承结构如图 6-18 所示，TimePicker 的类继承结构如图 6-19 所示。它们的类继承结构不同，所在的包也不同，但作用基本一样，用户在对话框中设置时间，也是使用的是 TimePicker 对象，主要区别是 TimePickerDialog 通过弹出对话框的方式，调用 OnTimeSetListener 方法，设置选择时间（单击 Done 按钮）。一般用户选择对话框视图时采用它。

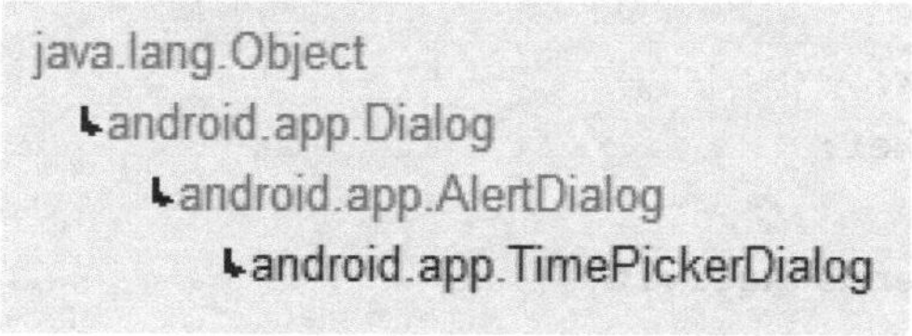

```
java.lang.Object
    ↳android.app.Dialog
        ↳android.app.AlertDialog
            ↳android.app.TimePickerDialog
```

```
java.lang.Object
    ↳android.view.View
        ↳android.view.ViewGroup
            ↳android.widget.FrameLayout
                ↳android.widget.TimePicker
```

图 6-18　TimePickerDialog 类继承关系 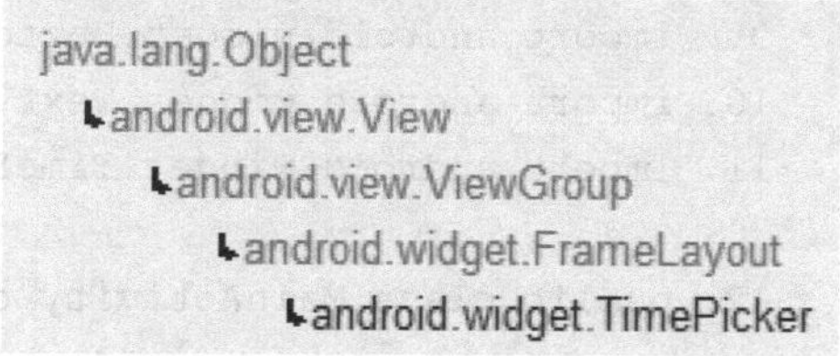　　　　图 6-19　TimePicker 类继承关系

上述简要介绍了 TimePickerDialog 和 TimePicker 两个类的不同与相同之处，下面通过一个 TimePickerDialog 应用案例介绍其应用。具体步骤如下。

（1）创建一个新的 Android 工程，工程名为 TimePickerDialogDemo，目标 API 选择 17（即 Android 4.2 版本），应用程序名为 TimePickerDialogDemo，包名为 com.bcpl.activity，创建的 Activity 的名字为 MainActivity。

（2）修改 res 目录下 layout 文件夹中的 activity_main.xml 文件，设置线性布局，添加一个 TextView 控件和 Button 按钮控件描述，并设置相关属性，代码如下所示。

```
1.   <?xml version="1.0" encoding="utf-8"?>
2.   <LinearLayout xmlns:android="http://schemas.android.com/apk/res/android"
3.       android:layout_width="wrap_content"
4.       android:layout_height="wrap_content"
5.       android:orientation="vertical">
6.       <TextView android:id="@+id/timeDisplay"
7.           android:layout_width="wrap_content"
8.           android:layout_height="wrap_content"
9.           android:text=""/>
10.      <Button android:id="@+id/pickTime"
11.          android:layout_width="wrap_content"
12.          android:layout_height="wrap_content"
13.          android:text="Change the time"/>
14.  </LinearLayout>
```

（3）修改 src 目录中 com.bcpl.activity 包下的 MainActivity.java 文件，代码如下。

```
1.   package com.bcpl.activity;

2.   import java.util.Calendar;

3.   import android.app.Activity;
4.   import android.app.Dialog;
```

```java
5.  import android.app.TimePickerDialog;
6.  import android.os.Bundle;
7.  import android.view.View;
8.  import android.view.View.OnClickListener;
9.  import android.widget.Button;
10. import android.widget.TextView;
11. import android.widget.TimePicker;

12. public class MainActivity extends Activity {

13.     private TextView mTimeDisplay;
14.     private Button mPickTime;
15.     private int mHour;
16.     private int mMinute;
17.     static final int TIME_DIALOG_ID = 0;

18.     @Override
19.     protected void onCreate(Bundle savedInstanceState) {
20.         // TODO Auto-generated method stub
21.         super.onCreate(savedInstanceState);
22.         this.setContentView(R.layout.main);

23.         mTimeDisplay = (TextView) findViewById(R.id.timeDisplay);
24.         mPickTime = (Button) findViewById(R.id.pickTime);
25.
26.         final Calendar c = Calendar.getInstance();
27.         mHour = c.get(Calendar.HOUR_OF_DAY);
28.         mMinute = c.get(Calendar.MINUTE);
29.
30.         updateDisplay();
31.
32.         this.mPickTime.setOnClickListener(new OnClickListener() {
33.
34.             @Override
35.             public void onClick(View v) {
36.
37.                 MainActivity.this.showDialog(TIME_DIALOG_ID);
38.             }
39.         });
40.
41.     }

42.     private void updateDisplay() {
43.         mTimeDisplay.setText(new StringBuilder().append(pad
```

```
                (mHour)).append(":")
44.                 .append(pad(mMinute)));
45.        }
46.
47.        private static String pad(int c) {
48.            if (c >= 10)
49.                return String.valueOf(c);
50.            else
51.                return "0" + String.valueOf(c);
52.        }
53.
54.        protected Dialog onCreateDialog(int id) {
55.            switch (id) {
56.            case TIME_DIALOG_ID:
57.                return new TimePickerDialog(this, mTimeSetListener,
                    mHour, mMinute,false);
58.            }
59.            return null;
60.        }
61.        private TimePickerDialog.OnTimeSetListener mTimeSetListener =
            new TimePickerDialog.OnTimeSetListener() {
62.            public void onTimeSet(TimePicker view, int hourOfDay, int
                minute) {
63.                mHour = hourOfDay;
64.                mMinute = minute;
65.                updateDisplay();
66.            }
67.        };

68. }
```

（4）部署运行 TimePickerDialogDemo 项目工程，程序运行结果如图 6-20 所示。

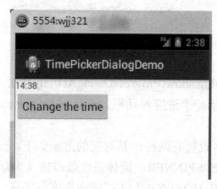

图 6-20　TimePickerDialogDemo 运行效果

单击 Change the time 按钮，出现如图 6-21 所示界面，可以修改时间，然后通过 Done 按钮设定。

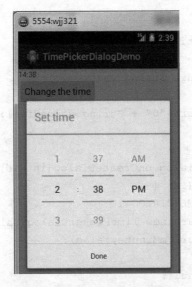

图 6-21　时间设置修改界面

6.2.5　进度对话框 ProgressDialog 及应用

1．进度对话框 ProgressDialog 简介

ProgressDialog 控件在 android.app 包中，继承自 android.app.AlertDialog，类继承结构如图 6-22 所示。它常用于显示载入进度、下载进度等。合理使用 ProgressDialog 能提高用户体验，让用户知道现在程序所处的状态。

```
java.lang.Object
  ↳android.app.Dialog
    ↳android.app.AlertDialog
      ↳android.app.ProgressDialog
```

图 6-22　ProgressDialog 类继承关系

使用代码 ProgressDialog.show(ProgressDialogActivity.this, "请稍等", "数据正在加载中...", true);可以创建并显示一个进度对话框。调用 setProgressStyle()方法，可以设置进度对话框风格。

ProgressDialog 的风格设置有两种，其对应的方法如下。

ProgressDialog.STYLE_SPINNER：旋体进度条风格（为默认风格）。

ProgressDialog.STYLE_HORIZONTAL：横向进度条风格。

ProgressDialog 控件的应用步骤主要如下。

（1）在布局上面加一个 Button，加入 OnClickListener。

（2）把 ProgressDialog 声明成全局的，并在 Button 的 OnClickListener 中创建，然后使用 show(Context context, CharSequence title, CharSequence message, boolean indeterminate)方法，显示 ProgressDialog。第一个参数为当前运行 Activity 的 Context，第二个参数是标题，第三个参数是内容，最后一个参数可选。

（3）在 Button 的 OnClickListener 中创建一个线程，让线程 run 的时候休眠 5s，然后使用 dismiss()方法，关闭刚才打开的 ProgressDialog 对话框。

2．ProgressDialog 应用

在应用中，人们经常可以看见程序加载中的对话框，一般在程序运行前加载数据可以使用 ProgressDialog 对话框，当后台程序运行完毕，需要用 dismiss()方法来关闭取得焦点，以免不能关闭 ProgressDialog 对话框。

（1）创建一个新的 Android 工程，工程名为 ProgressDialogDemo，目标 API 选择 17（即 Android 4.2 版本），应用程序名为 ProgressDialogDemo，包名为 com.bcpl.activity，创建的 Activity 的名字为 MainActivity，最小 SDK 版本根据选择的目标 API 会自动添加为 8，创建项目工程。

（2）修改 res 目录下 layout 文件夹中的 activity_main.xml 文件，设置线性布局，添加一个 Button 按钮控件描述，并设置相关属性，代码如下所示。

```
1.  <?xml version="1.0" encoding="utf-8"?>
2.  <LinearLayout xmlns:android="http://schemas.android.com/apk/res/android"
3.      android:orientation="vertical"
4.      android:layout_width="fill_parent"
5.      android:layout_height="fill_parent"
6.      >
7.      <Button android:text="启动程序"
8.          android:id="@+id/button1"
9.          android:layout_width="wrap_content"
10.         android:layout_height="wrap_content">
11.     </Button>
12. </LinearLayout>
```

（3）修改 src 目录中 com.bcpl.activity 包下的 MainActivity.java 文件，代码如下。

```
1.  package com.bcpl.activity;

2.  import android.app.Activity;
3.  import android.app.ProgressDialog;
4.  import android.os.Bundle;
5.  import android.view.View;
```

```
6.    import android.view.View.OnClickListener;
7.    import android.widget.Button;
8.    public class MainActivity extends Activity {
9.        private Button b1;
10.       private ProgressDialog dialog;
11.       private int currentValue = 0;
12.
13.       /** Called when the activity is first created. */
14.       @Override
15.       public void onCreate(Bundle savedInstanceState) {
16.           super.onCreate(savedInstanceState);
17.           setContentView(R.layout.main);
18.
19.           this.b1 = (Button) this.findViewById(R.id.button1);
20.           this.b1.setOnClickListener(new OnClickListener() {
21.
22.               @Override
23.               public void onClick(View v) {
24.
25.                   dialog = ProgressDialog.show(MainActivity.this, "",
26.                       "Loading. Please wait...", true);
27.                   dialog.setMax(100);
28.                   dialog.show();
29.                   new Thread(new ProgressDialogThread()).start();
30.
31.
32.               }
33.           });
34.       @Override
35.       public boolean onCreateOptionsMenu(Menu menu) {
36.           // Inflate the menu; this adds items to the action bar if it is present.
37.           getMenuInflater().inflate(R.menu.main, menu);
38.           return true;
39.       }
40.       }
41.
42.       class ProgressDialogThread implements Runnable{
43.
44.           @Override
45.           public void run() {
46.               while(currentValue <= 100){
```

```
47.             try {
48.                 Thread.sleep(100);
49.             } catch (InterruptedException e) {
50.                 // TODO Auto-generated catch block
51.                 e.printStackTrace();
52.             }
53.             currentValue += 2;
54.         }
55.
56.         dialog.dismiss();
57.         }
58.     }
59.     }
60. }
```

（4）部署运行 ProgressDialogDemo 项目工程，程序运行结果如图 6-23 所示。单击"启动程序"按钮，程序运行结果如图 6-24 所示。

图 6-23　ProgressDialogDemo 运行效果

图 6-24　程序运行效果

6.3　信息提示控件

6.3.1　Toast 控件及应用

Toast 控件位于 android.widget 包中，其类的继承结构如图 6-25 所示。Toast 是一种提供给用户快速、简短信息的视图。借助 Toast 类可以创建和显示该信息。

Toast 视图以浮于应用程序之上的视图形式呈现给用户。因为它并不获得焦点，即使用户正在输入什么也不会受到影响。它的目标是尽可能以不显眼的方式，使用户看到提供的信息。音量控制和信息设置、保存成功这两个例子就是使用 Toast。

```
java.lang.Object
  ↳android.widget.Toast
```

图 6-25　Toast 类继承结构

使用该类最简单的方法就是调用它的一个静态方法 makeText，让它来构造需要的一切并返回一个新的 Toast 对象，即生成一个从资源中取得的包含文本视图的标准 Toast 对象。

上面简述了 Toast 的功能和创建方法，下面通过一个 Toast 应用案例，简要介绍 Toast 显示快速、简短消息，以及显示图片的方法，具体步骤如下。

（1）创建一个新的 Android 工程，工程名为 ToastDemo，目标 API 选择 17（即 Android 4.2 版本），应用程序名为 ToastDemo，包名为 com.bcpl.activity，创建的 Activity 的名字为 MainActivity，最小 SDK 版本根据选择的目标 API 会自动添加为 8，创建项目工程。

（2）修改 res 目录下 layout 文件夹中的 activity_main.xml 文件，设置线性布局，添加一个 Button 按钮控件描述，并设置相关属性，代码如下所示。

```
1.  <?xml version="1.0" encoding="utf-8"?>
2.  <LinearLayout xmlns:android="http://schemas.android.com/apk/res/
    android"
3.      android:orientation="vertical"
4.      android:layout_width="fill_parent"
5.      android:layout_height="fill_parent"
6.      >
7.      <Button android:text="Toast 弹出"
8.          android:id="@+id/button1"
9.          android:layout_width="wrap_content"
10.         android:layout_height="wrap_content">
11.     </Button>
12. </LinearLayout>
```

（3）在 res 目录下 layout 文件夹中，新建 toast.xml 文件，设置线性布局，添加一个 ImageView 控件和 TextView 控件描述，并设置相关属性，代码如下所示。

```
1.  <?xml version="1.0" encoding="utf-8"?>
2.  <LinearLayout
3.    xmlns:android="http://schemas.android.com/apk/res/android"
4.    android:layout_width="match_parent"
5.    android:layout_height="match_parent" android:orientation=
      "horizontal">
6.     <ImageView android:src="@drawable/icon"
7.         android:layout_height="wrap_content"
8.         android:layout_width="wrap_content"
9.         android:id="@+id/imageView1">
10.    </ImageView>
11.    <TextView android:id="@+id/textView1"
```

```
12.            android:text="@string/info"
13.            android:layout_height="wrap_content"
14.            android:layout_width="wrap_content"
15.   android:textAppearance="?android:attr/textAppearanceMedium">
16.     </TextView>

17. </LinearLayout>
```

（4）修改 src 目录中 com.bcpl.activity 包下的 MainActivity.java 文件，代码如下。

```
1.  package com.bcpl.activity;
2.
3.  import android.app.Activity;
4.  import android.os.Bundle;
5.  import android.view.LayoutInflater;
6.  import android.view.View;
7.  import android.view.View.OnClickListener;
8.  import android.widget.Button;
9.  import android.widget.Toast;
10.
11. public class MainActivity extends Activity {
12.
13.     private Button b1;
14.
15.     /** Called when the activity is first created. */
16.     @Override
17.     public void onCreate(Bundle savedInstanceState) {
18.         super.onCreate(savedInstanceState);
19.         setContentView(R.layout.main);
20.
21.         this.b1 = (Button) this.findViewById(R.id.button1);
22.
23.         this.b1.setOnClickListener(new OnClickListener() {
24.
25.             @Override
26.             public void onClick(View v) {
27.
28.                 Toast t = Toast.makeText(MainActivity.this, "", Toast.
                       LENGTH_SHORT);
29.                 LayoutInflater inflater = LayoutInflater.from
                       (MainActivity.this);
30.                 View view = inflater.inflate(R.layout.toast, null);
31.                 t.setView(view);
32.                 t.show();
33.
```

```
34.            }
35.        });
36.
37.    }
38. }
```

（5）修改 res 目录下 values 文件夹中的 strings.xml 文件，代码如下所示。

```
1. <?xml version="1.0" encoding="utf-8"?>
2. <resources>
3.     <string name="hello">Hello World, MainActivity!</string>
4.     <string name="app_name">ToastDemo</string>
5.     <string name="info">短消息发送成功</string>
6. </resources>
```

（6）部署运行 ToastDemo 项目工程，程序运行结果如图 6-26 所示。
单击"toast 弹出"按钮，出现 Toast 提示信息，如图 6-27 所示。

图 6-26　ToastDemo 运行效果

图 6-27　Toast 提示信息

6.3.2　Notification 控件及应用

1．Notification 控件简介

Notification 控件位于 android.app 包下，类的继承结构如图 6-28 所示。Notification 也是 Android 系统中给用户消息提示的方式，位于屏幕最顶部的状态栏，通知的同时可以播放声音，以及振动提示用户，单击通知还可以返回指定的 Activity。通常它用来在状态栏中显示电池电量、信息强度等信息。按住状态栏，然后往下拖或者拉，可以打开状态栏并查看系统的提示信息。

通知的设置等操作相对比较简单，就是新建一个 Notification 对象，然后设置好通知的各项参数，然后使用系统后台运行的 NotificationManager 服务将通知发出来。可以通过 Notification.Builder 快速地创建构造 Notification 对象。

```
java.lang.Object
   ↳android.app.Notification
```

图 6-28　Notification 类继承关系

Notification 通常使用用法及步骤如下。

（1）得到 NotificationManager：

```
String ns = Context.NOTIFICATION_SERVICE;
NotificationManager    mNotificationManager    =    (NotificationManager)
getSystemService(ns);
```

（2）创建一个新的 Notification 对象：

```
Notification notification = new Notification();
notification.icon = R.drawable.notification_icon; //或者以复杂一些的方式创
建Notification
int icon = R.drawable.notification_icon; //通知图标
CharSequence tickerText = "Hello"; //状态栏(Status Bar)显示的通知文本提示
long when = System.currentTimeMillis(); //通知产生的时间，会在通知信息里显示
Notification notification = new Notification(icon, tickerText, when);
```

（3）填充 Notification 的各个属性：

```
Context context = getApplicationContext();
CharSequence contentTitle = "My notification";
CharSequence contentText = "Hello World!";
Intent notificationIntent = new Intent(this, MyClass.class);
PendingIntent   contentIntent   =   PendingIntent.getActivity(this,   0,
notificationIntent, 0);
notification.setLatestEventInfo(context,   contentTitle,   contentText,
contentIntent);
```

（4）发送通知：

```
private static final int ID_NOTIFICATION = 1;
mNotificationManager.notify(ID_NOTIFICATION, notification);
```

Notification 的手机提示方式有 4 种，具体如下。

（1）在状态栏(Status Bar)显示的通知文本提示，如：

```
notification.tickerText = "hello";
```

（2）发出提示音，如：

```
notification.defaults |= Notification.DEFAULT_SOUND;
```

```
notification.sound = Uri.parse("file:///sdcard/notification/ringer.
mp3");
notification.sound = Uri.withAppendedPath(Audio.Media.INTERNAL_CONTENT_
URI, "6");
```

(3) 手机振动,如:

```
notification.defaults |= Notification.DEFAULT_VIBRATE;
long[] vibrate = {0,100,200,300};
notification.vibrate = vibrate;
```

(4) LED 灯闪烁,如:

```
notification.defaults |= Notification.DEFAULT_LIGHTS;
notification.ledARGB = 0xff00ff00;
notification.ledOnMS = 300;
notification.ledOffMS = 1000;
notification.flags |= Notification.FLAG_SHOW_LIGHTS;
```

Notification 如果需要更新一个通知,只需要在设置好 notification 之后,再调用 setLatestEventInfo,然后重新发送一次通知即可。

2. Notification 应用

上面讲述了 Notification 通常用法及提示方式等基础知识,下面通过 Notification 案例,讲解如何向状态栏添加信息以及图片。

(1) 创建一个新的 Android 工程,工程名为 NotificationDemo,目标 API 选择 17(即 Android 4.2 版本),应用程序名为 NotificationDemo,包名为 com.bcpl.activity,创建的 Activity 的名字为 MainActivity,最小 SDK 版本根据选择的目标 API 会自动添加为 8,创建项目工程。

(2) 修改 res 目录下 layout 文件夹中的 activity_main.xml 文件,设置线性布局,添加两个 Button 按钮控件描述,并设置相关属性,代码如下所示。

```
1.   <?xml version="1.0" encoding="utf-8"?>
2.   <LinearLayout xmlns:android="http://schemas.android.com/apk/res/
     android"
3.       android:orientation="vertical"
4.       android:layout_width="fill_parent"
5.       android:layout_height="fill_parent"
6.       >
7.   <Button android:id="@+id/bt1"
8.        android:layout_height="wrap_content"
9.        android:layout_width="fill_parent"
10.       android:text="测试 Notification"
11.  />
```

```
12. <Button android:id="@+id/bt2"
13.     android:layout_height="wrap_content"
14.     android:layout_width="fill_parent"
15.     android:text="清除 Notification"
16. />
17. </LinearLayout>
```

（3）修改 src 目录中 com.bcpl.activity 包下的 MainActivity.java 文件，代码如下。

```
1.  package com.bcpl.activity;
2.  import android.app.Activity;
3.  import android.app.Notification;
4.  import android.app.NotificationManager;
5.  import android.app.PendingIntent;
6.  import android.content.Intent;
7.  import android.os.Bundle;
8.  import android.view.View;
9.  import android.view.View.OnClickListener;
10. import android.widget.Button;
11.
12. public class MainActivity extends Activity {
13.     /** Called when the activity is first created. */
14.     int notification_id=19172439;
15.     NotificationManager nm;
16.     @Override
17.     public void onCreate(Bundle savedInstanceState) {
18.         super.onCreate(savedInstanceState);
19.         setContentView(R.layout.main);
20.         nm=(NotificationManager)getSystemService(NOTIFICATION_
            SERVICE);
21.         Button bt1=(Button)findViewById(R.id.bt1);
22.         bt1.setOnClickListener(bt1lis);
23.         Button bt2=(Button)findViewById(R.id.bt2);
24.         bt2.setOnClickListener(bt2lis);
25.
26.     }
27.     OnClickListener bt1lis=new OnClickListener(){
28.
29.         @Override
30.         public void onClick(View v) {
31.             // TODO Auto-generated method stub
32.             showNotification(R.drawable.ic_launcher_home,"测试信息
                ","短信","北京政法内容");
33.         }
34.
```

```
35.        };
36.        OnClickListener bt2lis=new OnClickListener(){
37.
38.            @Override
39.            public void onClick(View v) {
40.                // TODO Auto-generated method stub
41.                //showNotification(R.drawable.home,"测试信息","短信","北
                   京政法信息技术系测试内容");
42.                nm.cancel(notification_id);
43.            }
44.
45.        };
46.        public void showNotification(int icon,String tickertext,String
                title,String content){
47.            //设置一个唯一的ID，随便设置
48.
49.            //Notification 管理器
50.            Notification notification=new Notification(icon,tickertext,
                System.currentTimeMillis());
51.            //后面的参数分别是显示在顶部通知栏的小图标，小图标旁的文字（短暂显示，自动
                消失）系统当前时间
52.            notification.defaults=Notification.DEFAULT_ALL;
53.            //这是设置通知时是否同时播放声音或振动，声音为 Notification.DEFAULT_
                SOUND
54.            //振动为 Notification.DEFAULT_VIBRATE;
55.            //Light 为 Notification.DEFAULT_LIGHTS
56.            //全部为 Notification.DEFAULT_ALL
57.            //如果是振动或者全部，必须在 AndroidManifest.xml 中加入振动权限
58.            PendingIntent pt=PendingIntent.getActivity(this, 0, new Intent
                (this,MainActivity.class), 0);
59.            //点击通知后的动作，这里是转回 MainActcity
60.            notification.setLatestEventInfo(this,title,content,pt);
61. //如果需要更新一个通知，在设置好 notification 之后，再调用
        setLatestEventInfo，然后重新发送一次通知即可。
62.            nm.notify(notification_id, notification);
63.
64.        }
65. }
```

（4）在 AndroidMainfest.xml 文件中添加振动器的权限，代码如下：

```
1. <uses-permission android:name="android.permission.VIBRATE"/>
```

（5）部署运行 NotificationDemo 项目工程，程序运行结果如图 6-29 所示。

单击"测试 Notification"按钮，状态栏显示如图 6-30 所示。

图 6-29　NotificationDemo 运行效果　　　　图 6-30　Notification 状态栏显示

然后按住状态栏，往下拖动，显示信息标题及内容，如图 6-31 所示。

图 6-31　信息标题及内容

如果需要更新一个通知，单击"测试 Notification"按钮，然后即可重新发送信息。

6.4　项目案例

学习目标：学习 Android 界面菜单、对话框、信息提示控件分类、方法、属性的设置等应用。

案例描述：使用 Android UI 菜单、对话框、信息提示控件，RelativeLayout 相对布局、Button 按钮，并设置相对父控件的位置、控件之间相对位置的属性，实现"妈咪宝贝"菜单选择界面。

案例要点：Menu、onOptionsItemSelected 方法、Toast 控件。

案例步骤：

（1）创建工程 Project_Chapter_6，选择 Android 4.2 作为目标平台。

（2）创建 activity_main.xml 文件，使用相对布局 RelativeLayout 和 Button 按钮，将创建的文件存放在 res/layout 下，代码如下。

```
1.  <RelativeLayout xmlns:android="http://schemas.android.com/apk/res/android"
2.      xmlns:tools="http://schemas.android.com/tools"
3.      android:layout_width="match_parent"
4.      android:layout_height="match_parent"
5.      android:background="@drawable/backthree"
6.
7.      android:paddingLeft="@dimen/activity_horizontal_margin"
8.      android:paddingRight="@dimen/activity_horizontal_margin"
9.
10.     tools:context=".MainActivity" >
11.
12.     <Button
13.         android:id="@+id/button1"
14.         android:layout_width="wrap_content"
15.         android:layout_height="wrap_content"
16.         android:layout_alignParentRight="true"
17.         android:layout_alignParentTop="true"
18.         android:layout_marginRight="70dp"
19.         android:layout_marginTop="31dp"
20.         android:background="@drawable/yun1"
21.         android:text="妈妈篇"
22.         android:textSize="30dp" />
23.
24.     <Button
25.         android:id="@+id/button2"
26.         android:layout_width="wrap_content"
27.         android:layout_height="wrap_content"
28.         android:layout_alignParentLeft="true"
29.         android:layout_below="@+id/button1"
30.         android:layout_marginTop="24dp"
31.         android:background="@drawable/yun2"
32.         android:text="宝宝篇"
33.         android:textSize="30dp" />
34.
35.     <Button
36.         android:id="@+id/daddy"
37.         android:layout_width="wrap_content"
38.         android:layout_height="wrap_content"
39.         android:layout_alignParentRight="true"
40.         android:layout_below="@+id/button2"
41.         android:layout_marginRight="34dp"
42.         android:background="@drawable/yun2"
43.         android:text="爸爸篇"
```

```
44.        android:textSize="30dp" />
45. </RelativeLayout>
```

(3) 在 src 目录下 com.bcpl.baby 包下，创建 MainActivity.java，代码如下。

```
1.  import com.bcpl.baby.daddy.DaddyActivity;
2.  import com.bcpl.baby.dear.DearActivity;
3.
4.  import com.bcpl.baby.mum.FenleiActivity;
5.  import com.bcpl.baby.mum.MumActivity;
6.  import com.bcpl.baby.mum.SearchActivity;
7.
8.  import android.os.Bundle;
9.
10. import android.app.Activity;
11. import android.content.Intent;
12. import android.graphics.Typeface;
13. import android.view.Menu;
14. import android.view.MenuItem;
15. import android.view.View;
16. import android.view.View.OnClickListener;
17. import android.widget.Button;
18. import android.widget.Toast;
19. public class MainActivity extends Activity implements OnClickListener {
20.     private Button b_MUM;
21.     private Button b_DEAR;
22.     private Button b_DAD;
23.     private Typeface typeFace;
24.
25.     @Override
26.     protected void onCreate(Bundle savedInstanceState) {
27.         super.onCreate(savedInstanceState);
28.         setContentView(R.layout.activity_main);
29.
30.         b_MUM = (Button) findViewById(R.id.button1);
31.         typeFace = Typeface.createFromAsset(getAssets(),
32.                 "fonts/DFPShaoNvW5-GB.ttf");
33.         b_MUM.setTypeface(typeFace);
34.
35.         b_DEAR = (Button) findViewById(R.id.button2);
36.         b_DEAR.setTypeface(typeFace);
37.
38.         b_DAD=(Button)findViewById(R.id.daddy);
39.         b_DAD.setTypeface(typeFace);
40.
```

```
41.        b_MUM.setOnClickListener(new OnClickListener() {
42.
43.            @Override
44.            public void onClick(View v) {
45.                // TODO Auto-generated method stub
46.                Intent intent1 = new Intent();
47.                intent1.setClass(MainActivity.this, MumActivity.class);
48.                startActivity(intent1);
49.            }
50.        });
51.        b_DEAR.setOnClickListener(new OnClickListener() {
52.
53.            @Override
54.            public void onClick(View arg0) {
55.                // TODO Auto-generated method stub
56.                Intent intent2 = new Intent();
57.                intent2.setClass(MainActivity.this, DearActivity.class);
58.                startActivity(intent2);
59.
60.                // mText.setTypeface(Typeface.createFromAsset(getAssets(),"fonts/DFPShaoNvW5-GB.ttf"));
61.            }
62.        });
63.        b_DAD.setOnClickListener(new OnClickListener() {
64.
65.            @Override
66.            public void onClick(View v) {
67.                // TODO Auto-generated method stub
68.                Intent intent3=new Intent(MainActivity.this, DaddyActivity.class);
69.                startActivity(intent3);
70.            }
71.        });
72.    }
73.
74.    // 设置菜单
75.    @Override
76.    public boolean onCreateOptionsMenu(Menu menu) {
77.        menu.add(0,0,0,"宝宝篇").setIcon(R.drawable.bao_bao);
78.        menu.add(0,1,1,"妈妈篇").setIcon(R.drawable.ma_mi);
79.        menu.add(0, 2, 2, "爸爸篇").setIcon(R.drawable.ba_ba);
80.        menu.add(0, 3, 3, "分类").setIcon(R.drawable.tb_recommend_
```

```
81.        menu.add(0,4,4,"搜索").setIcon(R.drawable.tb_seach);
82.        menu.add(0,5,5,"更多").setIcon(R.drawable.tb_more);
83.        return true;
84.    }
85.    @Override
86.    public boolean onOptionsItemSelected(MenuItem item){
87.        int id=item.getItemId();
88.        switch (id){
89.        case 0:
90.        Intent bb=new Intent(MainActivity.this,DearActivity.class);
91.        startActivity(bb);
92.        finish();
93.            break;
94.        case 1:
95.            Intent mm=new Intent(MainActivity.this,MumActivity.class);
96.            startActivity(mm);
97.            finish();
98.            break;
99.        case 2:
100.              Intent ba=new Intent(MainActivity.this,DaddyActivity.class);
101.              startActivity(ba);
102.              finish();
103.              break;
104.
105.        case 3:
106.              Intent fl=new Intent(MainActivity.this,FenleiActivity.class);
107.              startActivity(fl);
108.              finish();
109.              break;
110.        case 4:
111.              Intent ss=new Intent(MainActivity.this,SearchActivity.class);
112.              startActivity(ss);
113.              finish();
114.              break;
115.        case 5:
116.              Intent gd=new Intent(MainActivity.this,MoreActivity.class);
117.              startActivity(gd);
118.              finish();
```

```
119.                break;
120.            }
121.            return true;
122.        }
123.        @Override
124.        public void onClick(View arg0) {
125.            // TODO Auto-generated method stub
126.
127.        }
128.    }
```

（4）部署 Project_Chapter_6 工程，单击 menu 按钮，运行效果如图 6-32 所示。

图 6-32　Project_Chapter_6 运行效果

习　题

一、简答题

1. Android 系统 4.0 之后支持的菜单有哪些？在程序中创建菜单与以前版本有什么不同？

2. Android 系统 4.0 版本之后选项菜单可以分为哪些类别？它们的创建方法是什么？

3. 如何在 res 目录下 values 文件中 styles.xml 文件中自定义 MyDialog，然后自定义继承于 Dialog 的类 RegDialog，进行实例化 RegDialog 对象，并调用自定义风格？

4. Notification 控件常用于什么功能？

二、实训

要求：

单击"退出"按钮，实现退出 Dialog 功能，如图 6-33 所示。

图 6-33 退出提示

第 7 章

Android 组件消息通信与服务

学习目标

本章介绍了 Android 组件 Intent、Intent 对象包含的信息、使用 Intent 进行组件通信、Intent 广播消息、BroadcastReceiver 监听广播消息、Service 组件服务、Service 与 Activity 通信等。读者通过本章的学习，可掌握以下知识要点。

（1）Intent 常用类型、消息机制及启动方式。
（2）Android 组件 Activity、Service 和 BroadcastReceiver 及 Intent 的通信。
（3）Intent 启动 Activity 的方法及 Activity 返回值的获取。
（4）Intent 解析原理及常用的方法。
（5）BroadcastReceiver 监听广播消息过程及方法。
（6）Service 类组件服务应用。

7.1 Intent 消息通信

7.1.1 Intent 组件及通信

1. Intent 简介

Intent 提供了一种通用的消息系统，它允许在自己的应用程序与其他的应用程序间传递 Intent 来执行动作和产生事件。

Intent 负责对应用中一次操作的动作、动作涉及数据、附加数据进行描述，Android 则根据此 Intent 的描述，负责找到对应的组件，将 Intent 传递给调用的组件，并完成组件的调用。使用 Intent 可以激活 Android 应用的三个核心组件：活动、服务和广播接收器。

在 Android 系统中，Intent 的用途主要有以下三个。
（1）启动 Activity；
（2）启动 Service；
（3）在 Android 系统上发布广播消息（广播消息可以是接收到特定数据或消息，也可以是手机的信号变化或电池的电量过低等信息）。

通常 Intent 分为显式和隐式两类。显式的 Intent，就是指定了组件名字的，是由程序指定具体的目标组件来处理，即在构造 Intent 对象时就指定接收者，指定了一个明确的组件（setComponent 或 setClass）来使用处理 Intent。

```
Intent intent = new Intent(
    getApplicationContext() ,
    Test.class
);
startActivity(intent);
```

特别注意：被启动的 Activity 需要在 AndroidManifest.xml 中进行定义。

隐式的 Intent 就是没有指定 Intent 的组件名字，没有指定明确的组件来处理该 Intent。使用这种方式时，需要让 Intent 与应用中的 Intent Filter 描述表相匹配。需要 Android 根据 Intent 中的 Action、Data、Category 等来解析匹配。由系统接受调用并决定如何处理，即 Intent 的发送者在构造 Intent 对象时，并不知道也不关心接收者是谁，有利于降低发送者和接收者之间的耦合，如 startActivity(new Intent(Intent.ACTION_DIAL));。

```
Intent intent = new Intent();
intent.setAction("test.intent.IntentTest");
startActivity(intent);
```

目标组件（Activity、Service、BroadcastReceiver）是通过设置它们的 Intent Filter 来界定其处理的 Intent。如果一个组件没有定义 Intent Filter，那么它只能接收处理显式的 Intent，只有定义了 Intent Filter 的组件才能同时处理隐式和显式的 Intent。

一个 Intent 对象包含很多数据的信息，由以下 6 个部分组成。

（1）Action——要执行的动作。
（2）Data——执行动作要操作的数据。
（3）Category——被执行动作的附加信息。
（4）Extras——其他所有附加信息的集合。
（5）Type——显式指定 Intent 的数据类型（MIME）。
（6）Component——指定 Intent 的目标组件的类名称，比如要执行的动作、类别、数据、附加信息等。

下面就一个 Intent 中包含的信息进行简要介绍。

1）Action

使用 android:name 属性指定要为其服务的动作的名称，一个 Intent 的 Action 在很大程度上说明这个 Intent 要做什么，如查看（View）、删除（Delete）、编辑（Edit）等。Android 中预定义了很多 Action，可以参考 Intent 类查看，表 7-1 是 Android 文档中的几个动作。

此外，用户也可以自定义 Action，比如 com.flysnow.intent.ACTION_ADD。定义的 Action 最好能表明其所表示的意义以及要做什么，这样 Intent 中的数据才好填充。Intent 对象的 getAction()可以获取动作，使用 setAction()可以设置动作。

表 7-1 Action

Constant	Target component	Action
ACTION_CALL	activity	Initiate a phone call.
ACTION_EDIT	activity	Display data for the user to edit.
ACTION_MAIN	activity	Start up as the initial activity of a task, with no data input and no returned output.
ACTION_SYNC	activity	Synchronize data on a server with data on the mobile device.
ACTION_BATTERY_LOW	broadcast receiver	A warning that the battery is low.
ACTION_HEADSET_PLUG	broadcast receiver	A headset has been plugged into the device, or unplugged from it.
ACTION_SCREEN_ON	broadcast receiver	The screen has been turned on.
ACTION_TIMEZONE_CHANGED	broadcast receiver	The setting for the time zone has changed.

2）Data

Data 实质上是一个 URI，用于执行一个 Action 时所用到的数据的 URI 和 MIME。不同的 Action 有不同的数据规格，比如 ACTION_EDIT 动作，数据就可以包含一个用于编辑文档的 URI；如果是一个 ACTION_CALL 动作，那么数据就是一个包含类似 tel:6546541 的数据字段。所以上面提到自定义 Action 时要规范命名。数据的 URI 和类型对于 Intent 的匹配是很重要的，Android 往往根据数据的 URI 和 MIME 找到能处理该 Intent 的最佳目标组件。

3）Component（组件）

Component 指定 Intent 的目标组件的类名称。通常 Android 会根据 Intent 中包含的其他属性的信息，比如 action、data/type、category 进行查找，最终找到一个与之匹配的目标组件。

如果设置了 Intent 目标组件的名字，那么这个 Intent 就会被传递给特定的组件，而不再执行上述查找过程，指定了这个属性以后，Intent 的其他所有属性都是可选的。也就是显式 Intent。如果不设置，则是隐式的 Intent，Android 系统将根据 Intent Filter 中的信息进行匹配。

4）Category

Category 指定了用于处理 Intent 的组件的类型信息，一个 Intent 可以添加多个 Category，使用 addCategory()方法即可，使用 removeCategory()可删除一个已经添加的类别。Android 的 Intent 类里定义了很多常用的类别，可以参考使用。

5）Extras

Extras 用于处理 Intent 的目标组件的一些额外的信息。一个附加信息就是一个 key-value 的键值对（NVP），那么就可以通过 Intent 的 putExtra()方法把额外的信息加入到 Intent 对象中，然后用 getExtras 方法提取，供目标组件的使用。

此外，还有 PendingIntent 类提供了一种创建可由其他应用程序在稍晚的时间触发 Intent 的机制，需要使用的读者可以查阅相关资料。

2. 使用 Intent 进行组件通信

前面已经讲述了 Intent 的作用、分类及其包含的信息，从上述可以得知，Intent 就是一个动作的完整描述，包含动作的产生组件、接收组件和传递的数据信息。Intent 也可称为一个在不同组件之间传递的消息，这个消息在到达接收组件后，接收组件会执行相关的动作。Intent 为 Activity、Service 和 BroadcastReceiver 等组件提供了交互的能力，如图 7-1 所示。

对于 Activity、Service 和 BroadcastReceiver 这三个组件，它们都有自己独立的传递 Intent 的机制。

（1）Activity：对于 Activity 来说，它主要是通过 Context.startActivity() 或 Activity.startActivityForRestult() 来启动一个存在的 Activity 做一些事情。当使用 Activity.startActivityForResult() 启动一个 Activity 时，可以使用 Activity.setResult()返回一些结果信息，可以在 Activity.onActivityResult() 中得到返回的结果。

（2）Service：对于 Service 来说，它主要是通过 Context.startService() 初始化一个 Service 或者传递消息给正在运行的 Service。同样，也可以通过 Context.bindService() 建立一个调用组件和目标服务之间的连接。

（3）BroadcastReceiver：它可以通过 Context.sendBroadcast(),Context.sendOrderedBroadcast()以及 Context.sendStickyBroadcast()这些方法，传递 Intent 给感兴趣的广播。

消息之间的传递是没有重叠的，比如调用 startActivity()启动传播一个 Intent，只会传递给 Activity，而不会传递给 Service 和 BroadcastReceiver，反过来也是这样。

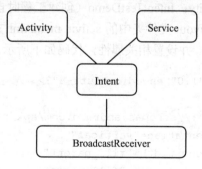

图 7-1　组件交互

7.1.2　使用 Intent 启动 Activity

在 Android 系统中，应用程序一般都有多个 Activity，Intent 可以实现不同 Activity 之间的切换和数据传递。

使用 Intent 启动 Activity 的方式主要有两种，分别是：显式启动和隐式启动。如前面章节所述一样，显式启动，必须在 Intent 中指明启动的 Activity 所在的类；而隐式启动，Android 系统根据 Intent 的动作和数据来决定启动哪一个 Activity，也就是说在隐式启动时，Intent 中只包含需要执行的动作和所包含的数据，并没有指明具体启动的 Activity，而是由 Android 系统和最终用户来决定。下面就显式和隐式启动 Activity 的通常用法进行介绍。

1. 显式启动 Activity 的通常用法

（1）新建一个 Intent。

（2）指定当前的应用程序上下文以及要启动的 Activity。

（3）把新建好的这个 Intent 作为参数传递给 startActivity()方法。

```
1.    Intent intent = new Intent(IntentTestDemo.this, NewActivity.class);
2.    startActivity(intent);
```

上述包含两个 Activity 类，分别是 IntentTestDemo 和 NewActivity，程序默认启动的是 IntentTestDemo。具体步骤如下。

（1）依照前面案例创建的步骤，新创建一个工程名为 IntentTestDemo 的工程，然后打开工程中的 AndroidManifest.xml 文件，在<application>根节点下添加<activity>标签，注册新添加的 activity，嵌套在<application>根节点标签下，添加代码如下：

```
1.    <activity android:name=".NewActivity"
2.    android:label="@string/app_name"
3.    /activity>
```

在 Android 应用程序中，用户使用的每个组件都必须在 AndroidManifest.xml 文件中的<application>节点内定义，<application>节点下共有两个<activity>节点，分别代表应用程序中所使用的两个 Activity：IntentTestDemo（创建工程时自动生成）和 NewActivity。

（2）修改 res 目录下 layout 文件夹中的 activit_main.xml 文件，设置线性布局，添加一个 Button 按钮控件描述，并设置相关属性，代码如下所示。

```
1.    <?xml version="1.0" encoding="utf-8"?>
2.    <LinearLayout xmlns:android="http://schemas.android.com/apk/res/android"
3.        android:orientation="vertical"
4.        android:layout_width="fill_parent"
5.        android:layout_height="fill_parent"
6.        >
7.    <Button android:id="@+id/bt1"
8.         android:layout_height="wrap_content"
9.         android:layout_width="fill_parent"
10.        android:text="测试显式 Intent"
11.   />
12.   </LinearLayout>
```

（3）修改 src 目录下 com.bcpl.activity 包下的 IntentTestDemoActivity.java 文件，添加显式使用 Intent 启动 Activity 的核心代码，代码如下。

```
1.  Button button = (Button)findViewById(R.id.bt1);
2.  button.setOnClickListener(new OnClickListener(){
3.      public void onClick(View view){
4.          Intent intent = new Intent(IntentTestDemoActivity.this,
              NewActivity.class);
5.          startActivity(intent);
6.      }
7.  });
```

在点击事件的处理函数中，Intent 构造函数的第 1 个参数是应用程序上下文，程序中的应用程序上下文就是 IntentTestDemo；第 2 个参数是接收 Intent 的目标组件，使用的是显式启动方式，直接指明了需要启动的 Activity。

（4）在 src 目录下 com.bcpl.activity 包下创建新的 NewActivity，在 res 目录下 layout 文件夹中创建 new_main.xml 文件，在 values 文件夹下的 strings.xml 文件中，添加 text 引用，让 NewActivity 界面显示：NewActivity application。

（5）部署运行程序，程序运行效果如图 7-2 所示，单击"测试显式 Intent"按钮，程序运行如图 7-3 所示。

图 7-2 IntentTestDemo 运行效果

图 7-3 测试显式 Intent 效果

2. 隐式启动 Activity 的通常用法

隐式启动 Activity 时，Android 系统在应用程序运行时解析 Intent，并根据一定的规则对 Intent 和 Activity 进行匹配，使 Intent 上的动作、数据与 Activity 完全匹配。

（1）在 AndroidManifest.xml 中注册声明需要匹配 Activity。

（2）程序代码中创建新的 Intent（可以向 Intent 中添加运行 Activity 所需要的附加信息）。

（3）将 Intent 传递给 startActivity()。

创建 Intent 时，在默认情况下，Android 系统会调用内置的 Web 浏览器，如：

```
Intent intent = new Intent(Intent.ACTION_VIEW, Uri.parse("http://www.
google.com"));
startActivity(intent);
```

上述代码中，Intent 的动作是 Intent.ACTION_VIEW，根据 URI 的数据类型来匹配动作，数据部分的 URI 是 Web 地址，使用 Uri.parse(urlString)方法，可以简单地把一个字符串解释成 Uri 对象。

创建 Intent 对象的语法如下：

```
Intent intent = new Intent(Intent.ACTION_VIEW, Uri.parse(urlString));
```

Intent 构造函数的第 1 个参数是 Intent 需要执行的动作，第 2 个参数是 URI，表示需要传递的数据。

Android 系统支持的常见动作字符串常量如表 7-2 所示。

表 7-2　Android 系统支持的常见动作字符串常量

动　作	说　明
ACTION_ANSWER	打开接听电话的 Activity，默认为 Android 内置的拨号盘界面
ACTION_CALL	打开拨号盘界面并拨打电话，使用 Uri 中的数字部分作为电话号码
ACTION_DELETE	打开一个 Activity，对所提供的数据进行删除操作
ACTION_DIAL	打开内置拨号盘界面，显示 Uri 中提供的电话号码
ACTION_EDIT	打开一个 Activity，对所提供的数据进行编辑操作
ACTION_INSERT	打开一个 Activity，在提供数据的当前位置插入新项
ACTION_PICK	启动一个子 Activity，从提供的数据列表中选取一项
ACTION_SEARCH	启动一个 Activity，执行搜索动作
ACTION_SENDTO	启动一个 Activity，向数据提供的联系人发送信息
ACTION_SEND	启动一个可以发送数据的 Activity
ACTION_VIEW	最常用的动作，对以 Uri 方式传送的数据，根据 Uri 协议部分以最佳方式启动相应的 Activity 进行处理。对于 http:address 将打开浏览器查看；对于 tel:address 将打开拨号呼叫指定的电话号码
ACTION_WEB_SEARCH	打开一个 Activity，对提供的数据进行 Web 搜索

隐式 Intent 应用具体步骤如下。

（1）同显式 Intent 一样，新创建一个工程，然后打开工程中的 AndroidManifest.xml 文件，在<application>根节点下添加<activity>标签，注册新添加的 activity，嵌套在<application>根节点标签下，添加代码如下。

```
1.    <activity
2.        android:name=".FirstActivity"
3.        android:label="First Activity">
4.        <intent-filter >
5.            <action android:name="com.android.activity.Me_Action"/>
6.            <category android:name="android.intent.category.DEFAULT"/>
7.        </intent-filter>
8.    </activity>
```

（2）修改 res 目录下 layout 文件夹中的 activity_main.xml 文件，设置线性布局，添加一个 Button 按钮控件描述，并设置相关属性，代码如下所示。

```
1.    <?xml version="1.0" encoding="utf-8"?>
```

```
2.  <LinearLayout xmlns:android="http://schemas.android.com/apk/res
    /android"
3.       android:orientation="vertical"
4.       android:layout_width="fill_parent"
5.       android:layout_height="fill_parent"
6.       >
7.  <Button android:id="@+id/bt1"
8.       android:layout_height="wrap_content"
9.       android:layout_width="fill_parent"
10.      android:text="测试隐式 Intent"
11. />
12. </LinearLayout>
```

（3）修改 src 目录下 com.bcpl.activity 包下的 IntentTestDemoActivity.java 文件，添加显式使用 Intent 启动 Activity 的核心代码，代码如下。

```
1.  Button button = (Button)findViewById(R.id.bt1);
2.  button.setOnClickListener(new OnClickListener(){
3.
4.
5.         @Override
6.         public void onClick(View view){
7.             Intent intent = new Intent();
8.             intent.setAction("com.android.activity.Me_Action");
9.             startActivity(intent);
10.        }
11. });
```

（4）在 src 目录下 com.bcpl.activity 包下的 FirstActivity.java 文件，代码如下。

```
1.  public class FirstActivity extends Activity {
2.      @Override
3.      protected void onCreate(Bundle savedInstanceState) {
4.          super.onCreate(savedInstanceState);
5.          setContentView(R.layout.second);
6.          Intent intent = new Intent(Intent.ACTION_VIEW, Uri.parse
              ("http://www.google.com"));
7.          startActivity(intent);
8.      }
9.  }
```

（5）修改 res 目录下 layout 文件夹中新创建的 second.xml 文件，并设置相关属性，代码如下所示。

```
1.  <?xml version="1.0" encoding="utf-8"?>
2.  <LinearLayout xmlns:android="http://schemas.android.com/apk/res/
```

```
    android"
3.      android:orientation="vertical"
4.      android:layout_width="fill_parent"
5.      android:layout_height="fill_parent"
6.      >
7.  <TextView
8.      android:layout_width="fill_parent"
9.      android:layout_height="wrap_content"
10.     android:text="@string/start"
11.     />
12. </LinearLayout>
```

(6) 修改 res 目录下 values 文件夹中的 strings.xml 文件,并设置相关属性,代码如下所示。

```
1.  <?xml version="1.0" encoding="utf-8"?>
2.  <resources>
3.      <string name="hello">Hello World, IntentTestDemoActivity!</string>
4.      <string name="app_name">IntentTestDemo</string>
5.      <string name="start">NewActivity application</string>
6.      <string name="app">NewActivity</string>
7.  </resources>
```

(7) 部署运行程序,程序运行效果如图 7-4 所示,单击"测试隐式 Intent"按钮,程序根据设定的网址生成一个 Intent,并以隐式启动的方式调用 Android 内置的 Web 浏览器,并打开指定的 Google 网站运行,如图 7-5 所示。

图 7-4 隐式 Intent 运行结果

图 7-5 Web 浏览

注意:

Android 本地的应用程序组件和第三方应用程序一样,都是 Intent 解析过程中的一部分。它们没有更高的优先度,可以被新的 Activity 完全代替,这些新的 Activity 宣告自己的 Intent Filter 能响应相同的动作请求。

隐式 Intent 与显式 Intent 相比，更有优势，它不需要指明需要启动哪一个 Activity，而由 Android 系统来决定，有利于使用第三方组件。此外，匹配的 Activity 可以是应用程序本身的，也可以是 Android 系统内置的，还可以是第三方应用程序提供的。因此，这种方式更加强调了 Android 应用程序中组件的可复用性。

在一个 Activity 中可以使用系统提供的 startActivity(Intent intent)方法打开新的 Activity。在打开新的 Activity 前，可以决定是否为新的 Activity 传递参数：

```
startActivity(new Intent(MainActivity.this, NewActivity.class));
```

Bundle 类用作携带数据，它类似于 Map，用于存放 key-value 名值对形式的值。相对于 Map，它提供了各种常用类型的 putXxx()/getXxx()方法，如 putString()/getString()和 putInt()/getInt()，putXxx()用于往 Bundle 对象中放入数据，getXxx()方法用于从 Bundle 对象里获取数据。Bundle 的内部实际上是使用了 HashMap<String, Object>类型的变量来存放 putXxx()方法放入的值。

启动 Activity 并传递数据：

```
1.    public final class Bundle implements Parcelable, Cloneable {
2.         ...
3.      Map<String, Object> mMap;
4.      public Bundle() {
5.          mMap = new HashMap<String, Object>();
6.          ...
7.      }
8.      public void putString(String key, String value) {
9.          mMap.put(key, value);
10.     }
11.     public String getString(String key) {
12.         Object o = mMap.get(key);
13.         return (String) o;
14.     }
15. }
```

在调用 Bundle 对象的 getXxx()方法时，方法内部会从该变量中获取数据，然后对数据进行类型转换，转换成什么类型由方法的 Xxx 决定，getXxx()方法会把转换后的值返回。

打开新的 Activity，并传递若干个参数给它：

```
1.  Intent intent = new Intent(MainActivity.this, NewActivity.class)
2.  Bundle bundle = new Bundle();//该类用作携带数据 bundle.putString("name",
    "lee");
3.  bundle.putInt("age", 4);
4.  intent.putExtras(bundle);//附带上额外的数据
5.  startActivity(intent);
```

在新的 Activity 中接收前面 Activity 传递过来的参数：

```
1.   public class NewActivity extends Activity
2.   {
3.          @Override
4.       protected void onCreate(Bundle savedInstanceState)
5.       {
6.           ...
7.           Bundle bundle = this.getIntent().getExtras();
8.           String name = bundle.getString("name");
9.           int age = bundle.getInt("age");
10.      }
11.  }
```

7.1.3 获取 Activity 返回值

在 Activity 中得到新打开的 Activity 关闭后返回的数据，则需要完成以下几步。

（1）在 Activity 中使用系统提供的 startActivityForResult(Intent intent, int requestCode) 方法来打开新的 Activity。

（2）在 Activity 中重写 onActivityResult(int requestCode, int resultCode, Intent data) 方法。

当新 Activity 关闭后，新 Activity 返回的数据通过 Intent 进行传递，Android 平台会调用前面 Activity 的 onActivityResult()方法，把存放了返回数据的 Intent 作为第三个输入参数传入，这样在 onActivityResult()方法中使用第三个输入参数可以取出新 Activity 返回的数据，代码如下。

```
1.   public class MainActivity extends Activity {
2.     @Override
3.     protected void onCreate(Bundle savedInstanceState) {
4.       ...
5.     Button button =(Button) this.findViewById(R.id.button);
6.     button.setOnClickListener(new View.OnClickListener(){
                                   //单击该按钮会打开一个新的Activity
7.        public void onClick(View v) {
8.          //第二个参数为请求码，可以根据需求自己编号
9.          startActivityForResult(new Intent(MainActivity.this, NewActivity.
            class), 1);
10.     }});
11.    }
12.    //第一个参数为请求码，即调用 startActivityForResult()传递过去的值
13.    //第二个参数为结果码，用于标识返回数据来自哪一个新 Activity
14.    @Override
15.    protected void onActivityResult(int requestCode, int resultCode,
```

```
          Intent data) {
16.       String result = data.getExtras().getString("result"));
                                  //得到新 Activity 关闭后返回的数据
17.    }
18. }
```

上面讲述了使用 startActivityForResult(Intent intent, int requestCode)方法打开新的 Activity，新 Activity 关闭前需要向前面的 Activity 返回数据，需要使用系统提供的 setResult(int resultCode, Intent data)方法实现，代码如下：

```
1.  public class NewActivity extends Activity {
2.     @Override protected void onCreate(Bundle savedInstanceState) {
3.        ...
4.        button.setOnClickListener(new View.OnClickListener(){
5.        public void onClick(View v) {
6.           Intent intent = new Intent();       //数据是使用 Intent 返回
7.           intent.putExtra("result", "返回的数据!");
                                        //把返回数据存入 Intent
8.           NewActivity.this.setResult(RESULT_CANCELED, intent);
                                        //设置返回数据
9.           NewActivity.this.finish();       //关闭 Activity
10.       }});
11.    }
12. }
```

setResult()方法的第一个参数值可以根据需要自己定义，上面代码中使用到的 RESULT_CANCELED 是系统 Activity 类定义的一个常量，值为 0，代码片断如下：

```
1.  public class android.app.Activity extends ...{
2.    public static final int RESULT_CANCELED = 0;
3.    public static final int RESULT_OK = -1;
4.    public static final int RESULT_FIRST_USER = 1;
5.  }
```

上述代码中请求码的作用主要在于：使用 startActivityForResult(Intent intent, int requestCode)方法打开新的 Activity。需要为 startActivityForResult()方法传入一个请求码（第二个参数）。请求码的值是根据业务需要由自己设定的，用于标识请求来源。

例如，一个 Activity 有两个 Button 按钮，单击这两个按钮都会打开同一个 Activity，不管是 button1 还是 button2 按钮打开新 Activity，当这个新 Activity 关闭后，系统都会调用前面 Activity 的 onActivityResult(int requestCode, int resultCode, Intent data)方法。在 onActivityResult()方法中，如果需要知道新 Activity 是由哪个按钮打开的，并且要做出相应的业务处理，则参考代码如下。

```
1.  public void onCreate(Bundle savedInstanceState) {
```

```
2.    ...
3.        button1.setOnClickListener(new View.OnClickListener(){
4.        public void onClick(View v) {
5.            startActivityForResult  (new  Intent(MainActivity.this,
              NewActivity.class), 1);
6.        }});
7.        button2.setOnClickListener(new View.OnClickListener(){
8.        public void onClick(View v) {
9.            startActivityForResult  (new  Intent(MainActivity.this,
              NewActivity.class), 2);
10.       }});
11.       @Override
12. protected void onActivityResult(int requestCode, int resultCode,
    Intent data) {
13.           switch(requestCode){
14.               case 1:
15.                   //来自按钮1的请求，做相应处理
16.               case 2:
17.                   //来自按钮2的请求，做相应处理
18.           }
19.       }
20. }
```

同样，上述代码中结果码的主要作用是：在一个 Activity 中，可能会使用 startActivityForResult()方法打开多个不同的 Activity 处理不同的业务，当这些新 Activity 关闭后，系统都会调用前面 Activity 的 onActivityResult(int requestCode, int resultCode, Intent data)方法。为了知道返回的数据来自于哪个新 Activity，在 onActivityResult()方法（假设 ResultActivity 和 NewActivity 为要打开的新 Activity）中处理代码参考如下。

```
1.  public class ResultActivity extends Activity {
2.      ...
3.      ResultActivity.this.setResult(1, intent);
4.      ResultActivity.this.finish();
5.  }
6.  public class NewActivity extends Activity {
7.      ...
8.      NewActivity.this.setResult(2, intent);
9.      NewActivity.this.finish();
10. }
11. public class MainActivity extends Activity { //在该 Activity 会打开
    //ResultActivity 和 NewActivity
12.         @Override
13. protected void onActivityResult(int requestCode, int resultCode,
    Intent data) {
```

```
14.        switch(resultCode){
15.            case 1:
16.                //ResultActivity 的返回数据
17.            case 2:
18.        //NewActivity 的返回数据
19.            }
20.        }
21. }
```

7.1.4 Intent Filter 原理与匹配机制

Intent Filter（Intent 过滤器）是一种根据 Intent 中的动作（Action）、类别（Category）和数据（Data）等内容，对适合接收该 Intent 的组件进行匹配和筛选的机制。

Intent 过滤器可以匹配数据类型、路径和协议，还包括可以用来确定多个匹配项顺序的优先级（Priority）。

应用程序的 Activity 组件、Service 组件和 BroadcastReceiver 都可以注册 Intent 过滤器，则这些组件在特定的数据格式上就可以产生相应的动作。

1. 注册 Intent Filter

（1）在 AndroidManifest.xml 文件的各个组件的节点下定义<intent-filter>节点，然后在<intent-filter>节点中声明该组件所支持的动作、执行的环境和数据格式等信息。

（2）在程序代码中动态地为组件设置 Intent 过滤器。

在上述（1）中，定义的<intent-filter>节点包含的标签有：<action>标签、<category>标签和<data>标签。

① <action>标签定义 Intent Filter 的"动作"。
② <category>标签定义 Intent Filter 的"类别"。
③ <data>标签定义 Intent Filter 的"数据"。

<intent-filter>节点支持的标签和属性，如表 7-3 所示。

表 7-3 <intent-filter>节点支持的标签和属性

标 签	属 性	说 明
<action>	android:name	指定组件所能响应的动作,用字符串表示,通常使用 Java 类名和包的完全限定名构成
<category>	android:category	指定以何种方式去服务 Intent 请求的动作
<data>	android:host	指定一个有效的主机名
	android:mimetype	指定组件能处理的数据类型
	android:path	有效的 URI 路径名
	android:port	主机的有效端口号
	android:scheme	所需要的特定的协议

<category>标签用来指定 Intent Filter 的服务方式,每个 Intent Filter 可以定义多个<category>标签,开发者可使用自定义的类别,或使用 Android 系统提供的类别,如表 7-4 所示。

表 7-4 Android 系统提供的类别

常 量 值	描 述
ALTERNATIVE	Intent 数据默认动作的一个可替换的执行方法
SELECTED_ALTERNATIVE	和 ALTERNATIVE 类似,但替换的执行方法不是指定的,而是被解析出来的
BROWSABLE	声明 Activity 可以由浏览器启动
DEFAULT	为 Intent 过滤器中定义的数据提供默认动作
HOME	设备启动后显示的第一个 Activity
LAUNCHER	在应用程序启动时首先被显示

AndroidManifest.xml 文件中的每个组件的<intent-filter>都被解析成一个 Intent Filter 对象。当应用程序安装到 Android 系统时,所有的组件和 Intent Filter 都会注册到 Android 系统中。这样,Android 系统便知道了如何将任意一个 Intent 请求通过 Intent Filter 映射到相应的组件上。

2. Intent 解析机制

当使用 startActivity 时,隐式 Intent 解析到一个单一的 Activity。如果存在多个 Activity 都有能够匹配在特定的数据上执行给定的动作,Android 会从这些中选择最好的一个进行启动。决定哪个 Activity 来运行的过程称为 Intent 解析。即 Intent 到 Intent Filter 的映射过程。

Intent 解析机制主要是通过查找已注册在 AndroidManifest.xml 中的所有 Intent Filter 及其中定义的 Intent,最终找到一个可以与请求的 Intent 达成最佳匹配的 Intent Filter。

Intent 解析的匹配规则如下。

(1) Android 系统把所有应用程序包中的 Intent 过滤器集合在一起,形成一个完整的 Intent 过滤器列表。

(2) 在 Intent 与 Intent 过滤器进行匹配时,Android 系统会将列表中所有 Intent 过滤器的"动作"和"类别"与 Intent 进行匹配,任何不匹配的 Intent 过滤器都将被过滤掉。没有指定"动作"的 Intent 过滤器可以匹配任何的 Intent,但是没有指定"类别"的 Intent 过滤器只能匹配没有"类别"的 Intent。

(3) 把 Intent 数据 Uri 的每个子部与 Intent 过滤器的<data>标签中的属性进行匹配,如果<data>标签指定了协议、主机名、路径名或 MIME 类型,那么这些属性都要与 Intent 的 Uri 数据部分进行匹配,任何不匹配的 Intent 过滤器均被过滤掉。

(4) 如果 Intent 过滤器的匹配结果多于一个,则可以根据在<intent-filter>标签中定义的优先级标签来对 Intent 过滤器进行排序,优先级最高的 Intent 过滤器将被选择。

在根据 Intent 解析匹配规则解析的过程中,Android 是通过 Intent 的 action、category、 data 这三个属性来进行判断的，判断方法如下。

（1）如果 Intent 指明了 action，则目标组件的 Intent Filter 的 action 列表中就必须包含这个 action，否则不能匹配。

（2）如果 Intent 没有提供 mimetype，系统将从 data 中得到数据类型。和 action 一样，目标组件的数据类型列表中必须包含 Intent 的数据类型，否则不能匹配。

（3）如果 Intent 中的数据不是 content：类型的 URI，而且 Intent 也没有明确指定它的 type，将根据 Intent 中数据的 scheme （比如 http: 或者 mailto:） 进行匹配。同上，Intent 的 scheme 必须出现在目标组件的 scheme 列表中。

（4）如果 Intent 指定了一个或多个 category，这些类别必须全部出现在组建的类别列表中。比如 Intent 中包含两个类别：LAUNCHER_CATEGORY 和 ALTERNATIVE_CATEGORY，解析得到的目标组件必须至少包含这两个类别。

一个 Intent 对象只能指定一个 action，而一个 Intent Filter 可以指定多个 action，action 的列表不能为空，否则它将组织所有的 Intent。

一个 Intent 对象的 action 必须和 Intent Filter 中的某一个 action 匹配，才能通过测试。如果 Intent Filter 的 action 列表为空，则不通过。如果 Intent 对象不指定 action，并且 Intent Filter 的 action 列表不为空，则通过测试。

下面针对 Intent 和 Intent Filter 中包含的子元素 Action（动作）、Data（数据）以及 Category（类别）进行比较检查的具体规则详细介绍。

1. 动作匹配测试

动作匹配指 Intent Filter 包含特定的动作或没有指定的动作。一个 Intent Filter 有一个或多个定义的动作，如果没有任何一个能与 Intent 指定的动作匹配，这个 Intent Filter 就算作是动作匹配检查失败。

<intent-filter>元素中可以包括子元素<action>，比如：

```
<intent-filter>
<action android:name="com.example.project.SHOW_CURRENT" />
<action android:name="com.example.project.SHOW_RECENT" />
<action android:name="com.example.project.SHOW_PENDING" />
</intent-filter>
```

一条<intent-filter>元素至少应该包含一个<action>，否则任何 Intent 请求都不能和该<intent-filter>匹配。如果 Intent 请求的 Action 和<intent-filter>中某一条<action>匹配，那么该 Intent 就通过了这条<intent-filter>的动作测试。如果 Intent 请求或<intent-filter>中没有说明具体的 Action 类型，那么会出现下面两种情况。

（1）如果<intent-filter>中没有包含任何 Action 类型，那么无论什么 Intent 请求都无法和这条<intent- filter>匹配；

（2）反之，如果 Intent 请求中没有设定 Action 类型，那么只要<intent-filter>中包含 Action 类型，这个 Intent 请求就将顺利地通过<intent-filter>的行为测试。

2. 类别匹配测试

Intent Filter 必须包含所有在解析的 Intent 中定义的种类。一个没有特定种类的 Intent Filter 只能与没有种类的 Intent 匹配。

<intent-filter>元素可以包含<category>子元素，比如：

```
<intent-filter…>
<category android:name="android.Intent.Category.DEFAULT" />
<category android:name="android.Intent.Category.BROWSABLE" />
</intent-filter>
```

只有当 Intent 请求中所有的 Category 与组件中某一个 Intent Filter 的<category>完全匹配时，才会让该 Intent 请求通过测试，Intent Filter 中多余的<category>声明并不会导致匹配失败。一个没有指定任何类别测试的 Intent Filter 只会匹配没有设置类别的 Intent 请求。

3. 数据匹配测试

Intent 的数据 URI 中的部分会与 Intent Filter 中的 data 标签比较。如果 Intent Filter 定义 scheme，host/authority，path 或 mimetype，这些值都会与 Intent 的 URI 比较。任何不匹配都会导致 Intent Filter 从列表中删除。

没有指定 data 值的 Intent Filter 会和所有的 Intent 数据匹配。

数据在<intent-filter>中的描述如下：

```
<intent-filter…>
<data android:type="video/mpeg" android:scheme="http"…/>
<data android:type="audio/mpeg" android:scheme="http"…/>
</intent-filter>
```

<data>元素指定了希望接收的 Intent 请求的数据 URI 和数据类型，URI 被分成三部分来进行匹配：scheme、authority 和 path。其中，用 setData()设定的 Intent 请求的 URI 数据类型和 scheme 必须与 Intent Filter 中所指定的一致。scheme 是 URI 部分的协议——例如 http:，mailto:，tel:。

若 Intent Filter 中还指定了 authority 或 path，它们也需要相匹配才会通过测试。

mimetype 是正在匹配的数据的数据类型。当匹配数据类型时，可以使用通配符来匹配子类型（如 bjzfs/*）。如果 Intent Filter 指定一个数据类型，它必须与 Intent 匹配；如果没有指定数据则全部匹配。

host-name 是介于 URI 中 scheme 和 path 之间的部分（如 www.google.com）。匹配主机名时，Intent Filter 的 scheme 也必须通过匹配。

path 是紧接在 host-name 后面的部分（如/ig）。path 只在 scheme 和 host-name 部分都匹配的情况下才匹配。

如果这个过程中多于一个组件解析出来，它们会以优先度来排序，可以在 Intent Filter

的节点里添加一个可选的标签。最高等级的组件会返回。

具体实例见隐式 Intent 启动 Activity。

7.2 Intent 广播消息

前面章节中已经讲述了 Intent 的用途，其中一个重要用途是发送广播消息。广播消息的内容可以是与应用程序密切相关的数据信息，也可以是 Android 的系统信息，例如网络连接变化、电池电量变化、接收到短信和系统设置变化等，应用程序和 Android 系统都可以使用 Intent 发送广播消息。如果应用程序注册了 BroadcastReceiver，则可以接收到指定的广播消息。

7.2.1 广播消息

使用 Intent 广播消息常用的方法如下。

（1）创建一个 Intent，在构造 Intent 时必须用一个全局唯一的字符串标识其要执行的动作，通常使用应用程序包的名称。

（2）调用 sendBroadcast()方法，就可把 Intent 携带的消息广播出去，如果要使用 Intent 传递额外数据，可以用 Intent 的 putExtra()方法。

利用 Intent 发送广播消息，并添加了额外的数据，然后调用 sendBroadcast()发生了广播消息的代码如下。

```
1.    String UNIQUE_STRING = "com.hisoft.BroadcastReceiverDemo";
2.    Intent intent = new Intent(UNIQUE_STRING);
3.    intent.putExtra("key1", "testValue1");
4.    intent.putExtra("key2", "testValue2");
5.    sendBroadcast(intent);
```

7.2.2 BroadcastReceiver 监听广播消息及应用

1. BroadcastReceiver 监听广播消息简介

BroadcastReceiver（广播接收者）位于 android.content 包下，其类的继承结构如图 7-6 所示，是用于接收 sendBroadcast()广播的 Intent，广播 Intent 的发送是通过调用 Context.sendBroadcast()、Context.sendOrderedBroadcast()来实现的。通常一个广播 Intent 可以被订阅了此 Intent 的多个广播接收者所接收。

广播是一种广泛运用在应用程序之间传输信息的机制，而 BroadcastReceiver 是对发送出来的广播进行过滤接收并响应的一类组件。

BroadcastReceiver 自身并不实现图形用户界面，但是当它收到某个通知后，BroadcastReceiver 可以启动 Activity 作为响应，或者通过 NotificationMananger 提醒用户，或者启动 Service 等。

BroadcastReceiver 为广播接收器，它和事件处理机制类似，只不过事件的处理机制是程序组件级别的，广播处理机制是系统级别的。它用于接收并处理广播通知，如由系统发起的地域变换、电量不足、来电来信等或者程序播放的广播。

BroadcastReceiver 通知用户的方式有多种，如启动 Activity、使用 NotificationManager、开启背景灯、振动设备、播放声音等，最典型的是在状态栏显示一个图标，用户通过点击它打开浏览通知内容。

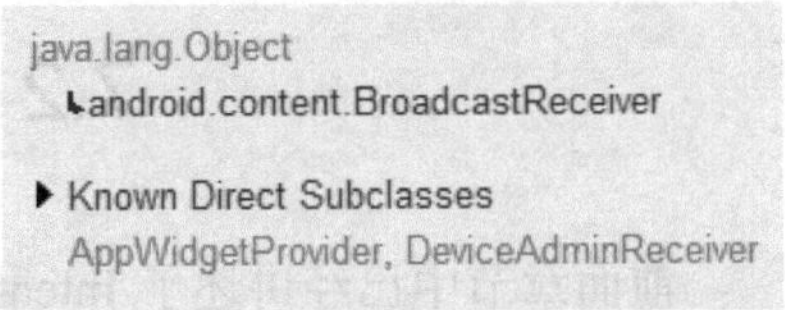

图 7-6 BroadcastReceiver 类继承关系

1）Broadcast Receiver 组件监听过程

使用 Broadcast Receiver 组件监听过滤接收的过程是：首先在需要发送信息的地方，把要发送的信息和用于过滤的信息（如 action、category）封装入一个 Intent 对象，然后通过调用 sendBroadcast()方法，把 Intent 对象以广播方式发送出去。当 Intent 发送以后，所有已在 AndroidManifest.xml 中或代码中注册的 BroadcastReceiver 会检查注册时的 Intent Filter 是否与发送的 Intent 相匹配，若匹配，则就会调用 BroadcastReceiver 的 onReceive()方法。所以在定义一个 BroadcastReceiver 时，需继承 BroadcastReceiver 类，并重载 onReceive()方法。代码如下。

```
1.  public class TestMeBroadcastReceiver extends BroadcastReceiver {
2.      @Override
3.      public void onReceive(Context context, Intent intent) {
4.          …
5.      }
6.  }
```

注册 BroadcastReceiver 的应用程序不需要一直运行，当 Android 系统接收到与之匹配的广播消息时，系统会自动启动此 BroadcastReceiver，在 BroadcastReceiver 接收到与之匹配的广播消息后，onReceive()方法会被调用。onReceive()方法必须要在 5s 内执行完毕，否则 Android 系统会认为该组件失去响应，并提示用户强行关闭该组件。

由于它的典型特征，BroadcastReceiver 通常适合用于做一些资源管理的工作。

2）BroadcastReceiver 用于监听实现方式

BroadcastReceiver 用于监听广播的 Intent。BroadcastReceiver 监听的运用，可以有以下两种方式来实现。

第一种方式是在 AndroidManifest.xml 文件中注册一个 BroadcastReceiver，并在其中使用 Intent Filter 指定要处理的广播消息 Intent。这是一种比较推荐的方法，因为它不需要手动注销广播（如果广播未注销，程序退出时可能会出错）。

第二种方式是直接在代码中实现，但需要手动注册注销。

它们通常的开发步骤如下。

（1）继承 BroadcastReceiver 类，实现自己的类，重写父类 BroadcastReceiver 中的 onReceive()方法。

(2) 在 AndroidManifest.xml 文件中为应用程序添加需要的权限。
(3) 在 AndroidManifest.xml 文件中或者程序代码中注册 BroadcastReceiver 对象。
(4) 等待接收广播，然后匹配。

方式 1 在 AndroidManifest.xml 文件中注册的实现过程如下。

(1) 在 AndroidManifest.xml 文件中注册一个 BroadcastReceiver，在<application>根节点标签下添加<receiver>标签和为应用程序添加需要的权限。

```
1.    < receiver android:name = ".MyBroadcastReceiver" >
2.        < intent-filter android:priority = "1000" >
3.            <action android:name = " android.provider.Telephony.
              SMS_RECEIVED" />
4.        </ intent-filter >
5.    </ receiver >
6.    < uses-permission android:name = "android.permission.RECEIVE_SMS" />
                                                            //添加权限
7.    < uses-permission android:name = "android.permission.SEND_SMS" />
```

(2) 在程序代码中调用 BroadcastReceiver 的 onReceive()方法。

```
1.    public class MyBroadcastReceiver extends BroadcastReceiver {
2.      //action 名称
3.      String SMS_RECEIVED = "android.provider.Telephony.SMS_RECEIVED" ;
4.       public void onReceive(Context context, Intent intent) {
5.         if (intent.getAction().equals( SMS_RECEIVED )) {
6.           //相关处理：地域变换、电量不足、来电来信；
7.         }
8.      }
9.    }
```

方式 2 在代码中注册的实现过程如下。

(1) 在程序代码中使用 registerReceiver 方法注册。

```
1.    IntentFilter intentFilter = new IntentFilter( "android.provider.
      Telephony.SMS_RECEIVED " );
2.    registerReceiver( mBatteryInfoReceiver , intentFilter);
```

(2) 在程序代码中调用 BroadcastReceiver 重写的 onReceive()方法。

```
1.    private BroadcastReceiver myBroadcastReceiver = new
      BroadcastReceiver() {
2.        @Override
3.        public void onReceive(Context context, Intent intent) {
4.            //相关处理，如收短信、监听电量变化信息
5.        }
6.    };
```

(3) 广播注销。

```
1.    //代码中注销广播
2.    unregisterReceiver(mBatteryInfoReceiver);
```

注意：在 Activity 中代码注销广播通常在 onPuase() 中注销，不在 Activity.onSaveInstanceState()中注销，因为这个方法是用来保存 Intent 状态的。

另外，因为 BroadcastReceiver 的生命周期很短，如果需要完成一项比较耗时的工作，应该通过发送 Intent 给 Service，由 Service 来完成。

3）BroadcastReceiver 广播发送方式

BroadcastReceiver 广播的发送方式有三种，分别是普通广播、异步广播、有序广播。

普通广播：发送一个广播，所以监听该广播的广播接收者都可以监听到该广播。

异步广播：当处理完之后的 Intent，依然存在，这时候 registerReceiver(BroadcastReceiver, Intent Filter)还能收到它的值，直到把它去掉，不能将处理结果传给下一个接收者，无法终止广播。

有序广播：按照接收者的优先级顺序接收广播，优先级别在 intent-filter 中的 priority 中声明，范围在–1000～1000 之间，值越大，优先级越高。可以终止广播意图的继续传播，接收者可以修改内容。

由于篇幅原因，此处不再详述，可参考相关文档详细了解它的收发及应用。

2. BroadcastReceiver 应用

前面介绍了 BroadcastReceiver 监听方式和使用方法，下面介绍一个 SMS（Short Message Service，短信）案例，通过 Emulator Control 向模拟器发送短信，模拟器收到短信将会提示，详细介绍 BroadcastReceiver 的应用。

(1) 创建一个新的 Android 工程，工程名为 NotificationDemo，目标 API 选择 17（即 Android 4.2 版本），应用程序名为 NotificationDemo，包名为 com.bcpl.activity，创建的 Activity 的名字为 MainActivity，最小 SDK 版本根据选择的目标 API 会自动添加为 8，创建项目工程。

(2) 修改 res 目录下 layout 文件夹中的 activity_main.xml 文件，设置线性布局，添加一个 TextView 控件描述，并设置相关属性，代码如下所示。

```
1.    <?xml version="1.0" encoding="utf-8"?>
2.    <LinearLayout xmlns:android="http://schemas.android.com/apk/res/android"
3.        android:orientation="vertical"
4.        android:layout_width="fill_parent"
5.        android:layout_height="fill_parent"
6.        >
7.    <TextView
8.        android:layout_width="fill_parent"
9.        android:layout_height="wrap_content"
```

```
10.        android:text="@string/hello"
11.        />
12. </LinearLayout>
```

(3) 修改 src 目录中 com.bcpl.activity 包下的 MainActivity.java 文件，代码如下。

```
1.  package com.bcpl.broadcast;
2.
3.  import android.content.BroadcastReceiver;
4.  import android.content.Context;
5.  import android.content.Intent;
6.  import android.os.Bundle;
7.  import android.telephony.SmsMessage;
8.  import android.widget.Toast;
9.
10.
11. public class SmsReceiver extends BroadcastReceiver
12. {
13.     //当接收到短信时被触发
14.     @Override
15.     public void onReceive(Context context, Intent intent)
16.     {
17.         //如果是接收到短信
18.         if (intent.getAction().equals(
19.             "android.provider.Telephony.SMS_RECEIVED"))
20.         {
21.             //abortBroadcast()方法是取消广播，将会让系统收不到短信，如果
                //不写或者注释掉，系统状态栏会有收到短信息提示，短信息收件箱会收
                //到发送的短信息)
22.             //abortBroadcast();
23.             StringBuilder sb = new StringBuilder();
24.             //接收由 SMS 传过来的数据
25.             Bundle bundle = intent.getExtras();
26.             //判断是否有数据
27.             if (bundle != null)
28.             {
29.                 //通过 pdus 可以获得接收到的所有短信消息
30.                 Object[] pdus = (Object[]) bundle.get("pdus");
31.                 //构建短信对象 array,并依据收到的对象长度来创建 array 的大小
32.                 SmsMessage[] messages = new SmsMessage[pdus.length];
33.                 for (int i = 0; i < pdus.length; i++)
34.                 {
35.                     messages[i] = SmsMessage
36.                         .createFromPdu((byte[]) pdus[i]);
37.                 }
38.                 //将送来的短信合并自定义信息于 StringBuilder 当中
39.                 for (SmsMessage message : messages)
40.                 {
```

```
41.                 sb.append("短信来源:");
42.                 //获得接收短信的电话号码
43.                 sb.append(message.getDisplayOriginatingAddress());
44.                 sb.append("\n------短信内容------\n");
45.                 //获得短信的内容
46.                 sb.append(message.getDisplayMessageBody());
47.             }
48.         }
49.         Toast.makeText(context, sb.toString()
50.                 , Toast.LENGTH_LONG).show();
51.     }
52. }
53. }
```

（4）修改 AndroidManifest.xml 文件，在 application 根节点下，添加配置 SMSReceiver 类，代码如下。

```
1. <receiver android:name=".SmsReceiver">
2.     <intent-filter android:priority="800">
3.         <action android:name="android.provider.Telephony.SMS_RECEIVED" />
4.     </intent-filter>
5. </receiver>
```

同时，在 AndroidManifest.xml 文件中 manifest 根节点下，添加设置应用程序接收短信的权限，以使应用程序可以成功地接收 SMS_RECEIVED 广播，代码如下。

```
1. <uses-permission android:name="android.permission.RECEIVE_SMS"/>
```

（5）测试发送短信息。打开 DDMS 视图，在 DDMS 视图中选择 Emulator Control 面板，在面板中选择 Telephone Actions 分组框，首先选择 SMS 选项，然后在 Incoming number 编辑框中输入接收短信息的手机号码，在 Message 多行文本框中输入内容，然后单击 Send 按钮发送短信息，如图 7-7 所示。

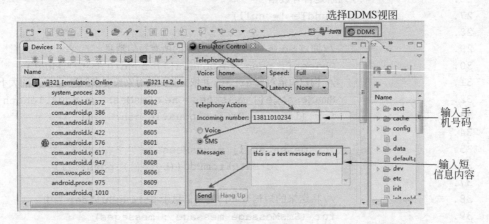

图 7-7 测试信息界面

（6）单击 Send 按钮发送短信息后，短信发送号码、内容信息出现，如图 7-8 所示，然后在界面顶部状态栏中出现有新短信息提示，按住鼠标往下拖动，显示如图 7-9 所示。

（7）单击选择短消息，信息显示如图 7-10 所示，可以选中信息进行查看，也可以单击 menu 按钮，在菜单中选择删除等操作。

图 7-8　发送显示　　　　　图 7-9　信息内容　　　　图 7-10　查看信息内容

在编写 SMSReceiver 类时需要注意如下 4 点。

（1）接收短信的 Broadcast Action 是 android.provider.Telephony.SMS_RECEIVED，因此，要在 onReceiver 方法的开始部分判断接收到的是否是接收短信的 Broadcast Action。

（2）需要通过 Bundle.get("pdus") 来获得接收到的短信消息。这个方法返回了一个表示短信内容的数组。每一个数组元素表示一条短信。这就意味着通过 Bundle.get("pdus") 可以返回多条系统接收到的短信内容。

（3）通过 Bundle.get("pdus") 返回的数组一般不能直接使用，需要使用 SmsMessage.createFromPdu 方法将这些数组元素转换成 SmsMessage 对象才可以使用。每一个 SmsMessage 对象表示一条短信。

（4）通过 SmsMessage 类的 getDisplayOriginatingAddress 方法可以获得发送短信的电话号码。通过 getDisplayMessageBody 方法可以获得短信的内容。

7.3　E-mail 邮件应用

1. 邮件协议 SMTP 简介

SMTP（Simple Mail Transfer Protocol，简单邮件传输协议）是一组用于由源地址到目的地址传送邮件的规则，或者说是由它来控制信件传输的一种中转方式。SMTP 属于

TCP/IP 协议族，它帮助每台计算机或移动终端设备在发送或中转信件时找到下一个目的地。通过 SMTP 所指定的服务器，可以把 E-mail 寄到收信人的服务器上。SMTP 服务器则是遵循 SMTP 的发送邮件服务器，用来发送或中转电子邮件。

在 Android 系统中，邮件的发送都通过内置的 Gmail 程序或者设置邮件服务器的方式完成，而系统平台的底层通信则是采用 SMTP 进行传输信息。

在 Android 系统里进行邮件客户端设计开发可以有两种方式：一种是调用 Android 系统自带的邮件服务；另外一种是采用 JavaMail 功能包的方式，下面分别进行介绍。

1）调用 Android 系统自带的邮件服务

```
//建立 Intent 对象
Intent intent = new Intent();
//设置对象动作
intent.setAction(Intent.ACTION_SEND);
//设置对方邮件地址
intent.putExtra(Intent.EXTRA_EMAIL, new String[]
{ "abc@com.cn","edf@com.cn" });
//设置标题内容
intent.putExtra(Intent.EXTRA_SUBJECT, "test");
//设置邮件文本内容
intent.putExtra(Intent.EXTRA_TEXT, "test mail");
//启动一个新的 ACTIVITY,"Sending mail..."是在启动这个
//ACTIVITY 的等待时间时所显示的文字
startActivity(Intent.createChooser(intent, "Sending
mail..."));
```

其优点是比较简单易用，而缺点是发送邮件的账号必须是 Gmail 账号。

只有上面的代码有可能还会出现异常，运行的时候会提示一个错误:no application can perform this action，这是由于没有在模拟器上配置 Gmail 邮箱，输入自己的 Gmail 账号和密码，默认使用的是 Gmail 账户发信。

2）采用 JavaMail 功能包

其优点是可以设置邮件服务器地址，不必局限于 Gmail 邮箱，而缺点是用法比较复杂。

在 Android 里使用 JavaMail 需要依赖三个包：activation.jar（JDK1.6 之前需要），additionnal.jar，mail.jar。

同时还要注意，在 AndroidManifest.xml 中添加访问互联网的权限，代码如下。

```
<uses-permission android:name="android.permission.INTERNET"></uses-permission>
```

对于 JavaMail，最基础的功能就是邮件的发送和接收。发送邮件主要包括三个部分：创建连接，创建邮件体，发送邮件。

JavaMail 中，是使用会话(Session)来管理连接的。创建一个连接，就需要创建一个

会话。在会话中，有两个重要的因素，一是会话的属性，二是会话的认证。在使用 Hotmail 等邮件工具的时候，就要设置"SMTP 服务器身份验证"，也就是会话的认证。

首先，创建一个连接属性。

```
Properties props = new Properties();
props.put("mail.smtp.host","smtp.126.com");
                            //设置 SMTP 的服务器地址是 smtp.126.com
props.put("mail.smtp.auth","true"); //设置 SMTP 服务器要身份验证。
```

再创建一个身份验证。身份验证稍微复杂一点儿，要创建一个 Authenticator 的子类，并重载 getPasswordAuthentication()方法，代码如下：

```
class PopupAuthenticator extends Authenticator {
    public PasswordAuthentication getPasswordAuthentication() {
    String username = "cqhcp"; //126 邮箱登录账号
    String pwd = "12345"; //登录密码
    return new PasswordAuthentication(username, pwd);
    }
}
```

创建身份验证的实例：

```
PopupAuthenticator auth = new PopupAuthenticator();
```

创建会话。关于会话的创建，有两种方法，具体请参看后续的文章，这里只简单使用一种。

```
Session session = Session.getInstance(props, auth);
```

定义邮件地址：

```
//发送人地址
Address addressFrom = new InternetAddress("cqhcp@126.com", "George Bush");
//收件人地址
Address addressTo = new InternetAddress("webmaster@javazy.com", "George Bush");
//抄送地址
Address addressCopy = new InternetAddress("haocongping@gmail.com", "George Bush");
```

创建邮件体：

```
message.setContent("Hello", "text/plain");
                            //或者使用 message.setText("Hello");
message.setSubject("Title");
message.setFrom(addressFrom);
```

```
message.addRecipient(Message.RecipientType.TO,addressTo);
message.addRecipient(Message.RecipientType.CC,addressCopy);
message.saveChanges();
```

发送邮件的过程:

```
Transport transport = session.getTransport("smtp");    //创建连接
transport.connect("smtp.126.com", "cqhcp", "12345");   //连接服务器
transport.send(message);                               //发送信息
transport.close();                                     //关闭连接
```

整体程序的代码如下:

```
class PopupAuthenticator extends Authenticator {
    public PasswordAuthentication getPasswordAuthentication() {
        String username = "cqhcp"; //163邮箱登录账号
        String pwd = "12345"; //登录密码
        return new PasswordAuthentication(username, pwd);
    }

}
Properties props = new Properties();
props.put("mail.smtp.host","smtp.126.com");
props.put("mail.smtp.auth","true");
PopupAuthenticator auth = new PopupAuthenticator();
Session session = Session.getInstance(props, auth);
MimeMessage message = new MimeMessage(session);
Address addressFrom = new InternetAddress("cqhcp@126.com", "George Bush");
Address addressTo = new InternetAddress("webmaster@javazy.com", "George Bush");
Address addressCopy = new InternetAddress("haocongping@gmail.com", "George Bush");
message.setContent("Hello", "text/plain");
                        //或者使用message.setText("Hello");
message.setSubject("Title");
message.setFrom(addressFrom);
message.addRecipient(Message.RecipientType.TO,addressTo);
message.addRecipient(Message.RecipientType.CC,addressCopy);
message.saveChanges();
Transport transport = session.getTransport("smtp");
transport.connect("smtp.126.com", "cqhcp", "12345");
transport.send(message);
transport.close();
```

若想在登录时判断输入的用户名和密码是否正确,正确时登录,不正确时提示出错

而不登录，只需像下面这样实现：

```
try {
        session.setDebug(true);
        Transport trans = session.getTransport("smtp");
        trans.connect("smtp.126.com",account, password);

} catch (AuthenticationFailedException ae) {
        ae.printStackTrace();
        DisplayToast("用户名或者密码错误！");
                        //其中DisplayToast是作者自己写的一个Toast

} catch (MessagingException mex) {
        mex.printStackTrace();
        Exception ex = null;
        if ((ex = mex.getNextException()) != null) {
            ex.printStackTrace();
        }
    }
```

2. 发送邮件应用

上面介绍了在 Android 中发送邮件常用的包及类和方法，下面通过一个使用发送邮件的应用案例，详细介绍邮件的应用。

（1）创建一个新的 Android 工程，工程名为 EMailDemo，目标 API 选择 17（即 Android 4.2 版本），应用程序名为 EMailDemo，包名为 com.bcpl.activity，创建的 Activity 的名字为 MainActivity，最小 SDK 版本根据选择的目标 API 会自动添加为 8，创建项目工程。

（2）修改 res 目录下 layout 文件夹中的 activity_main.xml 文件，设置线性布局并嵌套线性布局，添加 4 个 EditText 控件、4 个 TextView 和一个 Button 控件描述，并设置相关属性，代码如下所示。

```
1.   <?xml version="1.0" encoding="utf-8"?>
2.   <LinearLayout
xmlns:android="http://schemas.android.com/apk/res/android"
3.       android:orientation="vertical"
4.       android:layout_width="fill_parent"
5.       android:layout_height="fill_parent"
6.       >
7.      <LinearLayout
8.        android:layout_width="fill_parent"
9.        android:layout_height="wrap_content"
10.       android:orientation="horizontal">
11.         <TextView
```

```
12.        android:text="收件人地址："
13.        android:id="@+id/TextView01"
14.        android:textColor="@android:color/white"
15.        android:layout_width="wrap_content"
16.        android:layout_height="wrap_content">
17.        </TextView>
18.        <EditText
19.        android:text=""
20.        android:id="@+id/EditText01"
21.        android:textColor="#222222"
22.        android:layout_width="fill_parent"
23.        android:layout_height="wrap_content">
24.        </EditText>
25.    </LinearLayout>
26.    <LinearLayout
27.     android:layout_width="fill_parent"
28.     android:layout_height="wrap_content"
29.     android:orientation="horizontal">
30.        <TextView
31.        android:text="发件人地址："
32.        android:id="@+id/TextView04"
33.        android:textColor="@android:color/white"
34.        android:layout_width="wrap_content"
35.        android:layout_height="wrap_content">
36.        </TextView>
37.        <EditText
38.        android:text=""
39.        android:id="@+id/EditText04"
40.        android:textColor="#222222"
41.        android:layout_width="fill_parent"
42.        android:layout_height="wrap_content">
43.        </EditText>
44.    </LinearLayout>
45.    <LinearLayout
46.     android:layout_width="fill_parent"
47.     android:layout_height="wrap_content"
48.     android:orientation="horizontal">
49.        <TextView
50.        android:text="邮件主题："
51.        android:id="@+id/TextView02"
52.        android:textColor="@android:color/white"
53.        android:layout_width="wrap_content"
54.        android:layout_height="wrap_content">
55.        </TextView>
```

```
56.        <EditText
57.            android:id="@+id/EditText02"
58.            android:textColor="#222222"
59.            android:layout_width="fill_parent"
60.            android:layout_height="wrap_content">
61.        </EditText>
62.    </LinearLayout>
63.    <TextView
64.        android:text="邮件内容："
65.        android:textColor="@android:color/white"
66.        android:id="@+id/TextView03"
67.        android:layout_width="wrap_content"
68.        android:layout_height="wrap_content">
69.    </TextView>
70.    <EditText
71.        android:id="@+id/EditText03"
72.        android:textColor="#222222"
73.        android:layout_width="fill_parent"
74.        android:layout_height="100dip"
75.        android:gravity="top|left">
76.    </EditText>
77.    <Button
78.        android:text="发送"
79.        android:textColor="#222222"
80.        android:id="@+id/Button01"
81.        android:layout_width="wrap_content"
82.        android:layout_height="wrap_content">
83.    </Button>
84. </LinearLayout>
```

（3）修改 res 目录下 values 文件夹中的 strings.xml 文件，代码如下所示。

```
1. <?xml version="1.0" encoding="utf-8"?>
2. <resources>
3.     <string name="hello">Hello World, MainActivity!</string>
4.     <string name="app_name">EMailDemo</string>
5.     <string name="start">邮件发送中……</string>
6. </resources>
```

（4）修改 src 目录中 com.bcpl.activity 包下的 MainActivity.java 文件，代码如下。

```
1. package com.bcpl.activity;
2.
3. import android.app.Activity;
4. import android.content.Intent;
```

```
5.  import android.os.Bundle;
6.  import android.view.View;
7.  import android.view.View.OnClickListener;
8.  import android.widget.Button;
9.  import android.widget.EditText;
10. import android.widget.Toast;
11.
12. public class MainActivity extends Activity {
13.
14.     EditText etReceiver;//收件人
15.     EditText etSender;//发件人
16.     EditText etTheme;//主题
17.     EditText etMessage;//内容
18.     Button bSend;//"发送"按钮
19.     String strReceiver;//收件人信息
20.     String strSender;//发件人信息
21.     String strTheme;//主题信息
22.     String strMessage;//内容信息
23.
24.     @Override
25.     public void onCreate(Bundle savedInstanceState) {
26.         super.onCreate(savedInstanceState);
27.         setContentView(R.layout.main);
28.         etReceiver=(EditText)this.findViewById(R.id.EditText01);
                                                            //获取对象
29.         etSender=(EditText)this.findViewById(R.id.EditText04);
                                                            //获取对象
30.         etTheme=(EditText)this.findViewById(R.id.EditText02);
                                                            //获取对象
31.         etMessage=(EditText)this.findViewById(R.id.EditText03);
                                                            //获取对象
32.         bSend=(Button)this.findViewById(R.id.Button01);
                                                    //"发送"按钮
33.
34.
35.         bSend.setOnClickListener
36.         (
37.             new OnClickListener()
38.             {
39.                 @Override
40.                 public void onClick(View v) {
41.                     strReceiver=etReceiver.getText().toString().trim();
                                                            //获取收件人
42.                     strSender=etSender.getText().toString().trim();
```

```
43.            strTheme=etTheme.getText().toString().trim();
                                                        //获取主题
44.            strMessage=etMessage.getText().toString().trim();
                                                        //获取内容
45.            String parent="^[a-zA-Z][\\w\\.-]*[a-zA-Z0-9]@[a-zA-Z0-9]
               [\\w\\.-]*[a-zA-Z0-9]\\.[a-zA-Z][a-zA-Z\\.]*[a-zA-Z]$";
46.            if(!strReceiver.matches(parent))//查看收件人地址是否符合格式
47.            {
48.                Toast.makeText(MainActivity.this,"收件人地址格式
                   错误",Toast.LENGTH_SHORT).show();
49.            }else if(!strSender.matches(parent))
                                                //查看发件人地址是否符合格式
50.            {
51.                Toast.makeText(MainActivity.this,"发件
                   人地址格式错误",Toast.LENGTH_SHORT).show();
52.            }else//若都符合格式,则发送邮件
53.            {
54.                Intent intent=new Intent(android.content.
                   Intent.ACTION_SEND);//发送邮件功能
55.                intent.setType("plain/text");
56.                intent.putExtra(android.content.Intent.EXTRA_EMAIL,
                   strReceiver);
57.                intent.putExtra(android.content.Intent.EXTRA_CC,
                   strSender);///
58.                intent.putExtra(android.content.Intent.EXTRA_SUBJECT,
                   strTheme);
59.                intent.putExtra(android.content.Intent.EXTRA_TEXT,
                   strMessage);
60.                startActivity(Intent.createChooser(intent,
                   getResources().getString(R.string.start)));
61.            }
62.        }
63.    }
64.    );
65.  }
66.}
```

（5）部署 EMailDemo 项目工程，程序运行后如图 7-11 所示，填写发件人地址、收件人地址、邮件主题、邮件内容，然后单击"发送"按钮。如前面所讲，如果使用的是系统自带的 Gmail 程序，需要先配置好 Gmail 账户和密码，然后才能调用发送成功，否则会提示"No applications can perform this action"信息，如图 7-12 所示；如果使用的是

SMTP 服务器，需要设置好邮件服务器，才能正确发送成功。

图 7-11　EMailDemo 运行效果

图 7-12　发送邮件

7.4　手机短信发送应用

1．短信服务简介

短信服务是当前任何一款手机都不可缺少的应用程序之一，而且是用户手机使用频率最高的应用之一。在 Android 系统中，以前与短信应用相关的类主要位于包 android.telephony.gsm 中，android.telephony.gsm 包中包含的类有：GsmCellLocation、SmsManager、SmsMessage 和 SmsMessage.SubmitPdu，具体如表 7-5 所示。

表 7-5　android.telephony.gsm 包中的类

类　名	描　述
GsmCellLocation	表示 GSM 手机的位置
SmsManager	管理各种短信操作。这个类已经不再推荐使用，被 android.telephony.SmsManager 替代，以支持 GSM 和 CDMA
SmsMessage	表示具体的短信息，这个类已经不再推荐使用，被 android.telephony.SmsMessage 替代，来支持 GSM 和 CDMA
SmsMessage.SubmitPdu	这个类已经不再推荐使用，用 Use android.telephony.SmsMessage

现在与短信应用相关的类主要位于包 android.telephony 中，从表 7-5 中也可以看出，原来的一些位于 android.telephony.gsm 包中的类，已经不再推荐使用，被 android.telephony 包的类替代。

如上所述，在 Android 系统中 SmsManager 类管理短信息操作，用户利用它可以完成手机的短信发送与接收工作。其中，sendTextMessage() 方法需要传入 5 个值，依次是收件人地址（String）、发送人地址（String）、正文内容（String）、发送服务（PendingIntent）、送达服务（PendingIntent），其中收件人地址与正文内容是不能为 NULL 的参数。同打电话一样，涉及重要的信息必须在配置文件中分配权限，权限代码如下：<uses-permission android:name="android.permission.SEND_SMS"/>。

2. 短信发送与提示

下面通过一个使用 SmsManager 的应用案例，详细介绍短信发送与提示的应用。

（1）创建一个新的 Android 工程，工程名为 SendSMSDemo，目标 API 选择 17（即 Android 4.2 版本），应用程序名为 SendSMSDemo，包名为 com.bcpl.activity，创建的 Activity 的名字为 MainActivity，最小 SDK 版本根据选择的目标 API 会自动添加为 8，创建项目工程。

（2）修改 res 目录下 layout 文件夹中的 activity_main.xml 文件，设置线性布局，添加两个 EditText 控件、两个 TextView 和一个 Button 控件描述，并设置相关属性，EditText 属性设置只能输入数字，代码如下所示。

```
1.   <?xml version="1.0" encoding="utf-8"?>
2.   <LinearLayout xmlns:android="http://schemas.android.com/apk/res/android"
3.       android:orientation="vertical"
4.       android:layout_width="fill_parent"
5.       android:layout_height="fill_parent"
6.       >
7.       <TextView
8.           android:text="接收号码："
9.           android:id="@+id/TextView02"
10.          android:textSize="20dip"
11.          android:textStyle="bold"
12.          android:layout_width="wrap_content"
13.          android:layout_height="wrap_content"
14.          android:paddingLeft="5dip">
15.      </TextView>
16.      <EditText
17.          android:text=""
18.          android:id="@+id/dial_num"
19.          android:layout_width="fill_parent"
20.          android:layout_height="wrap_content">
21.      </EditText>
22.      <TextView
23.          android:text="短信内容："
24.          android:id="@+id/TextView01"
25.          android:layout_width="wrap_content"
26.          android:textSize="20dip"
27.          android:textStyle="bold"
28.          android:paddingLeft="5dip"
29.          android:layout_height="wrap_content">
30.      </TextView>
31.      <EditText
```

```
32.        android:text=""
33.        android:id="@+id/sms_content"
34.        android:layout_width="fill_parent"
35.        android:inputType="number"
36.        android:singleLine="false"
37.        android:gravity="top|left"
38.        android:layout_height="100dip">
39.    </EditText>
40.    <Button
41.        android:text="发送短信"
42.        android:id="@+id/send"
43.        android:textSize="20dip"
44.        android:layout_width="fill_parent"
45.        android:layout_height="wrap_content">
46.    </Button>
47. </LinearLayout>
```

(3) 修改 src 目录中 com.bcpl.activity 包下的 MainActivity.java 文件，代码如下。

```
1.  package com.bcpl.activity;
2.
3.  import android.app.Activity;
4.  import android.app.PendingIntent;
5.  import android.content.Intent;
6.  import android.os.Bundle;
7.  import android.telephony.PhoneNumberUtils;
8.  import android.telephony.SmsManager;
9.  import android.view.View;
10. import android.widget.Button;
11. import android.widget.EditText;
12. import android.widget.Toast;
13.
14. public class MainActivity extends Activity {
15.     /** Called when the activity is first created. */
16.     @Override
17.     public void onCreate(Bundle savedInstanceState) {
18.         super.onCreate(savedInstanceState);
19.         setContentView(R.layout.main);
20.
21.         Button bdial = (Button) this.findViewById(R.id.send);
22.         bdial.setOnClickListener(//为拨号按钮添加监听器
23.         //OnClickListener 为 View 的内部接口，其实现者负责监听点击事件
24.         new View.OnClickListener() {
25.             public void onClick(View v) {
26.                 //获取输入的电话号码
```

```
27.         EditText etTel = (EditText) findViewById(R.id.dial_
            num);
28.         String telStr = etTel.getText().toString();
29.
30.         //获取输入的短信内容
31.         EditText etSms = (EditText) findViewById(R.id.sms_
            content);
32.         String smsStr = etSms.getText().toString();
33.         //判断号码字符串是否合法
34.         if (PhoneNumberUtils.isGlobalPhoneNumber(telStr)) {
                //合法则发送短信
35.             v.setEnabled(false);
                //短信发送完成前将"发送"按钮设置为不可用
36.             sendSMS(telStr, smsStr, v);
37.         } else {//不合法则提示
38.             Toast.makeText(MainActivity.this, //上下文
39.                 "电话号码不符合格式!!! ", //提示内容
40.                 Toast.LENGTH_SHORT //信息显示时间
41.             ).show();
42.         }
43.         }
44.     });
45. }
46.
47. //自己开发的直接发送短信的方法
48. private void sendSMS(String telNo, String smsStr, View v) {
49.     PendingIntent pi = //创建 PendingIntent 对象
50.     PendingIntent.getActivity(this, 0,
51.         new Intent(this, MainActivity.class), 0);
52.     SmsManager sms = SmsManager.getDefault();
53.     sms.sendTextMessage(telNo, null, smsStr, pi, null);
        //收件人,发送人,正文,发送服务,送达服务,其中收件人和正文不可为空
54.     //短信发送成功给予提示
55.     Toast.makeText(MainActivity.this, //上下文
56.         "短信发送成功! ", //提示内容
57.         Toast.LENGTH_SHORT //信息显示时间
58.     ).show();
59.     v.setEnabled(true);//短信发送完成后恢复"发送"按钮的可用状态
60.     }
61. }
```

（4）部署 SendSMSDemo 项目工程，程序运行后，然后在"接受号码"编辑框中输入另外一个模拟器号码 5556，在"短消息内容"编辑框中输入如图 7-13 所示内容，单击"发送短信"按钮，信息发送成功，提示"短信发送成功！"，如图 7-14 所示，然后在

模拟器 5556 中查看接收的信息。在"接受号码"编辑框中不能输入不符合要求的手机号码，因 XML 中 EditText 已经设定属性只能输入数字，其他符号无法输入。

图 7-13　发送信息　　　　　　　　图 7-14　短信发送成功

　　然后如前面章节讲述步骤，在 Eclipse 中创建或者在命令行中使用命令创建 5556 模拟器，打开模拟器 5556 后，单击 Messaging 图标，即可查见刚才模拟器 5554 发送的短信息。

　　注意：如果短信内容过长，可以使用 SmsManager.divideMessage(String text)方法，把短信内容自动拆分成一个 ArrayList 数组，再根据数组长度循环发送，或者直接用 sendMultipartTextMessage 方法发送，参数与 sendTextMessage 类似，也就是短信内容转变为用 divideMessage 拆成的 ArrayList 数组。

　　3．短信发送状态查询

　　上面介绍了使用 SmsManager 发送短信息的应用，下面在上面案例的基础上来介绍短信息发送后其状态状况，帮助用户了解短信息的发送状态，以供用户决定下一步的短信息操作。

　　（1）创建一个新的 Android 工程，工程名为 GetSendSMSStateDemo，目标 API 选择 17（即 Android 4.2 版本），应用程序名为 GetSendSMSStateDemo，包名为 com.bcpl.activity，创建的 Activity 的名字为 MainActivity，最小 SDK 版本根据选择的目标 API 会自动添加为 8，创建项目工程。

　　（2）修改 res 目录下 layout 文件夹中的 activity_main.xml 文件，设置线性布局，添加两个 EditText 控件、一个 TextView 和一个 Button 控件描述，并设置相关属性，代码如下所示。

```xml
1.  <?xml version="1.0" encoding="utf-8"?>
2.  <LinearLayout xmlns:android="http://schemas.android.com/apk/res/android"
3.      android:orientation="vertical"
4.      android:layout_width="fill_parent"
5.      android:layout_height="fill_parent"
6.      >
7.      <TextView
8.          android:layout_width="fill_parent"
9.          android:layout_height="wrap_content"
10.         android:text="@string/SmsNumber"/>
11.     <EditText
12.         android:id="@+id/number"
13.         android:layout_width="fill_parent"
14.         android:layout_height="wrap_content"
15.         android:text="@string/SmstempNumber"/>
16.     <TextView
17.         android:layout_width="fill_parent"
18.         android:layout_height="wrap_content"
19.         android:text="@string/SmsBody"/>
20.     <EditText
21.         android:id="@+id/body"
22.         android:layout_width="fill_parent"
23.         android:layout_height="wrap_content"
24.         android:text="@string/SmstempBody"/>
25.     <Button
26.         android:id="@+id/send"
27.         android:layout_width="fill_parent"
28.         android:layout_height="wrap_content"
29.         android:text="@string/Smssend"/>
30.
31. </LinearLayout>
```

（3）修改 res 目录下 values 文件夹中的 strings.xml 文件，代码如下所示。

```xml
1.  <?xml version="1.0" encoding="utf-8"?>
2.  <resources>
3.      <string name="hello">Hello World, GetSendSMSStateDemoActivity!</string>
4.      <string name="app_name">GetSendSMSStateDemo</string>
5.      <string name="SmsNumber">收件人 </string>
6.      <string name="SmsBody">信息内容</string>
7.      <string name="SmstempNumber">5556</string>
8.      <string name="SmstempBody">测试短信发送状态！</string>
9.      <string name="Smssend">发送信息</string>
```

10. `</resources>`

（4）在 AndroidManifest.xml 文件中<manifest>根节点标签下，添加权限，代码如下。

1. `<uses-permission android:name="android.permission.SEND_SMS" />`

（5）修改 src 目录中 com.bcpl.activity 包下的 MainActivity.java 文件，代码如下。

```java
1.  package com.bcpl.activity;
2.  import android.app.Activity;
3.  import android.app.PendingIntent;
4.  import android.content.BroadcastReceiver;
5.  import android.content.Context;
6.  import android.content.Intent;
7.  import android.content.IntentFilter;
8.  import android.os.Bundle;
9.  import android.telephony.SmsManager;
10. import android.view.View;
11. import android.view.View.OnClickListener;
12. import android.widget.Button;
13. import android.widget.EditText;
14. import android.widget.Toast;
15. public class GetSendSMSStateDemoActivity extends Activity implements OnClickListener{
16.     EditText number;//电话号码
17.     EditText body;//短信内容
18.     Button send;//"发送"按钮
19.     @Override
20.     public void onCreate(Bundle savedInstanceState) {
                                                 //重写的 onCreate 方法
21.         super.onCreate(savedInstanceState);
22.         setContentView(R.layout.main);
23.         send = (Button) this.findViewById(R.id.send);
24.         number = (EditText) this.findViewById(R.id.number);
25.         body = (EditText) this.findViewById(R.id.body);
26.         send.setOnClickListener(this);//给按钮添加监听
27.         IntentFilter myIntentFilter = new IntentFilter("SMS_SEND_ACTION");                        //创建过滤器
28.         MySmsReceiver mySmsReceiver = new MySmsReceiver();
                                                 //创建广播接收
29.         registerReceiver(mySmsReceiver, myIntentFilter);
                                                 //注册广播接收
30.     }
31.     @Override
32.     public void onClick(View v) {//监听方法
```

```
33.        if(v == send){//如果是按下"发送"按钮
34.            send.setEnabled(false);    //设置按钮为不可用
35.            String strNumber = number.getText().toString();
                                         //得到发送电话号码
36.            String strBody = body.getText().toString();
                                         //得到需要发送的信息内容
37.            SmsManager smsManager = SmsManager.getDefault();
                                         //获取SmsManager对象
38.            Intent intentSend = new Intent("SMS_SEND_ACTION");
                                         //创建Intent
39.
40.            PendingIntent sendPI = PendingIntent.getBroadcast
               (getApplicationContext(), 0, intentSend, 0);
41.            smsManager.sendTextMessage(strNumber, null, strBody,
               sendPI, null);            //发送短信
42.            send.setEnabled(true);    //设置按钮为可用
43.        }
44.    }
45.    public class MySmsReceiver extends BroadcastReceiver{
                                         //自定义的广播接收类
46.        @Override
47.        public void onReceive(Context context, Intent intent) {
                                         //重写的onReceive方法
48.            switch(getResultCode()){
49.            case Activity.RESULT_OK://如发送成功
50.                Toast.makeText(context, "信息发送成功", Toast.LENGTH_
                   LONG).show();        //信息提示
51.                break;
52.            case SmsManager.RESULT_ERROR_GENERIC_FAILURE://发送失败
53.                Toast.makeText(context, "信息发送失败", Toast.LENGTH_
                   LONG).show();
54.                break;
55.            default://其他情况
56.                Toast.makeText(context, "状态未知", Toast.LENGTH_
                   LONG).show();
57.                break;
58.            }
59.        }
60.    }
61. }
```

（6）部署GetSendSMSStateDemo项目工程，程序运行后，在"收件人"编辑框中填入收件人的电话号码，在"信息内容"编辑框中写入需要发送的信息内容，单击"发送信息"按钮，如发送成功，结果如图7-15所示。

图 7-15 发送信息成功

7.5 网络访问及通信

本节介绍 Android 系统进行网络访问与数据处理的知识和方法，Android 系统中与访问网络相关的包，主要如表 7-6 所示。

表 7-6 Android 系统中与访问网络相关的包

序 号	包 名	包 描 述
1	java.net	包含访问网络有关的类，包括流和数据包 sockets、Internet 协议和常见 HTTP 处理。该包是一个多功能网络资源
2	org.apache.*	包含许多为 HTTP 通信提供精确控制和功能的包。可以将 Apache 视为流行的开源 Web 服务器
3	android.net	除核心 java.net.* 类以外，包含额外的网络访问 socket。该包包括 URI 类
4	android.net.http	包含处理 SSL 证书的类
5	android.net.wifi	包含在 Android 平台上管理有关 WiFi（802.11 无线 Ethernet）所有方面的类
6	android.telephony	包含用于管理和发送 SMS（文本）消息的类。支持 CDMA 或 android.telephony.cdma 等网络
7	java.io	该包中的类由其他 Java 包中提供的 socket 和连接使用。它们还用于与本地文件（在与网络进行交互时会经常出现）的交互
8	java.nio	包含表示特定数据类型的缓冲区的类。适合用于两个基于 Java 语言的端点之间的通信

Android 系统中访问网络资源的方式有 4 种，分别如下。

（1）使用 URL / HttpURLConnection 访问网络（URL / HttpURLConnection 位于

java.net 包中)。

```
//创建URL对象
URL url = new URL("http://www.google.com/");
//打开连接
HttpURLConnection http = (HttpURLConnection) url.openConnection();
```

(2) 使用 Socket 访问网络 (Socket 位于 java.net 包中)。

服务器端：
```
ServerSocket ser=new ServerSocket(8888);                    //设置监听端口号
    Socket socket=ser.accept();                             //获取连接的 socket 对象
    DataOutputStream sout=new DataOutputStream(socket.getOutputStream());
客户端：
    Socket socket=new Socket("192.168.1.14",8888);          //创建 Socket 对象
    DataInputStream sout=new DataInputStream(socket.getInputStream());
                                                            //获取输入流
```

(3) 使用 HttpClient 应用 Post 和 Get 方式访问网络 (HttpClient 位于 org.apache.http 包中)。

```
//DefaultHttpClient 表示默认属性
HttpClient httpClient = new DefaultHttpClient();
//使用 HttpGet 创建对象实例
HttpGet get = new HttpGet("http://www.google.com/");
HttpResponse rp = httpClient.execute(get);
```

(4) 使用 InetAddress 访问网络 (InetAddress 位于 android.net 包中)。

```
InetAddress inetAddress = InetAddress.getByName("192.168.1.1");
//端口
Socket client = new Socket(inetAddress,61203,true);
//取得数据
InputStream in = client.getInputStream();
OutputStream out = client.getOutputStream();
```

7.5.1 使用 URL 读取网络资源及应用

1. 读取 URL 引用的资源

在 Android 系统中，URL 类位于 java.net 包下，使用资源可以是简单的文件或目录，也可以是对更复杂的对象的引用。URL 可以由协议名、主机、端口和资源路径组成。

URL 的 openConnection()方法将返回一个 URLConnection 对象，该对象表示应用程序和 URL 之间的通信连接。程序可以通过 URLConnection 实例向该 URL 发送请求，读取 URL 引用的资源。

通常创建一个和 URL 的连接,并发送请求。读取此 URL 引用的资源需要如下几个步骤。

(1) 创建 URL 对象,如下。

```
URL myURL=new URL(HTTP://www.baidu.com/hello.txt);
```

(2) 通过调用 URL 对象 openConnection()方法来创建 URLConnection 对象,用类 URLConnection 表示一个打开的网络连接。

```
URLConnection ur=myURL.openConnection();
```

(3) 创建输入流,从网络上读到的数据用字节流的形式表示,如下。

```
InputStream is=ucon.getInputStream();
```

为了避免频繁读取字节流,提高读取效率,用 BufferedInputStream 缓存读到的字节流。

```
InputStream is=ur.getInputStream();
BufferedInputStream bis=new BufferedInputStream(is);
```

(4) 创建 BufferdInputStream 后,然后就可以用 read 方法读入网络数据。

```
ByteArrayBuffer baf=new ByteArrayBuffer(50);
    int current=0;
    while((current=bis.read())!=-1)
    {
     baf.append((byte)current);
    }
```

(5) 将字节流转换为可读取的字符串,并设置编码为 UTF-8。

```
myString=EncodingUtils.getString(baf.toByteArray(),"UTF-8");
```

如果读取的是.txt 等文件是 UTF-8 格式的,就需要对数据进行专门的转换。

(6) 在 AndroidManifest.xml 中加入访问因特网服务的权限:

```
<uses-permission android:name="android.permission.INTERNET" />
```

如果不加入,程序运行就会出现 permission denied 的异常。

2. 使用 URL 访问网络应用

上面介绍了 URL 读取网络资源常用的包及类和方法,下面通过一个访问网络资源的应用案例,详细介绍 URL 的应用。

(1) 创建一个新的 Android 工程,工程名为 NetWorkDemo,目标 API 选择 17(即 Android 4.2 版本),应用程序名为 NetWorkDemo,包名为 com.bcpl.activity,创建的 Activity

的名字为 MainActivity，最小 SDK 版本根据选择的目标 API 会自动添加为 8，创建项目工程。

（2）在 res 目录下 layout 文件夹中新建 mydata.xml 文件，设置线性布局，添加一个 ScrollView 控件、一个 TextView 描述，并设置相关属性，代码如下所示。

```
1.  <LinearLayout
2.  android:id="@+id/LinearLayout01"
3.  android:layout_width="fill_parent"
4.  android:layout_height="fill_parent"
5.  xmlns:android="http://schemas.android.com/apk/res/android">
6.  <ScrollView
7.  android:id="@+id/ScrollView01"
8.  android:layout_width="fill_parent"
9.  android:layout_height="wrap_content">
10. <TextView
11. android:id="@+id/TextView01"
12. android:layout_width="wrap_content"
13. android:layout_height="wrap_content"></TextView>
14. </ScrollView>
15. </LinearLayout>
```

（3）修改 src 目录中 com.bcpl.activity 包下的 NewActivity.java 文件，代码如下。

```
1.  package com.bcpl.activity;
2.  import java.io.BufferedInputStream;
3.  import java.io.InputStream;
4.  import java.net.URL;
5.  import java.net.URLConnection;
6.  import org.apache.http.util.ByteArrayBuffer;
7.  import android.app.Activity;
8.  import android.os.Bundle;
9.  import android.widget.TextView;
10. public class NewActivity extends Activity {
11. @Override
12. protected void onCreate(Bundle savedInstanceState) {
13. //TODO Auto-generated method stub
14. super.onCreate(savedInstanceState);
15. setContentView(R.layout.mydata);
16.
17. TextView tv=(TextView)findViewById(R.id.TextView01);
18.
19. String msg="";
20. try {
21. URL url=new URL("http://linux.chinaitlab.com/c/896411.html");
```

```
22. URLConnection con=url.openConnection();
23. InputStream is=con.getInputStream();
24. BufferedInputStream bis=new BufferedInputStream(is);
25.
26. ByteArrayBuffer baf=new ByteArrayBuffer(100);
27. int current=0;
28. while ((current=bis.read())!= -1) {
29. baf.append((byte)current);
30. }
31. msg=new String(baf.toByteArray(),"GBK");
32.
33. } catch (Exception e) {
34. msg=e.getMessage();
35. }
36. tv.setText(msg);
37. }
38. }
```

（4）在 AndroidManifest.xml 中<manifest>根节点标签下，添加访问网络的权限，代码如下。

```
1. <uses-permission android:name="android.permission.INTERNET" />
```

（5）部署 NetWorkDemo 项目工程，程序运行后，获取 URL 指定地址的网页资源，并显示出来，如图 7-16 所示。

图 7-16　NetWorkDemo 提取网页内容

7.5.2 使用 HTTP 访问网络资源及应用

Android 中使用 HTTP 访问网络资源主要是通过 POST 方式和 GET 方式进行网络请求。HTTP 访问通信中的 POST 和 GET 请求方式不同。GET 可以获得静态页面，也可以把参数放在 URL 字符串后面，传递给服务器。而 POST 方法的参数是放在 HTTP 请求中。因此，在编程之前，应当首先明确使用的请求方法，然后再根据所使用的方式选择相应的编程方式。

HttpURLConnection 位于 java.net 包下，继承于 URLConnection 类，二者都是抽象类。其对象主要通过 URL 的 openConnection 方法获得。与上述的 URLConnection 方法相比较，HttpURLConnection 用于预先不知道数据长度的数据的接收与发送。

（1）HttpURLConnection 使用 POST 方式请求访问资源的通常用法，代码如下所示。

```
1.   URL url = new URL("http://www.zfjsjx.cn/ ");
2.   HttpURLConnection urlConn=(HttpURLConnection)url.openConnection();
3.   //通过以下方法可以对请求的属性进行一些设置
4.   //设置输入和输出流
5.   urlConn.setDoOutput(true);
6.   urlConn.setDoInput(true);
7.   //设置请求方式为 POST
8.   urlConn.setRequestMethod("POST");
9.   //POST 请求不能使用缓存
10.  urlConn.setUseCaches(false);
11.  //关闭连接
12.  urlConn.disConnection();
```

（2）HttpURLConnection 默认使用 GET 方式请求访问资源，代码如下所示。

```
1.   //使用 HttpURLConnection 打开连接
2.            HttpURLConnection urlConn = (HttpURLConnection)
              url.openConnection();
3.            //得到读取的内容(流)
4.            InputStreamReader in = new InputStreamReader
              (urlConn.getInputStream());
5.            //为输出创建 BufferedReader
6.            BufferedReader buffer = new BufferedReader(in);
7.            String inputLine = null;
8.            //使用循环来读取获得的数据
9.            while (((inputLine = buffer.readLine()) != null))
10.           {
11.               //在每一行后面加上一个"\n"来换行
12.               resultData += inputLine + "\n";
```

```
13.         }
14.         //关闭 InputStreamReader
15.         in.close();
16.         //关闭 HTTP 连接
17.         urlConn.disconnect();
```

如果需要使用 POST 方式,则需要 setRequestMethod 设置,代码如下。

```
1.  String httpUrl = "http://www.zfjsjx.cn:8080/a.jsp";
2.      //获得的数据
3.      String resultData = "";
4.  URL url = null;
5.  try
6.  {
7.      //构造一个 URL 对象
8.      url = new URL(httpUrl);
9.  }
10. catch (MalformedURLException e)
11. {
12.     Log.e(DEBUG_TAG, "MalformedURLException");
13. }
14. if (url != null)
15. {
16.     try
17.     {
18.         //使用 HttpURLConnection 打开连接
19.         HttpURLConnection urlConn = (HttpURLConnection)
            url.openConnection();
20.         //因为这个是 POST 请求,需要设置为 true
21.         urlConn.setDoOutput(true);
22.         urlConn.setDoInput(true);
23.         //设置以 POST 方式
24.         urlConn.setRequestMethod("POST");
25.         //POST 请求不能使用缓存
26.         urlConn.setUseCaches(false);
27.         urlConn.setInstanceFollowRedirects(true);
28.         //配置本次连接的 Content-type,配置为 application/x-www-
            form-urlencoded 的
29.         urlConn.setRequestProperty("Content-Type","application/
            x-www-form-urlencoded");
30.         //连接,从 postUrl.openConnection()至此的配置必须要在
            connect 之前完成
31.         //要注意的是 connection.getOutputStream 会隐含地进行
            connect
32.         urlConn.connect();
```

```
33.         //DataOutputStream流
34.         DataOutputStream out = new DataOutputStream(urlConn.
            getOutputStream());
35.         //要上传的参数
36.         String content = "par=" + URLEncoder.encode("ABCDEFG",
            "gb2312");
37.         //将要上传的内容写入流中
38.         out.writeBytes(content);
39.         //刷新、关闭
40.         out.flush();
41.         out.close();
```

上面介绍了 URL 读取网络资源常用的包及类和方法,下面通过一个访问网络资源的应用案例,详细介绍 URL 的应用。

(1)创建一个新的 Android 工程,工程名为 HttpURLConnectionDemo,目标 API 选择 17(即 Android 4.2 版本),应用程序名为 HttpURLConnectionDemo,包名为 com.bcpl.activity,创建的 Activity 的名字为 HttpURLConnectionDemoActivity,最小 SDK 版本根据选择的目标 API 会自动添加为 8,创建项目工程。

(2)修改 res 目录下 layout 文件夹中的 activity_main.xml 文件,设置线性布局,添加一个 EditText 控件、一个 TextView 和一个 Button 控件描述,并设置相关属性,代码如下所示。

```
1.  <?xml version="1.0" encoding="utf-8"?>
2.  <LinearLayout
    xmlns:android="http://schemas.android.com/apk/res/android"
3.      android:orientation="vertical"
4.      android:layout_width="fill_parent"
5.      android:layout_height="fill_parent"
6.      >
7.      <TextView
8.          android:layout_width="fill_parent"
9.          android:layout_height="wrap_content"
10.         android:text="@string/resource"
11.         />
12.     <EditText
13.         android:id="@+id/url"
14.         android:layout_width="fill_parent"
15.         android:layout_height="wrap_content"
16.         android:text="@string/surl"
17.         />
18.     <TextView
19.         android:layout_width="fill_parent"
20.         android:layout_height="wrap_content"
```

```
21.        android:text="@string/goalfile"
22.        />
23.    <EditText
24.        android:id="@+id/target"
25.        android:layout_width="fill_parent"
26.        android:layout_height="wrap_content"
27.        android:text="/mnt/sdcard/bjzf.rar"
28.        />
29.    <Button
30.        android:id="@+id/down"
31.        android:layout_width="fill_parent"
32.        android:layout_height="wrap_content"
33.        android:text="@string/down"
34.        />
35.    <!-- 定义一个水平进度条,用于显示下载进度 -->
36.    <ProgressBar
37.        android:id="@+id/bar"
38.        android:layout_width="fill_parent"
39.        android:layout_height="wrap_content"
40.        android:max="100"
41.        style="@android:style/Widget.ProgressBar.Horizontal"
42.        />
43. </LinearLayout>
```

(3) 修改 res 目录下 values 文件中的 strings.xml 文件,代码如下。

```
1.  <?xml version="1.0" encoding="utf-8"?>
2.  <resources>
3.      <string name="hello">Hello World, HttpURLConnectionDemoActivity!
        </string>
4.      <string name="app_name">HttpURLConnectionDemo</string>
5.      <string name="down">点击下载网络资源</string>
6.      <string name="surl">http://rsdownload.rising.com.cn/for_down/
        rsfree2011/ravf/set1137225.exe</string>
7.      <string name="resource">网络资源 URL 路径:</string>
8.      <string name="goalfile">目标保存文件:</string>
9.  </resources>
```

(4) 修改 AndroidManifest.xml 文件,在<manifest>根节点下添加访问网络权限、SD 卡写入数据权限、SD 卡创建删除文件权限,代码如下。

```
1.  <!-- 在 SD 卡中创建与删除文件权限 -->
2.  <uses-permission android:name="android.permission.MOUNT_UNMOUNT_
    FILESYSTEMS"/>
3.  <!-- 向 SD 卡读写入数据权限 -->
```

4. `<uses-permission android:name="android.permission.WRITE_EXTERNAL_STORAGE"/>`
5. `<uses-permission android:name="android.permission.READ_EXTERNAL_STORAGE"/>`
6. `<!-- 授权访问网络 -->`
7. `<uses-permission android:name="android.permission.INTERNET"/>`

（5）在 src 目录中 com.bcpl.activity 包下创建 AccessUtil.java 文件，代码如下。

```java
1.  package com.bcpl.activity;
2.  import java.io.InputStream;
3.  import java.io.RandomAccessFile;
4.  import java.net.HttpURLConnection;
5.  import java.net.URL;
6.
7.  public class AccessUtil
8.  {
9.      //定义下载资源的路径
10.     private String path;
11.     //指定所下载的文件的保存位置
12.     private String targetFile;
13.     //定义需要使用多少线程下载资源
14.     private int threadNum;
15.     //定义下载的线程对象
16.     private DownloadThread[] threads;
17.     //定义下载的文件的总大小
18.     private int fileSize;
19.
20.     public AccessUtil(String path, String targetFile, int threadNum)
21.     {
22.         this.path = path;
23.         this.threadNum = threadNum;
24.         //初始化 threads 数组
25.         threads = new DownloadThread[threadNum];
26.         this.targetFile = targetFile;
27.     }
28.
29.     public void download() throws Exception
30.     {
31.         URL url = new URL(path);
32.         HttpURLConnection conn = (HttpURLConnection) url.openConnection();
33.         conn.setConnectTimeout(5 * 1000);
34.         conn.setRequestMethod("GET");
35.         conn.setRequestProperty(
```

```
36.            "Accept",
37.            "image/gif, image/jpeg, image/pjpeg, image/pjpeg,
               application/x-shockwave-flash, application/xaml+xml,
                  application/vnd.ms-xpsdocument,
application/x-ms-xbap,          application/x-ms-application,
application/vnd.ms-excel,          application/vnd.ms-powerpoint,
application/msword, */*");
38.            conn.setRequestProperty("Accept-Language", "zh-CN");
39.            conn.setRequestProperty("Charset", "UTF-8");
40.            conn.setRequestProperty(
41.               "User-Agent",
42.               "Mozilla/4.0 (compatible; MSIE 7.0; Windows NT 5.2; Trident/
                  4.0; .NET CLR 1.1.4322; .NET CLR 2.0.50727; .NET CLR
                  3.0.04506.30; .NET CLR 3.0.4506.2152; .NET CLR 3.5.30729)");
43.            conn.setRequestProperty("Connection", "Keep-Alive");
44.            //得到文件大小
45.            fileSize = conn.getContentLength();
46.            conn.disconnect();
47.            int currentPartSize = fileSize / threadNum + 1;
48.            RandomAccessFile file = new RandomAccessFile(targetFile,
               "rw");
49.            //设置本地文件的大小
50.            file.setLength(fileSize);
51.            file.close();
52.            for (int i = 0; i < threadNum; i++)
53.            {
54.               //计算每条线程的下载的开始位置
55.               int startPos = i * currentPartSize;
56.               //每个线程使用一个RandomAccessFile进行下载
57.               RandomAccessFile currentPart = new RandomAccessFile
                  (targetFile,
58.                  "rw");
59.               //定位该线程的下载位置
60.               currentPart.seek(startPos);
61.               //创建下载线程
62.               threads[i] = new DownloadThread(startPos, currentPartSize,
63.                  currentPart);
64.               //启动下载线程
65.               threads[i].start();
66.            }
67.         }
68.
69.         //获取下载的完成百分比
70.         public double getCompleteRate()
71.         {
```

```java
72.        //统计多条线程已经下载的总大小
73.        int sumSize = 0;
74.        for (int i = 0; i < threadNum; i++)
75.        {
76.            sumSize += threads[i].length;
77.        }
78.        //返回已经完成的百分比
79.        return sumSize * 1.0 / fileSize;
80.    }
81.
82.    private class DownloadThread extends Thread
83.    {
84.        //当前线程的下载位置
85.        private int startPos;
86.        //定义当前线程负责下载的文件大小
87.        private int currentPartSize;
88.        //当前线程需要下载的文件块
89.        private RandomAccessFile currentPart;
90.        //定义该线程已下载的字节数
91.        public int length;
92.
93.        public DownloadThread(int startPos, int currentPartSize,
94.            RandomAccessFile currentPart)
95.        {
96.            this.startPos = startPos;
97.            this.currentPartSize = currentPartSize;
98.            this.currentPart = currentPart;
99.        }
100.
101.        @Override
102.        public void run()
103.        {
104.            try
105.            {
106.                URL url = new URL(path);
107.                HttpURLConnection conn = (HttpURLConnection) url
108.                    .openConnection();
109.                conn.setConnectTimeout(5 * 1000);
110.                conn.setRequestMethod("GET");
111.                conn.setRequestProperty(
112.                    "Accept", "image/gif, image/jpeg, image/pjpeg, image/pjpeg," +
113.                        " application/x-shockwave-flash, application/xaml+xml, application/vnd.
```

```
                         ms-xpsdocument, application/x-ms-xbap,
                         application/x-ms-application,
                         application/vnd.ms-excel,
                         application/vnd.ms-powerpoint,
                         application/msword, */*");
114.            conn.setRequestProperty("Accept-Language", "zh-CN");
115.            conn.setRequestProperty("Charset", "UTF-8");
116.            InputStream inStream = conn.getInputStream();
117.            //跳过 startPos 个字节，表明该线程只下载自己负责那部分文件。
118.            inStream.skip(this.startPos);
119.            byte[] buffer = new byte[1024];
120.            int hasRead = 0;
121.            //读取网络数据，并写入本地文件
122.            while (length < currentPartSize
123.                && (hasRead = inStream.read(buffer)) != -1)
124.            {
125.                currentPart.write(buffer, 0, hasRead);
126.                //累计该线程下载的总大小
127.                length += hasRead;
128.            }
129.            currentPart.close();
130.            inStream.close();
131.        }
132.        catch (Exception e)
133.        {
134.            e.printStackTrace();
135.        }
136.    }
137. }
138. }
```

（6）修改 src 目录中 com.bcpl.activity 包下的 HttpURLConnectionDemoActivity.java 文件，代码如下。

```
1.  package com.bcpl.activity;
2.  import java.util.Timer;
3.  import java.util.TimerTask;
4.
5.  import android.app.Activity;
6.  import android.os.Bundle;
7.  import android.os.Handler;
8.  import android.os.Message;
9.  import android.view.View;
10. import android.view.View.OnClickListener;
```

```java
11. import android.widget.Button;
12. import android.widget.EditText;
13. import android.widget.ProgressBar;
14.
15. public class HttpURLConnectionDemoActivity extends Activity
16. {
17.     EditText url;
18.     EditText target;
19.     Button downBn;
20.     ProgressBar bar;
21.     AccessUtil downUtil;
22.     private int mDownStatus;
23.
24.     @Override
25.     public void onCreate(Bundle savedInstanceState)
26.     {
27.         super.onCreate(savedInstanceState);
28.         setContentView(R.layout.activity_main);
29.         //获取程序界面中的三个界面控件
30.         url = (EditText) findViewById(R.id.url);
31.         target = (EditText) findViewById(R.id.target);
32.         downBn = (Button) findViewById(R.id.down);
33.         bar = (ProgressBar) findViewById(R.id.bar);
34.
35.         StrictMode.ThreadPolicy
36.         policy=new StrictMode.ThreadPolicy.Builder().permitAll().build();
37.         StrictMode.setThreadPolicy(policy);
38.         //创建一个Handler对象
39.         final Handler handler = new Handler()
40.         {
41.             @Override
42.             public void handleMessage(Message msg)
43.             {
44.                 if (msg.what == 0x123)
45.                 {
46.                     bar.setProgress(mDownStatus);
47.                 }
48.             }
49.         };
50.         downBn.setOnClickListener(new OnClickListener()
51.         {
52.             @Override
53.             public void onClick(View v)
```

```
54.         {
55.             //初始化 DownUtil 对象
56.             downUtil = new AccessUtil(url.getText().toString(),
57.                 target.getText().toString(), 4);
58.             try
59.             {
60.                 //开始下载
61.                 downUtil.download();
62.             }
63.             catch (Exception e)
64.             {
65.                 e.printStackTrace();
66.             }
67.             //定义每秒调度获取一次系统的完成进度
68.             final Timer timer = new Timer();
69.             timer.schedule(new TimerTask()
70.             {
71.                 @Override
72.                 public void run()
73.                 {
74.                     //获取下载任务的完成比率
75.                     double completeRate = downUtil.getCompleteRate();
76.                     mDownStatus = (int) (completeRate * 100);
77.                     //发送消息通知界面更新进度条
78.                     handler.sendEmptyMessage(0x123);
79.                     //下载完全后取消任务调度
80.                     if (mDownStatus >= 100)
81.                     {
82.                         timer.cancel();
83.                     }
84.                 }
85.             }, 0, 100);
86.         }
87.     });
88.   }
89. }
```

注意：Android 3.0 版本开始就强制开发者不能在主线程中访问网络，要求访问网络需放在独立的线程中。在开发中，为了防止访问网络阻塞主线程，一般都要把访问网络放在独立线程中或者异步线程 AsyncTask 中，否则会报以下错误：

android.os.NetworkOnMainThreadException
at android.os.StrictMode$AndroidBlockGuardPolicy.onNetwork(StrictMode.java:1117)

但是由于某些原因，开发者想要忽略这些强制策略问题，可以在 onCreate()方法里面加上以下代码：

```
StrictMode.ThreadPolicy policy=new StrictMode.ThreadPolicy.Builder().
permitAll().build();
StrictMode.setThreadPolicy(policy);
```

同时在 AndroidManifest.xml 中<uses-sdk android:minSdkVersion="11" />的最小 SDK 版本需为 11 以上即可解决上述错误问题。

（7）部署 HttpURLConnectionDemo 项目工程，程序运行如图 7-17 所示，单击"点击下载网络资源"按钮，开始下载，下载完成后存储路径，如图 7-18 和图 7-19 所示。

图 7-17　HttpURLConnectionDemo 运行效果 　　图 7-18　下载网络资源

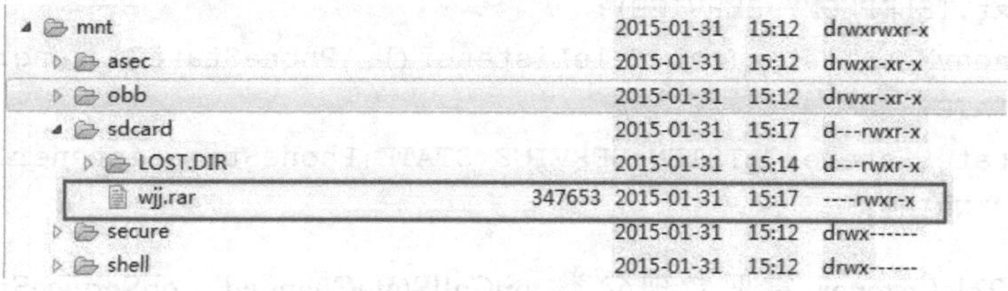

图 7-19　下载文件存储路径

7.6　电话拨打服务及应用

1．电话服务简介

Android 系统中，主要通过 Radio Interface Layer (RIL)来提供电话服务以及各个相关硬件之间的抽象层等。RIL 负责数据的可靠传输、AT 命令的发送以及 Response 的解析。应用处理器通过 AT 命令集与带 GPRS 功能的无线通信模块通信。

在 Android 系统中，与电话应用相关的类主要位于 android.telephony 包下，具体如表 7-7 所示。

表 7-7 android.telephony 包中与电话应用相关的类

类 名	描 述
CellLocation	抽象类，表示设备的位置
PhoneNumberFormattingTextWatcher	监听一个 TextView 控件，如果有电话号码进入，则调用 formatNumber()方法处理电话号码
PhoneNumberUtils	处理电话号码字符串的各种工具
PhoneStateListener	监听手机设备中电话状态变化的监听类，监听的内容包含服务的状态、信号的强弱、短信息的等待提示等
TelephonyManager	提供对手机设备中电话服务信息的访问

使用 Android 系统调用电话拨打，代码如下：

```
Intent intent = new Intent("android.intent.action.DIAL", Uri.parse("tel:10086") );
startActivity(intent);
```

在执行上述代码后，即可进入拨打电话呼叫界面。

在 Android 系统中，监听电话状态的功能如下。

首先建立一个继承于 PhoneStateListener 的电话监听类（如 TeleListener），并调用 TelephonyManager 监听它，关键代码如下：

```
TelephonyManager mTelephonyMgr = (TelephonyManager) getSystemService
(Context.TELEPHONY_SERVICE);
mTelephonyMgr.listen(new TeleListener(), PhoneStateListener.LISTEN_
CALL_STATE|
PhoneStateListener.LISTEN_SERVICE_STATE|PhoneStateListener.LISTEN_SIG
NAL_STRENGTH);
```

此外，TeleListener 需要实现父类 onCallStateChanged、onServiceStateChanged、onSignalStrengthChanged 几个方法，根据方法名便可知其功能，这里就不再详述。

2．接打电话

上面介绍了电话服务常用的包及类，下面通过一个使用电话服务拨打电话的应用案例，详细介绍电话服务的应用。

（1）创建一个新的 Android 工程，工程名为 PhoneCallDemo，目标 API 选择 17（即 Android 4.2 版本），应用程序名为 PhoneCallDemo，包名为 com.bcpl.activity，创建的 Activity 的名字为 MainActivity，最小 SDK 版本根据选择的目标 API 会自动添加为 8，创建项目工程。

（2）修改 res 目录下 layout 文件夹中的 activity_main.xml 文件，设置线性布局，添加一个 EditText 控件、一个 TextView 和一个 Button 控件描述，并设置相关属性，代码如下所示。

```xml
1.  <?xml version="1.0" encoding="utf-8"?>
2.  <LinearLayout
    xmlns:android="http://schemas.android.com/apk/res/android"
3.      android:orientation="vertical"
4.      android:layout_width="fill_parent"
5.      android:layout_height="fill_parent"
6.      >
7.      <TextView
8.       android:layout_width="fill_parent"
9.       android:layout_height="wrap_content"
10.      android:text="请您输入电话号码："
11.     />
12.     <EditText
13.      android:id="@+id/phone_num"
14.      android:layout_width="fill_parent"
15.      android:layout_height="wrap_content">
16.     </EditText>
17.
18.     <Button
19.      android:text="拨打"
20.      android:id="@+id/call"
21.      android:layout_width="wrap_content"
22.      android:layout_height="wrap_content">
23.     </Button>
24. </LinearLayout>
```

（3）修改 src 目录中 com.bcpl.activity 包下的 MainActivity.java 文件，代码如下。

```java
1.  package com.bcpl.activity;
2.  import java.util.regex.Matcher;
3.  import java.util.regex.Pattern;
4.  import android.app.Activity;
5.  import android.content.Intent;
6.  import android.net.Uri;
7.  import android.os.Bundle;
8.  import android.view.View;
9.  import android.view.View.OnClickListener;
10. import android.widget.Button;
11. import android.widget.EditText;
12. import android.widget.Toast;
13.
14. public class MainActivity extends Activity {
15.
```

```java
16.     private EditText phone_num;
17.     private Button call;
18.
19.     @Override
20.     public void onCreate(Bundle savedInstanceState) {
21.         super.onCreate(savedInstanceState);
22.         setContentView(R.layout.main);
23.
24.         this.phone_num=(EditText)this.findViewById(R.id.phone_num);
25.         this.call=(Button)this.findViewById(R.id.call);
26.
27.         this.call.setOnClickListener
28.         (
29.             new OnClickListener()
30.             {
31.                 @Override
32.                 public void onClick(View v) {
33.                     String number=phone_num.getText().toString().trim();//获取输入的手机号码
34.                     boolean flag=phoneNumber(number);
35.                     if(flag)
36.                     {
37.                         Intent intent=new Intent("android.intent.action.CALL",Uri.parse("tel:"+number));
38.                         startActivity(intent);
39.                         phone_num.setText("");
40.                     }else
41.                     {
42.                         Toast.makeText(MainActivity.this, "您输入的电话号码格式不正确",Toast.LENGTH_SHORT ).show();
43.                         phone_num.setText("");         //将EditText字符设为空
44.                     }
45.                 }
46.
47.             }
48.         );
49.     }
50.     public boolean phoneNumber(String number)
51.     {
52.     boolean flag=false;
53.     String pare="\\d{11}";//11个整数的手机号码正则式
54.     String pare2="\\d{12}";//12个整数的座机号码正则式
```

```
55.     CharSequence num=number;//获取电话号码
56.     Pattern pattern=Pattern.compile(pare);//判断是否为手机号码
57.     Matcher matcher=pattern.matcher(num);
58.     Pattern pattern2=Pattern.compile(pare2);//判断是否为座机号码
59.     Matcher matcher2=pattern2.matcher(num);
60.     if(matcher.matches()||matcher2.matches())//如果符合格式
61.     {
62.         flag=true;//标志位设为true
63.     }
64.     return flag;
65.     }
66. }
```

（4）部署 PhoneCallDemo 项目工程，程序运行后如图 7-20 所示，在编辑框中输入电话号码，如果输入的电话号码格式不符合要求，单击"拨打"按钮，会显示"您输入的电话号码格式不正确"提示信息，然后编辑框中的内容会置空，等待重新输入；如果输入座机号，长度符合 12 位条件要求，如"010828903214"，单击"拨打"按钮，则调用系统电话程序，拨打电话，如图 7-21 所示。

图 7-20　PhoneCallDemo 运行效果　　　　图 7-21　拨打电话界面

7.7　Service 组件服务

Service 位于 android.app 包下，它是一个不能与用户交互的，不能自己启动的，长期运行在后台的应用组件，每一个 Service 都必须在 AndroidMainfest.xml 文件中使用 <service> 标签在 <application> 标签根节点下进行声明，Service 的启动可以通过 Context.startService() 或 Context.bindService() 完成。

1．Service 生命周期

Android Service 的生命周期相对简单，它只继承了 onCreate()、onStart()、onDestroy()

三个方法。第一次启动 Service 时，先后调用了 onCreate()、onStart()这两个方法，当停止 Service 时，则执行 onDestroy()方法。这里需要注意的是，如果 Service 已经启动了，当再次启动 Service 时，不会再执行 onCreate()方法，而是直接执行 onStart()方法，具体的应用见后续的案例。

2. Service 与 Activity 通信

Service 后端的数据最终是需要呈现在前端 Activity 之上的，因为启动 Service 时，系统会重新开启一个新的进程，这就涉及不同进程间通信的问题了，当我们想获取启动的 Service 实例时，可以用到 bindService 和 onBindService 方法，它们分别执行了 Service 中的 IBinder()和 onUnbind()方法，具体应用可参见前面项目案例。

7.8 项目案例

学习目标：Android 组件 Intent 对象、Handler 及 Socket 等方法及应用。

案例描述：使用客户端和服务端交互的模式，通过 Handler 类、handleMessage、Socket、OutputStreamWriter 等方法，实现用户注册、用户登录、用户验证等操作。此外，本案例添加了数据库的操作，限于篇幅不再介绍。

案例要点：Socket、Handler、Bundle、OutputStreamWriter、Thread 等相关方法。

案例步骤：

（1）创建一个新的 Android 工程，工程名为 BabyLogin，目标 API 选择 17（即 Android 4.2 版本）。

（2）在 res 目录下 layout 文件夹中创建 client.xml 文件，设置线性布局嵌套，添加 TextView、Button、EditText 按钮控件描述，并设置相关属性，代码如下所示。

```
1.  <?xml version="1.0" encoding="utf-8"?>
2.  <LinearLayout
    xmlns:android="http://schemas.android.com/apk/res/android"
3.      android:layout_width="match_parent"
4.      android:layout_height="match_parent"
5.      android:orientation="vertical" >
6.      <LinearLayout android:id="@+id/FrameLayout01"
7.          android:layout_height="wrap_content"
8.          android:layout_width="fill_parent">
9.
10.         <TextView android:id="@+id/textView2"
11.             android:layout_height="wrap_content"
12.             android:layout_width="wrap_content"
13.             android:layout_weight="1"></TextView>
14.         <EditText android:id="@+id/EditText02"
15.             android:layout_height="wrap_content"
```

```
16.        android:text="msg"
17.        android:layout_width="wrap_content"
18.        android:layout_weight="10"/>
19.    <Button android:id="@+id/Button01"
20.        android:layout_height="wrap_content"
21.        android:text="say"
22.        android:layout_gravity="right"
23.        android:layout_width="wrap_content"
24.        android:layout_weight="1"></Button>
25.
26.  </LinearLayout>
27.  <TextView
28.      android:id="@+id/TextView01"
29.      android:layout_width="fill_parent"
30.      android:layout_height="wrap_content"
31.      android:text=""
32.      />
33.
34. </LinearLayout>
```

（3）在 res 目录下 layout 文件夹中创建 register.xml 文件，设置 LinearLayout 和 TableLayout 嵌套，添加 TextView、Button、EditText 按钮控件描述，并设置相关属性，代码如下所示。

```
1.  <?xml version="1.0" encoding="utf-8"?>
2.  <LinearLayout xmlns:android="http://schemas.android.com/apk/res/android"
3.      android:layout_width="match_parent"
4.      android:layout_height="match_parent"
5.      android:orientation="vertical" >
6.
7.      <TextView
8.          android:layout_width="fill_parent"
9.          android:layout_height="wrap_content"
10.         android:gravity="center"
11.         android:text="注册" />
12.
13.     <TableLayout
14.         android:layout_width="fill_parent"
15.         android:layout_height="wrap_content"
16.         android:stretchColumns="1" >
17.
18.         <TableRow>
19.
```

```
20.        <TextView
21.            android:layout_width="wrap_content"
22.            android:layout_height="wrap_content"
23.            android:gravity="center"
24.            android:text="用户名: " />
25.
26.        <EditText
27.            android:id="@+id/usernameRegister"
28.            android:layout_width="fill_parent"
29.            android:layout_height="wrap_content"
30.            android:hint="请输入用户名!" />
31.    </TableRow>
32.
33.    <TableRow>
34.
35.        <TextView
36.            android:layout_width="wrap_content"
37.            android:layout_height="wrap_content"
38.            android:layout_gravity="center"
39.            android:text="密码:" />
40.
41.        <EditText
42.            android:id="@+id/passwordRegister"
43.            android:layout_width="fill_parent"
44.            android:layout_height="wrap_content"
45.            android:password="true" />
46.    </TableRow>
47.    <TableRow>
48.        <TextView />
49.        <LinearLayout>
50.            <Button
51.                android:id="@+id/Register"
52.                android:layout_width="150dp"
53.                android:layout_height="wrap_content"
54.                android:text="注册" />
55.        </LinearLayout>
56.    </TableRow>
57.    </TableLayout>
58. </LinearLayout>
```

（4）在 src 目录中 com.bcpl.service 包下创建 RegisterActivity.java 文件，关于 www.bcpl.service 包下 User.java、UserService.java、DatabaseHelper.java 代码限于篇幅不再介绍，RegisterActivity.java 代码如下。

```
1.  import www.bcpl.service.User;
2.  import www.bcpl.service.UserService;
3.
4.  import android.app.Activity;
5.  import android.content.Intent;
6.  import android.os.Bundle;
7.  import android.util.Log;
8.  import android.view.View;
9.  import android.view.View.OnClickListener;
10. import android.widget.Button;
11. import android.widget.EditText;
12. import android.widget.Toast;
13.
14. public class RegisterActivity extends Activity {
15.
16.     EditText username;
17.     EditText password;
18.     Button register;
19.
20.     protected void onCreate(Bundle savedInstanceState)
21.     {
22.         super.onCreate(savedInstanceState);
23.         setContentView(R.layout.register);
24.         initViews();
25.     }
26.
27.     private void initViews()
28.     {
29.         username=(EditText) findViewById(R.id.usernameRegister);
30.         password=(EditText) findViewById(R.id.passwordRegister);
31.         register=(Button) findViewById(R.id.Register);
32.
33.         register.setOnClickListener(new OnClickListener() {
34.
35.             @Override
36.             public void onClick(View v) {
37.                 if(null!=username&&null!=password)
38.                 {
39.                     String name=username.getText().toString();
40.                     String pass=password.getText().toString();
41.                     if("".equals(name.trim())&&"".equals(pass.trim()))
42.                     {
43.                         Log.i("TAG","注册失败");
```

```
44.                    Toast.makeText(RegisterActivity.this, "请输
                       入用户名或密码", Toast.LENGTH_LONG).show();
45.                    return;
46.                }
47.                UserService userService=new UserService
                   (RegisterActivity.this);
48.                User user=new User(name,pass);
49.                boolean flag=userService.register(user);
50.                if(flag)
51.                {
52.                    Log.i("TAG", "注册成功");
53. //                 Toast.makeText(RegisterActivity.this, "注册
                       成功", Toast.LENGTH_LONG).show();
54.                    Intent i=new Intent();
55.                    i.setClass(RegisterActivity.this,
                       DBActivity.class);
56.                    startActivity(i);
57.                }
58.            }
59.        });
60.    }
61. }
```

（5）在 src 目录下创建 com.bcpl.activity 包中的 ClientActivity.java 文件，设置连接的服务器端 IP 地址和端口，代码如下。

```
1.  package www.bcpl.activity;
2.  import java.io.BufferedReader;
3.  import java.io.BufferedWriter;
4.  import java.io.IOException;
5.  import java.io.InputStreamReader;
6.  import java.io.OutputStreamWriter;
7.  import java.net.InetAddress;
8.  import java.net.Socket;
9.
10. import android.app.Activity;
11. import android.os.Bundle;
12. import android.os.Handler;
13. import android.view.View;
14. import android.widget.Button;
15. import android.widget.EditText;
16. import android.widget.TextView;
17. import android.widget.Toast;
```

```
18.
19. public class ClientActivity extends Activity {
20.
21.     public static Handler mHandler=new Handler();
22.     TextView textView,textView2;
23.     EditText editText2;
24.     String tmp;
25.     Socket clientSocket;
26.
27.     @Override
28.     public void onCreate(Bundle savedInstanceState)
29.     {
30.         super.onCreate(savedInstanceState);
31.         setContentView(R.layout.client);
32.
33.         textView=(TextView) findViewById(R.id.TextView01);
34.         textView2=(TextView) findViewById(R.id.textView2);
35.         editText2=(EditText) findViewById(R.id.EditText02);
36.
37.         Thread t=new Thread(readData);
38.
39.         t.start();
40.
41.         Bundle bu=this.getIntent().getExtras();
42.         textView2.setText(bu.getString("name"));
43.         Button button1=(Button) findViewById(R.id.Button01);
44.
45.         button1.setOnClickListener(new Button.OnClickListener() {
46.
47.             @Override
48.             public void onClick(View v) {
49.
50.                 try {
51.                     if(clientSocket.isConnected())
52.                     {
53.                         BufferedWriter bw;
54.                         try{
55.                             bw=new BufferedWriter(new OutputStreamWriter
                                 (clientSocket.getOutputStream()));
56.
57.                             bw.write(textView2.getText()+":"+
                                 editText2.getText()+"\n");
58.
59.                             bw.flush();
```

```
60.            }catch(IOException e){
61.                Toast.makeText(ClientActivity.this, "客
                   户端输入输出错误~", Toast.LENGTH_LONG).show();
62.            }
63.                editText2.setText("");
64.            }
65.        } catch (NullPointerException e) {
66.            Toast.makeText(ClientActivity.this, "连接服
                   务器端失败，请确认服务器已开启！", Toast.LENGTH_
                   LONG).show();
67.        }
68.            }
69.        });
70.    }
71.
72.    private Runnable updateText=new Runnable(){
73.        public void run(){
74.            textView.append(tmp+"\n");
75.        }
76.    };
77.
78.    private Runnable readData=new Runnable(){
79.        public void run(){
80.            try{
81.                int serverPort=5050;
82.                clientSocket=new Socket("10.22.53.151",serverPort);
83.
84.                BufferedReader br=new BufferedReader(new
                       InputStreamReader(clientSocket.getInputStream()));
85.
86.                while(clientSocket.isConnected()){
87.                    tmp=br.readLine();
88.                    if(null!=tmp)
89.                    {
90.                        mHandler.post(updateText);
91.                    }
92.                }
93.            }catch(IOException e){
94.                Toast.makeText(ClientActivity.this, "客户端输入输出错
                       误~", Toast.LENGTH_LONG).show();
95.            }
96.        }
97.    };
98. }
```

（6）另外创建 Java 工程，创建服务器端代码，具体如下。

```java
1.   package www.bcpl;
2.   import java.io.BufferedReader;
3.   import java.io.BufferedWriter;
4.   import java.io.IOException;
5.   import java.io.InputStreamReader;
6.   import java.io.OutputStreamWriter;
7.   import java.net.ServerSocket;
8.   import java.net.Socket;
9.   import java.util.ArrayList;
10.
11.  public class SocketServer {
12.
13.      private static int serverport = 5050;
14.      private static ServerSocket serverSocket;
15.
16.      //用串列来存储每一个 client
17.      private static ArrayList<Socket> players=new ArrayList<Socket>();
18.
19.      //程序入口点
20.      public static void main(String[] args) {
21.          try {
22.              serverSocket = new ServerSocket(serverport);
23.              System.out.println("Server is start.");
24.
25.              //当 Server 连接时
26.              while (!serverSocket.isClosed()) {
27.                  //终端显示等待客户端连接
28.                  System.out.println("Wait new client connect");
29.                  //呼叫等待接受客户端连接
30.                  waitNewPlayer();
31.              }
32.
33.          } catch (IOException e) {
34.              System.out.println("Server Socket ERROR");
35.          }
36.
37.      }
38.
39.      //等待接受客户端连接
40.      public static void waitNewPlayer() {
41.          try {
42.              Socket socket = serverSocket.accept();
```

```
43.
44.            //创建新的使用者
45.            createNewPlayer(socket);
46.        } catch (IOException e) {
47.
48.        }
49.
50.    }
51.
52.    //创建新的使用者
53.    public static void createNewPlayer(final Socket socket) {
54.
55.        //以新的线程来执行
56.        Thread t = new Thread(new Runnable() {
57.            @Override
58.            public void run() {
59.                try {
60.                    //增加新的使用者
61.                    players.add(socket);
62.
63.                    //取得网络数据流
64.                    BufferedReader br = new BufferedReader(
65.                            new InputStreamReader(socket.
                                getInputStream()));
66.
67.                    //当Socket已经连接时执行
68.                    while (socket.isConnected()) {
69.                        //取得网络流信息
70.                        String msg= br.readLine();
71.
72.                        //输出信息
73.                        System.out.println(msg);
74.
75.                        //广播信息给其他的客户端
76.                        castMsg(msg);
77.                    }
78.
79.                } catch (IOException e) {
80.
81.                }
82.
83.                //移除客户端
84.                players.remove(socket);
85.            }
```

```
86.        });
87.
88.        //启动执行线程
89.        t.start();
90.    }
91.
92.    //广播信息给其他的客户端
93.    public static void castMsg(String Msg){
94.        //创建socket数组
95.        Socket[] ps=new Socket[players.size()];
96.
97.        //将players转为数组存入ps
98.        players.toArray(ps);
99.
100.       //循环取出ps中的每一个元素
101.       for (Socket socket :ps ) {
102.           try {
103.               //输出信息流
104.               BufferedWriter bw;
105.               bw = new BufferedWriter( new OutputStreamWriter
                   (socket.getOutputStream()));
106.               //写入信息到输出流
107.               bw.write(Msg+"\n");
108.               //立即发送
109.               bw.flush();
110.           } catch (IOException e) {
111.
112.           }
113.       }
114.    }
115.  }
```

（7）在AndroidManifest.xml文件中添加网络访问权限，代码如下。

```
1.  <uses-permission android:name="android.permission.INTERNET" />
```

然后在<application>标签下添加注册启动的<activity>，代码如下。

```
1.  <activity android:name=".RegisterActivity"></activity>
2.  <activity android:name=".ClientActivity"></activity>
3.  <activity android:name=".DBActivity"></activity>
```

（8）部署运行，服务器端结果如图7-22所示，"妈咪宝贝"客户端界面如图7-23所示，用户注册成功界面如图7-24所示，给服务器端发送信息和服务器端显示信息如图7-25和图7-26所示。

图 7-22　服务器端运行

图 7-23　"妈咪宝贝"登录客户端

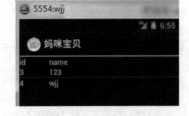

图 7-24　注册成功界面

图 7-25　给服务器端发信息

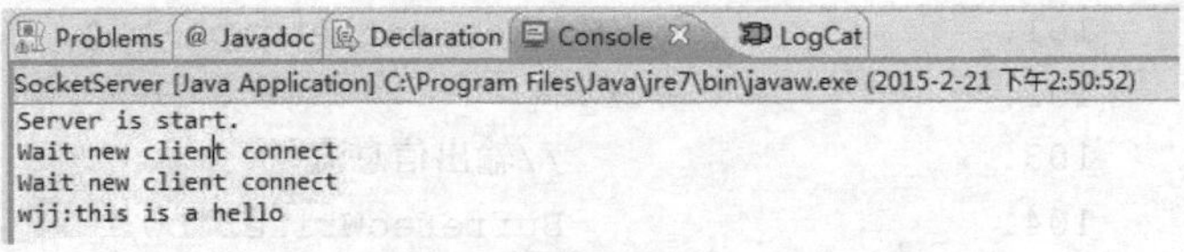

图 7-26　服务器端接收到的信息显示

习　　题

一、简答题

1．Intent 主要类型有哪些？它们之间的区别是什么？

2．通常 Intent 对象由几个部分构成？它主要用于哪些程序功能？

3．Android 系统 3.0 版本之后，在程序中访问网络资源与以前版本有什么不同？

二、实训

要求：

单击宝宝篇中的"宝宝学动物"，会出现动物图片，单击右侧的"中"会中文发音，单击左侧的 E 会英文发音，单击中间的"文字框"会出现童谣，如图 7-27 所示。

图 7-27　宝宝学动物界面显示

第 8 章

Android 数据存储及应用

学习目标

本章介绍了 Android 数据存储与访问的常用方式、SD 卡存储与访问方法、SQLite 数据库(创建、操作、管理及应用)、数据共享(Uri、UriMatcher 和 ContentUris、ContentResolver 操作数据)、网络存储应用等。读者通过本章的学习，能够掌握以下知识要点。

（1）数据存储与访问的通常用法。
（2）创建、访问 SD 卡及应用。
（3）SQLite 数据库创建、操作、管理及应用。
（4）Uri、UriMatcher 和 ContentUris。
（5）创建 ContentProvider、ContentResolver 操作数据。
（6）网络存储应用。

在 Android 系统中，数据存储和使用与通常的数据操作有很大的不同，首先，Android 中所有的应用程序数据，都为自己应用程序所有，其他应用程序如果共享、访问别的应用程序数据，必须通过 Android 系统提供的方式，才能访问或者暴露自己的私有数据供其他应用程序使用。Android 平台中实现数据存储的方式有 5 种，分别是：使用 SharedPreferences 存储数据、文件存储数据、SQLite 数据库存储数据、使用 ContentProvider 存储数据和网络存储数据。

1．SharedPreferences

SharedPreferences 功能类似于 Windows 系统中的 ini 配置文件，主要用于系统的配置信息的保存，比如保留界面设置的颜色、保留登录用户名等，以便下次登录时使用。

2．文件（Files）存储

Android 移动操作系统基于 Linux 核心，所以在 Android 系统中，文件也是 Linux 形式的文件系统。文件保存在设备的内部存储器上，在 Linux 系统下的/data/data/<package name>/files 目录中。

3．数据库

在 Android 系统中，数据存储、管理使用的数据库是轻便型的数据库 SQLite，SQLite

是一个开源的嵌入式关系数据库,与普通关系型数据库一样,也具有 ACID 的特性。

4．ContentProvider

ContentProvider（数据提供者）是在应用程序间共享数据的一种接口机制。ContentProvider 提供了更为高级的数据共享方法,应用程序可以指定需要共享的数据,而其他应用程序则可以在不知道数据来源、路径的情况下,对共享数据进行查询、添加、删除和更新等操作。

5．网络存储

前面介绍的几种存储都是将数据存储在本地设备上,除此之外,还有一种存储（获取）数据的方式,通过网络来实现数据的存储和获取。通过调用 WebService 返回的数据或是解析 HTTP 实现网络数据交互。

在 Android 系统中,按数据的共享方式可以分为:应用程序内自用和数据被其他应用程序共享两种。

1．本应用程序内使用

通常应用中需要的数据一般都是只能为本应用程序使用,使用 SharedPreferences、文件存储、SQLite 数据库存储方式创建的应用程序,默认为本程序使用,其他程序无法获取数据操作。

在 Android 中,可以在控制台使用 adb 命令查看本程序使用的数据:

```
adb shell  //进入手机的文件系统
cd /data/data //进入目录
```

使用 ls 命令查看,可以发现在系统中安装的每个应用程序在对应目录下都有一个文件夹,再次进入应用程序后,使用 ls 命令查看,会出现 shared_prefs、files、databases 几个目录,它们其实就是存放应用程序内自用的数据,内容分别由 SharedPreferences、文件存储、SQLite 数据库存储这三种方式创建。当然如果没有创建过,这个目录可能不存在。

2．应用程序数据共享

这类数据通常是一些共用数据,很多程序都会来调用,比如电话簿数据等。在 Android 系统中,由文件存储、数据库存储、SharedPreferences 创建的数据都可以通过系统提供的特定方式访问,实现数据共享。

8.1 SharedPreferences 存储及访问

8.1.1 SharedPreferences 简介

前面已经讲述了在 Android 系统中,可以使用 SharedPreferences 类来保存一些系统的配置信息、窗口的状态等,SharedPreferences 接口位于 android.content 包下。它是一个轻量级的存储类，特别适合用于保存软件配置参数。

使用 SharedPreferences 保存数据，最终是以 XML 文件存放数据，是基于 XML 文件存储键值对（Name/Value Pair，NVP）数据。XML 处理时 Dalvik 会通过自带底层的本地 XML Parser 解析，比如 XMLpull 方式。SharedPreferences 保存数据的文件存放在目录 /data/data/<package name>/shared_prefs 下。SharedPreferences 不仅能够保存数据，还能够实现不同应用程序间的数据共享。

由于 SharedPreferences 对用户完全屏蔽对文件系统的操作过程，在开发中 SharedPreferences 对象本身只能获取数据而不支持存储和修改，存储修改是通过 Editor 对象实现的。

SharedPreferences 支持各种基本数据类型，包括整型、布尔型、浮点型和长型等。

1. SharedPreferences 访问模式

Android 系统中，SharedPreferences 分为许多权限，其支持的访问模式有三种，分别是：私有、全局读和全局写。

私有（Context.MODE_PRIVATE）：为默认操作模式，代表该文件是私有数据，只能被应用本身访问，在该模式下，写入的内容会覆盖原文件的内容。

全局读（Context.MODE_WORLD_READABLE）：不仅创建程序可以对其进行读取或写入，其他应用程序也有读取操作的权限，但没有写入操作的权限。

全局写（Context.MODE_WORLD_WRITEABLE）：创建程序和其他程序都可以对其进行写入操作，但没有读取的权限。

2. 使用 SharedPreferences 实现存储

1）定义 SharedPreferences 的访问模式

在使用 SharedPreferences 前，先定义 SharedPreferences 的访问模式。

如将访问模式定义为私有模式：

public static int MODE = Context.MODE_PRIVATE;

也可以将 SharedPreferences 的访问模式设定为既可以全局读，也可以全局写，设定如下：

```
public static int MODE = Context.MODE_WORLD_READABLE + Context.MODE_WORLD_WRITEABLE;
```

2）定义 SharedPreferences 的名称

SharedPreferences 的名称与在 Android 文件系统中保存的文件同名。因此，只要具有相同的 SharedPreferences 名称的 NVP 内容，都会保存在同一个文件中，如：

```
public static final String PR_NAME = "SaveFile";
```

3）获取 SharedPreferences 对象

使用 SharedPreferences，需要将上述定义的访问模式和 SharedPreferences 名称作为参数，传递到 getSharedPreferences()函数，并获取到 SharedPreferences 对象：

```
SharedPreferences sharedPreferences = getSharedPreferences(PR_NAME, MODE);
```

4）利用 edit()方法获取 Editor 对象

在获取到 SharedPreferences 对象后，则可以通过 SharedPreferences.Editor 类对 SharedPreferences 进行修改。

```
Editor editor= sharedPreferences.edit();
```

5）通过 Editor 对象存储键值对数据

```
editor.putString("Name", "John");
editor.putInt("Age",28);
editor.putFloat("Height", 1.77);
```

6）通过 commit()方法提交数据

```
editor.commit();
```

完成上述步骤后，如果需要从已经保存的 SharedPreferences 中读取数据，同样是调用 getSharedPreferences()方法，并在方法的第一个参数中指明需要访问的 SharedPreferences 名称，然后通过 get<Type>()方法获取保存在 SharedPreferences 中的 NVP。

```
1.  SharedPreferences sharedPreferences = getSharedPreferences(PR_NAME,
    MODE);
2.  String name = sharedPreferences.getString("Name","name");
3.  int age = sharedPreferences.getInt("Age", 20);
4.  float height = sharedPreferences.getFloat("Height",);
```

上述代码中、get<Type>()方法中的第一个参数是 NVP 的名称。

第二个参数是在无法获取到数值的时候使用的默认值，如 getFloat()第二个参数为默认值，如果 preference 中不存在该 key，将返回默认值。

3．访问其他应用程序数据的 SharedPreferences

如果需要创建访问其他应用程序数据的 SharedPreferences，其前提条件如下。

在 SharedPreferences 对象创建时，为其指定 Context.MODE_WORLD_READABLE 或者 Context.MODE_WORLD_WRITEABLE 权限：

```
Context otherApps = createPackageContext("com.hisoft.sharedpreferences",
Context.CONTEXT_IGNORE_SECURITY);
SharedPreferences sharedPreferences = otherApps.getSharedPreferences
("testApp", Context.MODE_WORLD_READABLE);
String name = sharedPreferences.getString("name", "");
int age = sharedPreferences.getInt("age", 1);
```

如果想采用读取 XML 文件方式，直接访问其他应用 SharedPreferences 对应的 XML 文件，代码如下。

```
File sfx = new File("/data/data/<package name>/shared_prefs/
mypreferences.xml");
//<package name>应替换成应用的包名
```

4. 访问资源文件

（1）访问存储在 res 目录下的文件，如 res/raw 目录下：

```
InputStream ismp3 = getResources().openRawResource(R.raw.testVideo);
                                                    //存放声音文件
```

（2）访问存储在 assets 目录下的文件：

```
InputStream anyFile =getAssets().open(name);//存放数据文件
```

注意：存储文件的大小有限制。

SharedPreferences 对象与后续讲解的 SQLite 数据库相比，省略了创建数据库、创建表、写 SQL 语句等诸多操作，相对而言更加方便、简洁。但是 SharedPreferences 也有其自身缺陷，比如其只能存储 boolean、int、float、long 和 String 5 种简单的数据类型，无法进行条件查询等。所以不论 SharedPreferences 的数据存储操作如何简单，它也只能是存储方式的一种补充，而无法完全替代如 SQLite 数据库这样的其他数据存储方式。

8.1.2 访问本程序数据

前面简单介绍了 SharedPreferences 的基础知识和存储访问应用方法，下面通过一个案例，详细介绍 SharedPreferences 访问本程序数据的应用。

（1）创建一个新的 Android 工程，工程名为 SharedPreferencesDemo，目标 API 选择 17（即 Android 4.2 版本），应用程序名为 SharedPreferencesDemo，包名为 com.bcpl.sharedpreferences，创建的 Activity 的名字为 MainActivity，最小 SDK 版本根据选择的目标 API 会自动添加为 8，创建项目工程。

（2）修改 res 目录下 layout 文件夹中的 activity_main.xml 文件，设置线性布局，添加一个 TextView 控件和两个 EditText 控件描述，并设置相关属性，代码如下所示。

```
1.  <?xml version="1.0" encoding="utf-8"?>
2.  <LinearLayout xmlns:android="http://schemas.android.com/apk/res/android"
3.      android:orientation="vertical"
4.      android:layout_width="fill_parent"
5.      android:layout_height="fill_parent"
6.      >
7.      <TextView
```

```
8.      android:layout_width="fill_parent"
9.      android:layout_height="wrap_content"
10.     android:text="@string/inputname"
11.    />
12. <EditText android:layout_width="match_parent"
13.           android:layout_height="wrap_content"
14.           android:id="@+id/username">
15.     <requestFocus></requestFocus>
16. </EditText>
17. </LinearLayout>
```

（3）修改 res 目录下 values 文件夹中的 strings.xml 文件，代码如下所示。

```
1. <?xml version="1.0" encoding="utf-8"?>
2. <resources>
3.     <string name="hello">Hello World, MainActivity!</string>
4.     <string name="app_name">SharedPreferencesDemo</string>
5.     <string name="inputname">请输入用户名：</string>
6. </resources>
```

（4）修改 src 目录中 com.bcpl.sharedpreferences 包下的 MainActivity.java 文件，代码如下。

```
1. package com.bcpl.sharedpreferences;
2. import android.app.Activity;
3. import android.content.SharedPreferences;
4. import android.content.SharedPreferences.Editor;
5. import android.os.Bundle;
6. import android.view.Menu;
7. import android.view.MenuItem;
8. import android.widget.EditText;
9.
10. public class MainActivity extends Activity {
11.
12.     private EditText et_name;
13.     private static final String NAME = "name";
14.     private static final int EXIT = 1;
15.
16.     /** Called when the activity is first created. */
17.     @Override
18.     public void onCreate(Bundle savedInstanceState) {
19.         super.onCreate(savedInstanceState);
20.         setContentView(R.layout.activity_main);
21.
22.         this.et_name = (EditText) this.findViewById(R.id.username);
23.
```

```
24.         SharedPreferences sp = this.getSharedPreferences
            ("mypreference", MODE_WORLD_READABLE);
25.         String username = sp.getString(NAME, "");
26.
27.         this.et_name.setText(username);
28.
29.     }
30.
31.     @Override
32.     protected void onDestroy() {
33.      super.onDestroy();
34.     SharedPreferences sp = this.getSharedPreferences("mypreference",
        MODE_WORLD_READABLE);
35.      SharedPreferences.Editor edit = sp.edit().putString(NAME, this.
         et_name.getText().toString());
36.      edit.commit();
37.     }
38.
39.     @Override
40.     public boolean onCreateOptionsMenu(Menu menu) {
41.         menu.add(0, EXIT, 0, "退出程序");
42.         return true;
43.     }
44.
45.     @Override
46.     public boolean onOptionsItemSelected(MenuItem item) {
47.
48.         if(item.getItemId() == EXIT){
49.             this.finish();
50.         }
51.         return super.onOptionsItemSelected(item);
52.     }
53. }
```

（5）部署运行程序，SharedPreferencesDemo 工程运行结果如图 8-1 所示。

输入用户名"Gavin"，下次程序启动后，会自动读取用户名到编辑框中，如图 8-2 所示。

数据存储在路径 data/data/com.bcpl.sharepreferences/shared_prefs/目录下，通过单击 Eclipse 菜单 Window→Show View→Other 命令，在对话框中展开 android 文件夹，选择下面的 File Explorer 视图，然后在 File Explorer 视图中展开，如图 8-3 所示，名称为 mypreferences.xml 文件，单击 Pull a file from a device 按钮，导出文件，文件内容以下代码所示。

图 8-1 SharedPreferencesDemo 运行效果

图 8-2 自动读取用户名

图 8-3 数据存储路径

```
1.  <?xml version='1.0' encoding='utf-8' standalone='yes' ?>
2.  <map>
3.      <string name="name">Gavin</string>
4.  </map>
```

8.1.3 读取其他应用程序数据

上述简单介绍了 SharedPreferences 访问其他应用程序数据的条件和方法,下面通过一个案例,详细介绍 SharedPreferences 访问其他应用程序数据的应用。

(1) 创建一个新的 Android 工程,工程名为 OtherSharedPreferencesDemo,目标 API 选择 17(即 Android 4.2 版本),应用程序名为 OtherSharedPreferencesDemo,包名为 com.bcpl.activity,创建的 Activity 的名字为 MainActivity,最小 SDK 版本根据选择的目标 API 会自动添加为 8,创建项目工程。

(2) 修改 res 目录下 layout 文件夹中的 activity_main.xml 文件,设置线性布局,添加一个 TextView 控件描述,并设置相关属性,代码如下所示。

```
1.  <?xml version="1.0" encoding="utf-8"?>
2.  <LinearLayout xmlns:android="http://schemas.android.com/apk/res/
    android"
3.      android:orientation="vertical"
4.      android:layout_width="fill_parent"
5.      android:layout_height="fill_parent"
6.      >
7.      <TextView android:id="@+id/textview1"
```

```
8.      android:layout_width="fill_parent"
9.      android:layout_height="wrap_content"
10.     android:text=""
11.  />
12. </LinearLayout>
```

（3）修改 src 目录中 com.bcpl.activity 包下的 MainActivity.java 文件，代码如下。

```
1.  package com.bcpl.activity;
2.
3.  import android.app.Activity;
4.  import android.content.Context;
5.  import android.content.SharedPreferences;
6.  import android.content.pm.PackageManager.NameNotFoundException;
7.  import android.os.Bundle;
8.  import android.widget.TextView;
9.
10. public class MainActivity extends Activity {
11.
12.     private TextView tv;
13.
14.     /** Called when the activity is first created. */
15.     @Override
16.     public void onCreate(Bundle savedInstanceState) {
17.         super.onCreate(savedInstanceState);
18.         setContentView(R.layout.activity_main);
19.
20.         Context ctx = null;
21.         try
22.         {
23.             //获取其他程序所对应的Context
24.             ctx = createPackageContext("com.bcpl.
                    sharedpreferences",
25.                 Context.CONTEXT_IGNORE_SECURITY);
26.         }
27.         catch (NameNotFoundException e)
28.         {
29.             e.printStackTrace();
30.         }
31.         //使用其他程序的Context获取对应的SharedPreferences
32.         SharedPreferences prefs = ctx.getSharedPreferences
                ("mypreference",
33.             Context.MODE_WORLD_READABLE);
34.         //读取数据
35.         String name = prefs.getString("name", "");
```

```
36.         this.tv = (TextView) findViewById(R.id.textview1);
37.         //显示读取的数据内容
38.         this.tv.setText("被其他应用程序写入的name的值:" + name);
39.     }
40. }
```

(4)部署运行 OtherSharedPreferencesDemo 工程,读取上一案例存储的 name 值,程序运行结果如图 8-4 所示。

图 8-4 程序运行结果

8.2 SQLite 数据库存储及操作

8.2.1 SQLite 数据库简介

SQLite 是 2000 年由 D. Richard Hipp 发布的轻量级嵌入式关系型数据库,它支持 SQL,是一个开源的项目,在 Android 系统平台中集成了嵌入式关系型数据库(SQLite)。

1. SQLite 数据库体系结构

SQLite 数据库由 SQL 编译器、内核、后端以及附件 4 部分组成。SQLite 通过利用虚拟机和虚拟数据库引擎(VDBE),使调试、修改和扩展 SQLite 的内核变得更加方便。SQLite 数据库体系结构如图 8-5 所示。

2. SQLite 数据库和其他数据库的区别

SQLite 和其他数据库最大的不同就是对数据类型的支持,SQLite3 支持 NULL、INTEGER、REAL(浮点数字)、TEXT(字符串文本)和 BLOB(二进制对象)数据类型。虽然它支持的类型只有 5 种,但实际上 SQLite3 也接受 varchar(n)、char(n)、decimal(p,s) 等数据类型,只不过在运算或保存时

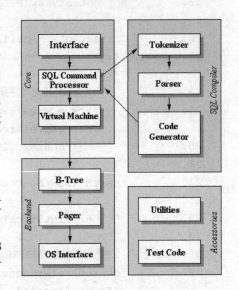

图 8-5 SQLite 数据库体系结构

会转成对应的 5 种数据类型。创建一个表时,可以在 CREATE TABLE 语句中指定某列

的数据类型，但是可以把任何数据类型放入任何列中。当某个值插入数据库时，SQLite 将检查它的类型。如果该类型与关联的列不匹配，则 SQLite 会尝试将该值转换成该列的类型。如果不能转换，则该值将作为其本身具有的类型存储。比如可以把一个字符串（String）放入 INTEGER 列。SQLite 称其为"弱类型"。此外，SQLite 不支持一些标准的 SQL 功能，特别是外键约束，嵌套 transcaction 和 RIGHT OUTER JOIN 和 FULL OUTER JOIN，还有一些 ALTER TABLE 功能。除了上述功能外，SQLite 是一个完整的 SQL 系统，拥有完整的触发器、交易等。

注意：定义为 INTEGER PRIMARY KEY 的字段只能存储 64 位整数，当向这种字段中保存除整数以外的数据时，将会产生错误。

另外，SQLite 在解析 CREATE TABLE 语句时，会忽略 CREATE TABLE 语句中跟在字段名后面的数据类型信息。

8.2.2 创建 SQLite 数据库方式

创建 SQLite 数据库的方式有两种，分别是使用 SQLite3 工具命令行方式和使用程序编码方式，下面就分别介绍它们创建数据库的过程。

1. SQLite3 工具命令行方式

SQLite3 是 SQLite 数据库自带的一个基于命令行的 SQL 命令执行工具，并可以显示命令执行结果。SQLite3 工具被集成在 Android 系统中，用户在 Linux 的命令行界面中输入"sqlite3"可启动 SQLite3 工具，并得到工具的版本信息。在命令窗口中输入"adb shell"命令可以启动 Linux 的命令行界面，过程如下所示：

（1）首先用命令或在 Eclipse 中启动模拟器，然后在命令窗口中输入命令"adb shell"进入设备 Linux 控制台，出现提示符"#"后，输入命令"sqlite3"，如图 8-6 所示。

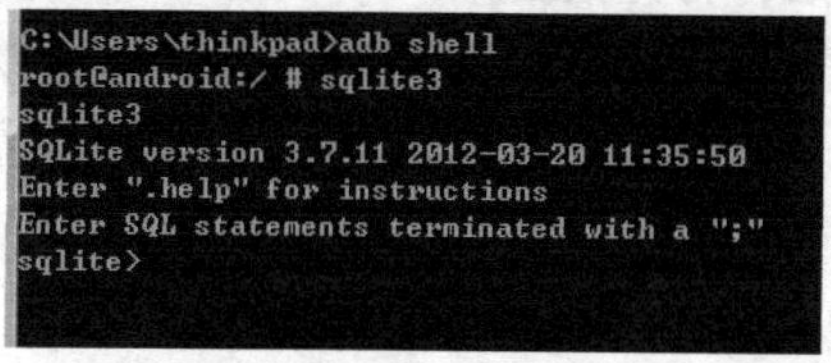

图 8-6　进入 SQLite

在启动 SQLite3 工具后，提示符从"#"变为"sqlite>"，表示命令行界面进入与 SQLite 数据库的交互模式，此时可以输入命令建立、删除或修改数据库中的内容。

正确退出 SQLite3 工具的方法是使用命令".exit"，如图 8-7 所示。

（2）命令行方式手动创建 SQLite 数据库，步骤如下。

① 在命令窗口中输入命令"adb shell"进入设备 Linux 控制台。

② ＃cd/data/data，进入应用 data 目录，如图 8-8 所示。

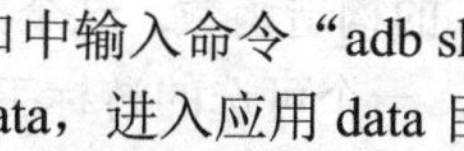

图 8-7　SQLite3 退出命令 　　　图 8-8　进入 data 目录

③ # ls，列表目录，查看文件，如图 8-9 所示。

找到自己的项目包目录并进入，如图 8-10 所示。

图 8-9　查看目录　　　　　　　　图 8-10　进入项目包目录

④ 使用 ls 命令查看有无 databases 目录，如果没有，则创建一个，命令如图 8-11 所示。

⑤ ls 命令查看列表目录会看到有一个文件为 mydb.db，即 SQLite 数据库，如图 8-12 所示。

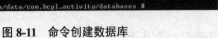

图 8-11　命令创建数据库　　　　　　图 8-12　查看 SQLite 数据库文件

2. 使用程序编码方式

在程序代码中动态建立 SQLite 数据库是比较常用的方法。在程序运行过程中，当需要进行数据库操作时，应用程序会首先尝试打开数据库，此时如果数据库不存在，程序则会自动建立数据库，然后再打开数据库。

在 Android 应用程序中创建使用 SQLite 数据库有两种方式：一种是自定义类继承 SQLiteOpenHelper；另外一种是调用 openOrCreateDatabases() 方法创建数据库。下面就对它们分别进行介绍。

1）自定义类继承 SQLiteOpenHelper 创建数据库

在 Android 应用程序中使用 SQLite，必须自己创建数据库，然后创建表、索引，填充数据。Android 提供了 SQLiteOpenHelper 帮助创建一个数据库，只要继承 SQLiteOpenHelper 类，就可以轻松创建数据库。SQLiteOpenHelper 类根据开发应用程序的需要，封装了创建和更新数据库使用的逻辑。

创建 SQLiteOpenHelper 的子类，至少需要实现以下三个方法。

（1）构造函数，调用父类 SQLiteOpenHelper 的构造函数。这个方法需要 4 个参数：上下文环境（例如，一个 Activity），数据库名字，一个可选的游标工厂（通常是 Null），一个代表正在使用的数据库模型版本的整数。

（2）onCreate() 方法，它需要一个 SQLiteDatabase 对象作为参数，根据需要对这个对象填充表和初始化数据。

(3) onUpgrade() 方法，它需要三个参数：一个 SQLiteDatabase 对象，一个旧的版本号和一个新的版本号。这样就可以知道如何把一个数据库从旧的模型转变到新的模型。

应用程序编码创建 SQLite 数据库的通常步骤如下。

(1) 创建自己的类 DatabaseHelper 继承 SQLiteOpenHelper，并实现上述三个方法，代码如下。

```
1.  public class DatabaseHelper extends SQLiteOpenHelper {
2.    DatabaseHelper(Context context, String name, CursorFactory
      cursorFactory, int version)
3.    {
4.      super(context, name, cursorFactory, version);
5.    }
6.
7.    @Override
8.    public void onCreate(SQLiteDatabase db) {
9.      //TODO 创建数据库后，对数据库的操作
10.   }
11.
12.   @Override
13.   public void onUpgrade(SQLiteDatabase db, int oldVersion, int
      newVersion) {
14.     //TODO 更改数据库版本的操作
15.   }
16.
17.   @Override
18.   public void onOpen(SQLiteDatabase db) {
19.     super.onOpen(db);
20.     //TODO 每次成功打开数据库后首先被执行
21.   }
22. }
```

(2) 获取 SQLiteDatabase 类对象实例。

根据需要改变数据库的内容，决定是调用 getReadableDatabase() 或 getWriteableDatabase() 方法，获取 SQLiteDatabase 实例，如：

```
db=(new DatabaseHelper(getContext())).getWritableDatabase();
```

上面这段代码会返回一个 SQLiteDatabase 类的实例，使用这个对象，就可以查询或者修改数据库。当完成了对数据库的操作（如 Activity 已经关闭）时，需要调用 SQLiteDatabase 的 Close() 方法来释放掉数据库连接。

2) 调用 openOrCreateDatabase ()方法创建数据库

android.content.Context 中提供了方法 openOrCreateDatabase () 来创建数据库。
```
db = context .openOrCreateDatabase( String DATABASE_NAME, int
Context. MODE_PRIVATE, null );
```

8.2.3 SQLite 数据库操作

在编程实现时，一般将所有对数据库的操作都封装在一个类中，因此只要调用这个类，就可以完成对数据库的添加、更新、删除和查询等操作，前面已经讲述了如何创建数据库，下面就对在数据库中创建表、索引、给表添加数据等操作进行介绍。

1）创建表和索引

为了创建表和索引，需要调用 SQLiteDatabase 的 execSQL() 方法来执行 DDL 语句。如果没有异常，这个方法没有返回值。

```
db.execSQL("CREATE TABLE mytable (_id INTEGER PRIMARY KEY  AUTOINCREMENT,
title TEXT, value REAL);");
```

上述语句创建表名为 mytable，表有一个列名为 _id，并且是主键，列值是会自动增长的整数，另外还有两列：title（字符）和 value（浮点数）。SQLite 会自动为主键列创建索引。通常情况下，第一次创建数据库时创建了表和索引。

另外，SQLiteDatabase 类提供了一个重载后的 execSQL(String sql, Object[] bindArgs) 方法。

使用这个方法支持使用占位符参数(?)。使用例子如下：

```
SQLiteDatabase db =…;
db.execSQL("insert into person(name, age) values(?,?)",new Object[]
{"Tom", 4});
db.close();
```

一个参数为 SQL 语句，第二个参数为 SQL 语句中占位符参数的值，参数值在数组中的顺序要和占位符的位置对应。

如果不需要改变表的 schema，不需要删除表和索引。删除表和索引，需要使用 execSQL() 方法调用 DROP INDEX 和 DROP TABLE 语句。

2．给表添加数据

给数据库中的表添加数据有以下两种方法。

（1）使用 execSQL() 方法执行 INSERT, UPDATE, DELETE 等语句来更新表的数据。execSQL() 方法适用于所有不返回结果的 SQL 语句。

例如：db.execSQL("INSERT INTO widgets (name, inventory)"+ "VALUES ('Sprocket', 5)");

（2）使用 SQLiteDatabase 对象的 insert(), update(), delete() 方法。这些方法把 SQL 语句的一部分作为参数。

① insert()方法

insert()方法用于添加数据，各个字段的数据使用 ContentValues 进行存放。

ContentValues 类似于 MAP，相对于 MAP，它提供了存取数据对应的 put(String key, Xxx value)和 getAsXxx(String key)方法，key 为字段名称，value 为字段值。

例如：

```
SQLiteDatabase db = databaseHelper.getWritableDatabase();
ContentValues values = new ContentValues();
values.put("name", "Tom");
values.put("age", 4);
long rowid = db.insert("person", null, values);
                            //返回新添记录的行号，与主键id无关
```

不管第三个参数是否包含数据，执行 insert()方法必然会添加一条记录，如果第三个参数为空，会添加一条除主键之外其他字段值为 Null 的记录。

② update()方法

update()方法有 4 个参数，分别是表名，表示列名和值的 ContentValues 对象，可选的 WHERE 条件和可选的填充 WHERE 语句的字符串，这些字符串会替换 WHERE 条件中的"?"标记。

update() 根据条件，更新指定列的值，所以用 execSQL() 方法可以达到同样的目的。WHERE 条件及其参数和用过的其他 SQL APIs 类似。

例如：

```
String[] parms=new String[] {"this is a string"};
db.update("widgets", replacements, "name=?", parms);
```

③ delete()方法

delete() 方法的使用和 update() 类似，使用表名，可选的 WHERE 条件和相应的填充 WHERE 条件的字符串。例如：

```
db.delete("person", "personid<?", new String[]{"2"});
db.close();
```

3. 查询数据库

在 Android 系统中，数据库查询结果的返回值并不是数据集合的完整拷贝，而是返回数据集的指针，这个指针就是 Cursor 类。

Cursor 类支持在查询的数据集合中以多种方式移动，并能够获取数据集合的属性名称和序号。

查询数据库使用 SELECT 语句从 SQLite 数据库中检索数据有两种方法，分别是：使用 rawQuery()直接调用 SELECT 语句；使用 query()方法构建一个查询。

1）使用 rawQuery()直接调用 SELECT 语句

调用 SQLiteDatabase 类的 rawQuery()方法，用于执行 select 语句。

例如：Cursor c=db.rawQuery("SELECT name FROM sqlite_master WHERE type='table' AND name='mytable'", null);

rawQuery()方法的第一个参数为 SELECT 语句；第二个参数为 SELECT 语句中占位

符参数的值,如果 SELECT 语句没有使用占位符,该参数可以设置为 null。

带占位符参数的 SELECT 语句使用示例如下:

```
Cursor cursor = db.rawQuery("select * from person where name like ? and
age=?", new String[]{"%Tom%", "4"});
```

在上面的例子中,查询 SQLite 系统表(sqlite_master)检查 table 表是否存在。返回值是一个 cursor 对象,这个对象的方法可以迭代查询结果。如果查询是动态的,使用这个方法就会非常复杂。

例如,当需要查询的列在程序编译的时候不能确定,这时候使用 query() 方法会方便很多。

2)使用 query()方法构建一个查询

调用 SQLiteDatabase 类的 query()函数,query()函数的语法如下:

```
Cursor android.database.sqlite.SQLiteDatabase.query(String table, String[]
columns, String selection, String[] selectionArgs, String groupBy, String
having, String orderBy,String limit)
```

query()函数的参数说明如表 8-1 所示。

表 8-1 query()函数的参数说明

位置	类型+名称	说明
1	String table	表名称
2	String[] columns	返回的属性列名称
3	String selection	查询条件子句
4	String[] selectionArgs	如果在查询条件中使用问号,则需要定义替换符的具体内容
5	String groupBy	分组方式
6	String having	定义组的过滤器
7	String limit	指定偏移量和获取的记录数

例如:

```
1.    SQLiteDatabase db = databaseHelper.getWritableDatabase();
2. Cursor cursor = db.query("person", new String[]{"personid,name,age"},
"name like ?", new String[]{"%Tom%"}, null, null, "personid desc", "1,2");
3.    while (cursor.moveToNext()) {
4.        int personid = cursor.getInt(0);
                            //获取第一列的值,第一列的索引从 0 开始
5.        String name = cursor.getString(1);    //获取第二列的值
6.        int age = cursor.getInt(2);           //获取第三列的值
7.    }
8.    cursor.close();
9.    db.close();
```

在 Android 的 SQLite 数据库中使用游标,不论如何执行查询,都会返回一个

Cursor 对象。

Cursor 类常用方法和说明，如表 8-2 所示。

表 8-2　Cursor 类常用方法和说明

方　　法	说　　明
moveToFirst	将指针移动到第一条数据上
moveToNext	将指针移动到下一条数据上
moveToPrevious	将指针移动到上一条数据上
getCount	获取集合的数据数量
getColumnIndexOrThrow	返回指定属性名称的序号，如果属性不存在则产生异常
getColumnName	返回指定序号的属性名称
getColumnNames	返回属性名称的字符串数组
getColumnIndex	根据属性名称返回序号
moveToPosition	将指针移动到指定的数据上
getPosition	返回当前指针的位置
getString，getInt 等	获取给定字段当前记录的值
requery	重新执行查询得到游标
close	释放游标资源

8.2.4　SQLite 数据库管理及应用

在 Android 系统中，针对 SQLite 数据库的查看和管理有两种方式，一种是使用 Eclipse 插件 DDMS 查看和管理；另外一种是使用 Android 工具包中的 adb 工具来查看和管理。如前所述，Android 项目中 SQLite 数据库位置为：/data/data/<package-name>/databases/。

1. 使用 Eclipse 插件 DDMS 查看和管理 SQLite 数据库

（1）在 Eclipse 中打开 DDMS 视图，如图 8-13 所示。

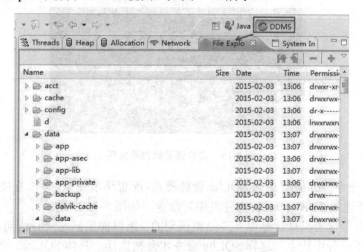

图 8-13　DDMS 视图下查看 data

注意：如果是模拟器进行项目调试，必须先启动模拟器，打开 DDMS 视图，才能有内容。

（2）选择 File Explorer 窗口，然后在 /data/data/<package-name>/. 目录下，打开 databases 文件，即可看见 SQLite 数据库文件。单击 Pull a file from the device 按钮，可以导出 SQLite 数据库文件，然后可以选择 SQLite 界面管理工具，如 sqlite man、sqlite administrator 等打开操作，如图 8-14 所示。

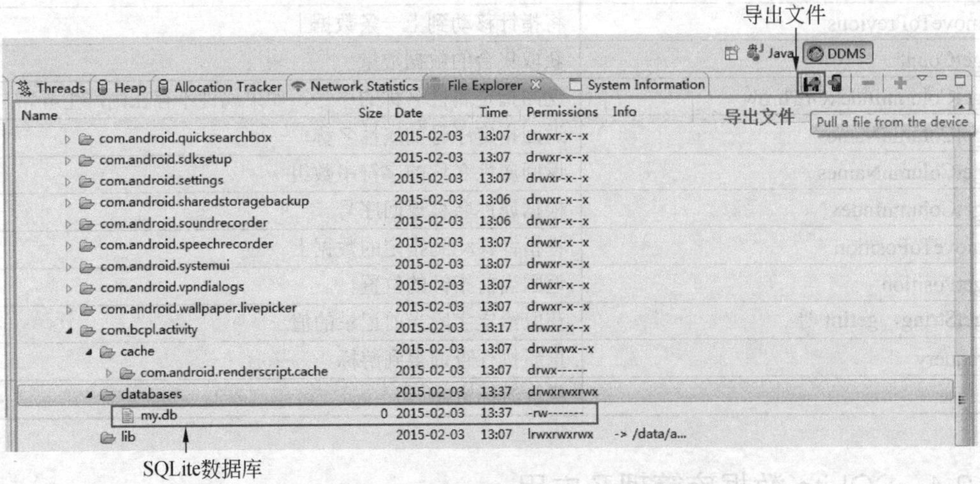

图 8-14　SQLite 数据库文件及导出数据库文件

2. 使用 adb 工具管理 SQLite 数据库

（1）在控制台窗口中（运行中输入"cmd"）输入命令"adb shell"进入设备 Linux 控制台，出现提示符"#"后，然后输入命令"cd /data/data/<package-name>/databases"进入目录。

（2）使用 ls 命令查看数据库文件是否存在，如图 8-15 所示。

图 8-15　命令查看数据库文件

（3）输入命令"sqlite3,"进入 SQLite 管理模式（配置环境变量，可以直接使用 SQLite3 工具，否则要进入 SDK 下 tools 文件夹中）命令，如图 8-16 所示。

SQLite 命令行工具默认是以";"结束语句的。所以如果只是一行语句，要在末尾加";"，或者在下一行中输入，这样 SQLite 命令才会被执行。具体 SQLite3 的命令如表 8-3 所示。

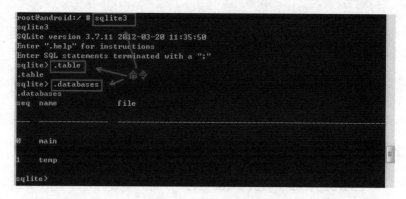

图 8-16　SQLite 工具命令

表 8-3　SQLite3 的命令列表

编　号	命　令	说　明
1	.bail ON|OFF	遇到错误时停止，默认为 OFF
2	.databases	显示数据库名称和文件位置
3	.dump ?TABLE? ...	将数据库以 SQL 文本形式导出
4	.echo ON|OFF	开启和关闭回显
5	.exit	退出
6	.explain ON|OFF	开启或关闭适当输出模式，如果开启模式将更改为 column，并自动设置宽度
7	.tables	查看数据库的表列表

其他命令可随时以.help 命令查看帮助。SQL 命令直接在此命令行上执行即可。

上面介绍了 SQLite 数据库创建方式、操作、管理及访问方法，下面通过一个案例，详细介绍访问 SQLite 数据库的应用。

（1）创建一个新的 Android 工程，工程名为 SQLiteDemo，目标 API 选择 17（即 Android 4.2 版本），应用程序名为 SQLiteDemo，包名为 com.bcpl.activity，创建的 Activity 的名字为 MainActivity，最小 SDK 版本根据选择的目标 API 会自动添加为 8，创建项目工程。

（2）修改 res 目录下 layout 文件夹中的 activity_main.xml 文件，设置线性布局，添加一个 EditText 控件、6 个 Button 控件和一个 ScrollView 控件描述，并设置相关属性，代码如下所示。

```
1.  <LinearLayout xmlns:android="http://schemas.android.com/apk/res/android"
2.      xmlns:tools="http://schemas.android.com/tools"
3.      android:layout_width="fill_parent"
4.      android:layout_height="fill_parent"
5.      android:orientation="vertical"
6.      >
7.
```

```xml
8.      <TextView
9.          android:layout_width="wrap_content"
10.         android:layout_height="wrap_content"
11.         android:text="@string/hello_world" />
12.     <Button
13.         android:id="@+id/create_open"
14.         android:layout_width="120dip"
15.         android:layout_height="wrap_content"
16.         android:text="@string/open_create"
17.         android:textColor="#00ffff"
18.         />
19.     <Button
20.         android:id="@+id/close"
21.         android:layout_width="120dip"
22.         android:layout_height="wrap_content"
23.         android:text="@string/close"
24.         />
25.     <Button
26.         android:id="@+id/insert"
27.         android:layout_width="wrap_content"
28.         android:layout_height="wrap_content"
29.         android:text="@string/insert"
30.         />
31.     <Button
32.         android:id="@+id/delete"
33.         android:layout_width="wrap_content"
34.         android:layout_height="wrap_content"
35.         android:text="@string/delete"
36.         />
37.     <Button
38.         android:id="@+id/update"
39.         android:layout_width="wrap_content"
40.         android:layout_height="wrap_content"
41.         android:text="@string/update"
42.         />
43.     <Button
44.         android:id="@+id/query"
45.         android:layout_width="wrap_content"
46.         android:layout_height="wrap_content"
47.         android:text="@string/query"
48.         />
49.     <ScrollView
50.         android:id="@+id/ScrollView01"
51.         android:layout_width="wrap_content"
```

```
52.         android:layout_height="wrap_content"
53.         >
54.     <EditText
55.         android:id="@+id/edit"
56.         android:layout_width="fill_parent"
57.         android:layout_height="wrap_content"
58.         />
59.     </ScrollView>
60. </LinearLayout>
```

（3）修改 src 目录中 com.bcpl.activity 包下的 MainActivity.java 文件，代码如下。

```
1.  package com.bcpl.activity;
2.
3.  import android.app.Activity;
4.  import android.database.Cursor;
5.  import android.database.sqlite.SQLiteDatabase;
6.  import android.os.Bundle;
7.  import android.view.View;
8.  import android.view.View.OnClickListener;
9.  import android.widget.Button;
10. import android.widget.EditText;
11. import android.widget.Toast;
12.
13. public class MainActivity extends Activity {
14.
15.     private SQLiteDatabase sld;
16.     private Button create_open, close, insert, delete, update, query;
17.
18.     @Override
19.     public void onCreate(Bundle savedInstanceState) {
20.         super.onCreate(savedInstanceState);
21.         setContentView(R.layout.main);
22.
23.         //初始化"创建数据库"按钮
24.         this.create_open=(Button)this.findViewById(R.id.create_open);
25.         this.create_open.setOnClickListener(
26.           new OnClickListener()
27.           {
28.                @Override
29.                public void onClick(View v) {
30.                    createOrOpenDatabase();
31.                }
32.           }
33.         );
```

```
34.
35.     //初始化"关闭数据库"按钮
36.     this.close=(Button)this.findViewById(R.id.close);
37.     this.close.setOnClickListener(
38.       new OnClickListener()
39.       {
40.           @Override
41.           public void onClick(View v) {
42.               closeDatabase();
43.           }
44.       }
45.     );
46.
47.     //初始化"添加记录"按钮
48.     this.insert=(Button)this.findViewById(R.id.insert);
49.     this.insert.setOnClickListener(
50.       new OnClickListener()
51.       {
52.           @Override
53.           public void onClick(View v) {
54.               insert();
55.           }
56.       }
57.     );
58.
59.     //初始化"删除记录"按钮
60.     this.delete=(Button)this.findViewById(R.id.delete);
61.     this.delete.setOnClickListener(
62.       new OnClickListener()
63.       {
64.           @Override
65.           public void onClick(View v) {
66.               delete();
67.           }
68.       }
69.     );
70.
71.     //初始化"修改记录"按钮
72.     this.update=(Button)this.findViewById(R.id.update);
73.     this.update.setOnClickListener(
74.       new OnClickListener()
75.       {
76.           @Override
77.           public void onClick(View v) {
```

```
78.              update();
79.          }
80.       }
81.    );
82.
83.    //初始化"查询记录"按钮
84.    this.query=(Button)this.findViewById(R.id.query);
85.    this.query.setOnClickListener(
86.       new OnClickListener()
87.       {
88.          @Override
89.          public void onClick(View v) {
90.             query();
91.          }
92.       }
93.    );
94. }
95.
96. //创建或打开数据库的方法
97. public void createOrOpenDatabase()
98. {
99.    try
100.      {
101.         sld=SQLiteDatabase.openDatabase
102.         (
103.             "/data/data/com.hisoft.activity/mydb",
                                                        //数据库所在路径
104.             null,                                   //CursorFactory
105.             SQLiteDatabase.OPEN_READWRITE|SQLiteDatabase.
                 CREATE_IF_NECESSARY              //读写、若不存在则创建
106.         );
107.         appendMessage("数据库已经成功打开! ");
108.         String sql="create table if not exists student(stuno
             char(5),stuname varchar(20),stuage integer,stuclass char(5))";
109.         sld.execSQL(sql);
110.         appendMessage("student 已经成功创建! ");
111.      }
112.      catch(Exception e)
113.      {
114.         Toast.makeText(this, "数据库错误: "+e.toString(), Toast.
             LENGTH_SHORT).show();
115.      }
116.   }
117.
```

```
118.        //关闭数据库的方法
119.        public void closeDatabase()
120.        {
121.            try
122.            {
123.                sld.close();
124.                appendMessage("数据库已经成功关闭！");
125.            }
126.            catch(Exception e)
127.            {
128.                Toast.makeText(this, "数据库错误："+e.toString(),
                    Toast.LENGTH_SHORT).show();;
129.            }
130.        }
131.
132.        //插入记录的方法
133.        public void insert()
134.        {
135.            try
136.            {
137.                String sql="insert into student values('10001','张三
                    ',10,'11010')";
138.                sld.execSQL(sql);
139.                appendMessage("成功插入一条记录！");
140.            }
141.            catch(Exception e)
142.            {
143.                Toast.makeText(this, "数据库错误："+e.toString(),
                    Toast.LENGTH_SHORT).show();;
144.            }
145.        }
146.
147.        //删除记录的方法
148.        public void delete()
149.        {
150.            try
151.            {
152.                String sql="delete from student;";
153.                sld.execSQL(sql);
154.                appendMessage("成功删除所有记录！");
155.            }
156.            catch(Exception e)
157.            {
158.                Toast.makeText(this, "数据库错误："+e.toString(),
```

```
                    Toast.LENGTH_SHORT).show();;
159.            }
160.        }
161.
162.        //修改记录的方法
163.        public void update()
164.        {
165.            try
166.            {
167.                String sql="update student set stuname='李四'";
168.                sld.execSQL(sql);
169.                appendMessage("成功更新记录！");
170.            }
171.            catch(Exception e)
172.            {
173.                Toast.makeText(this, "数据库错误："+e.toString(),
                    Toast.LENGTH_SHORT).show();;
174.            }
175.        }
176.
177.        //查询的方法
178.        public void query()
179.        {
180.            try
181.            {
182.                String sql="select * from student where stuage>?";
183.                Cursor cur=sld.rawQuery(sql, new String[]{"5"});
184.                appendMessage("学号\t\t姓名\t\t年龄\t班级");
185.                while(cur.moveToNext())
186.                {
187.                    String sno=cur.getString(0);
188.                    String sname=cur.getString(1);
189.                    int sage=cur.getInt(2);
190.                    String sclass=cur.getString(3);
191.                    appendMessage(sno+"\t"+sname+"\t\t"+sage+"\t"+sclass);
192.                }
193.                cur.close();
194.            }
195.            catch(Exception e)
196.            {
197.                Toast.makeText(this, "数据库错误："+e.toString(),
                    Toast.LENGTH_SHORT).show();;
198.            }
199.        }
```

```
200.
201.        //向文本区中添加文本
202.        public void appendMessage(String msg)
203.        {
204.            EditText et=(EditText)this.findViewById(R.id.EditText01);
205.            et.append(msg+"\n");
206.        }
207.    }
```

（4）在 src 目录中 com.bcpl.activity 包下创建 MyContentProvider.java 文件，代码如下。

```
1.   package com.bcpl.activity;
2.   import android.content.ContentProvider;
3.   import android.content.ContentValues;
4.   import android.content.UriMatcher;
5.   import android.database.Cursor;
6.   import android.database.sqlite.SQLiteDatabase;
7.   import android.net.Uri;
8.   public class MyContentProvider extends ContentProvider {
9.       private static final UriMatcher um;
10.      static
11.      {
12.          um=new UriMatcher(UriMatcher.NO_MATCH);
13.          um.addURI("com.bcpl.provider.student", "stu", 1);
14.      }
15.
16.      SQLiteDatabase sld;
17.
18.      @Override
19.      public String getType(Uri uri) {
20.          return null;
21.      }
22.
23.      @Override
24.      public Cursor query(Uri uri, String[] projection, String selection,
25.              String[] selectionArgs, String sortOrder) {
26.
27.          switch(um.match(uri))
28.          {
29.              case 1:
30.
31.                  Cursor cur=sld.query
32.                  (
33.                          "student",
```

```
34.                    projection,
35.                    selection,
36.                    selectionArgs,
37.                    null,
38.                    null,
39.                    sortOrder
40.            );
41.            return cur;
42.        }
43.        return null;
44.    }
45.
46.    @Override
47.    public int delete(Uri arg0, String arg1, String[] arg2) {
48.        //TODO Auto-generated method stub
49.        return 0;
50.    }
51.
52.    @Override
53.    public Uri insert(Uri uri, ContentValues values) {
54.        //TODO Auto-generated method stub
55.        return null;
56.    }
57.
58.    @Override
59.    public boolean onCreate() {
60.
61.        sld=SQLiteDatabase.openDatabase
62.    (
63.            "/data/data/com.bcpl.activity/mydb",  //数据库所在路径
64.            null,                                 //CursorFactory
65.            SQLiteDatabase.OPEN_READWRITE|SQLiteDatabase.CREATE_
                IF_NECESSARY  //读写、若不存在则创建
66.    );
67.
68.        return false;
69.    }
70.
71.    @Override
72.    public int update(Uri uri, ContentValues values, String selection,
73.            String[] selectionArgs) {
74.        //TODO Auto-generated method stub
75.        return 0;
76.    }
77. }
```

（5）在 res 目录下的 values 文件中，修改 strings.xml 文件，添加按钮的引用文字如下。

```xml
<?xml version="1.0" encoding="utf-8"?>
<resources>
    <string name="app_name">SQLiteDemo</string>
    <string name="action_settings">Settings</string>
    <string name="hello_world">SQLite 数据库操作</string>
    <string name="open_create">创建/打开数据库</string>
    <string name="close">关闭数据库</string>
    <string name="insert">添加记录</string>
    <string name="delete">删除记录</string>
    <string name="update">修改记录</string>
    <string name="query">查询记录</string>
</resources>
```

（6）在 AndroidManifest.xml 文件中，<application>根节点下添加<provider>节点标签，添加权限为后续的案例应用提供数据接口，暴露数据，代码如下。

```xml
<provider
    android:name=".MyContentProvider"
    android:authorities="com.bcpl.provider.student"
/>
```

（7）部署 SQLiteDemo 工程，程序运行结果如图 8-17 所示。

单击"创建/打开数据库"按钮，如果数据库存在，则打开数据库；如果数据库不存在，则创建数据库，并同时在数据库中创建 student 表，运行效果如图 8-18 所示。

图 8-17　SQLiteDemo 运行效果　　图 8-18　打开数据库并创建数据库表

单击"添加记录"按钮，程序中代码默认设置的 SQL 语句添加一条学生编号为 10001 的记录，如果添加成功，显示"成功插入一条记录！"信息，如图 8-19 所示。

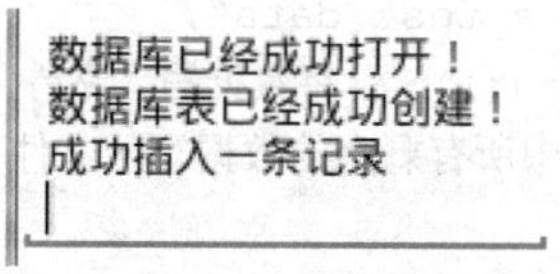

```
数据库已经成功打开！
数据库表已经成功创建！
成功插入一条记录
```

图 8-19　插入记录

然后单击"查询记录"按钮、"修改记录"按钮、"删除记录"按钮、"关闭数据库"按钮执行相关操作。

8.3　文件存储及读写

Android 系统使用的是基于 Linux 的文件系统，应用程序开发人员可以建立和访问程序自身的私有文件，也可以访问保存在资源目录中的原始文件和 XML 文件。此外，还可以在 SD 卡等外部存储设备中保存文件信息等。

8.3.1　文件存储及应用

在 Android 系统中，允许应用程序创建仅能够自身访问的私有文件，文件保存在设备的内部存储器上，文件默认保存路径位置是：/data/data/<package name>/ files/目录下。

Android 系统不仅支持标准 Java 的 IO 类和方法，还提供了能够简化读写流式文件过程的方法，关于文件存储，Activity 提供了 openFileOutput()方法和 openFileInput()方法。

openFileOutput()方法可以用于把数据输出到文件中。

openFileInput()方法为打开应用程序私有文件读取数据。

具体的实现过程与在 J2SE 环境中保存数据到文件中是一样的。文件可用来存放大量数据，如文本、图片、音频等。

1．openFileOutput()方法用法

openFileOutput()方法为打开应用程序私有文件写入数据，如果指定的文件不存在，则创建一个新的文件。

（1）openFileOutput()方法的语法声明：

```
public FileOutputStream openFileOutput(String name, int mode)
```

第一个参数是文件名称，这个参数不能包含路径分隔符"/"。
第二个参数是文件操作模式。
（2）使用 openFileOutput()方法创建新文件，代码如下：

```
1.   String NAME = "test.txt";//定义了创建文件的名称 test.txt
2.   FileOutputStream fos = openFileOutput(NAME,Context.MODE_PRIVATE);
     //使用 openFileOutput()方法以私有模式建立文件
```

```
3.    String ts= "This is a test data";
4.    fos.write(ts.getBytes());//将数据写入文件
5.    fos.flush();//将缓存中所有剩余的数据写入文件
6.    fos.close();//关闭流
```

2. openFileInput()方法用法

如果要打开存放在/data/data/<package name>/files 目录下应用程序私有的文件，可以使用 openFileInput()方法。

(1) openFileInput()方法的语法声明：

```
public FileInputStream openFileInput (String name)
```

方法参数也是文件名称，字符串中不能包含路径分隔符"/"。

(2) 使用 openFileInput ()方法打开已有文件，代码如下。

```
1.    FileInputStream inStream = this.getContext().openFileInput("test.txt");
2.    try {
3.        ByteArrayOutputStream outStream =
4.            new ByteArrayOutputStream();
5.        byte[] buffer = new byte[1024];
6.        int length = -1;
7.        while((length = inStream.read(buffer)) != -1 ){
8.            outStream.write(buffer, 0, length);
9.        }
10.       outStream.close();
11.       inStream.close();
12.       return outStream.toString();
13.   } catch (IOException e) {
14.       Log.i("FileTest", e.getMessage());
15.   }
```

如果想直接使用文件的绝对路径，可以使用如下代码：

```
1.    File file = new File("/data/data/ com.bcpl.activity /files/test.txt");
2.    FileInputStream inStream = new FileInputStream(file);
```

上面第 1 行文件路径中的"com.hisoft.activity"为应用所在包。

对于私有文件只能被创建该文件的应用访问，如果希望文件能被其他应用读和写，可以在创建文件时，指定 Context.MODE_WORLD_READABLE 和 Context.MODE_WORLD_WRITEABLE 权限。

Activity 还提供了 getCacheDir()和 getFilesDir()方法：

getCacheDir()方法用于获取/data/data/<package name>/cache 目录；

getFilesDir()方法用于获取/data/data/<package name>/files 目录。

注意：

（1）openFileOutput()和 openFileInput()方法使用时，必须使用 try{}…catch{}捕获异常。

（2）创建的文件保存在/data/data/<package name>/files 目录下，如/data/data/com.hisoft.activity/files/test.txt。

同前面讲述的一样，通过 File Explorer 视图，在 File Explorer 视图中展开/data/data/<package name>/files 目录，即可看到该文件。Android 系统支持 4 种文件操作模式，如表 8-4 所示。

表 8-4 Android 系统支持 4 种文件操作模式

文件操作模式	值	描述
MODE_PRIVATE	0	私有模式，缺陷模式，文件仅能够被文件创建程序访问，或具有相同 UID 的程序访问。为默认操作模式，代表该文件是私有数据，只能被应用本身访问，在该模式下，写入的内容会覆盖原文件的内容，如果想把新写入的内容追加到原文件中。可以使用 Context.MODE_APPEND
MODE_APPEND	32 768	追加模式，模式会检查文件是否存在，存在就往文件中追加内容，否则就创建新文件
MODE_WORLD_READABLE	1	全局读模式，允许任何程序读取私有文件
MODE_WORLD_WRITEABLE	2	全局写模式，允许任何程序写入私有文件

注意： 在使用上述模式时，可以用"+"来选择多种模式，比如：

openFileOutput(FILENAME, Context.MODE_PRIVATE + MODE_WORLD_READABLE);

上述介绍了文件存储访问方式及访问方法，下面通过一个文件存储案例，详细介绍访问 File 的应用。

（1）创建一个新的 Android 工程，工程名为 FileWriteAndReadDemo，目标 API 选择 17（即 Android 4.2 版本），应用程序名为 FileWriteAndReadDemo，包名为 com.bcpl.activity，创建的 Activity 的名字为 MainActivity，最小 SDK 版本根据选择的目标 API 会自动添加为 8，创建项目工程。

（2）修改 res 目录下 layout 文件夹中的 activity_main.xml 文件，添加 EditText、TextView、Button 控件，代码如下。

```
1.   <?xml version="1.0" encoding="utf-8"?>
2.   <LinearLayout xmlns:android="http://schemas.android.com/apk/
     res/android"
3.       android:orientation="vertical"
4.       android:layout_width="fill_parent"
5.       android:layout_height="fill_parent"
6.       >
7.
8.       <TextView
9.           android:layout_width="fill_parent"
```

```
10.            android:layout_height="wrap_content"
11.            android:text="@string/filename"
12.        />
13.
14.        <EditText
15.            android:layout_width="fill_parent"
16.            android:layout_height="wrap_content"
17.            android:id="@+id/filename"
18.        />
19.
20.        <TextView
21.            android:layout_width="fill_parent"
22.            android:layout_height="wrap_content"
23.            android:text="@string/content"
24.        />
25.
26.        <EditText
27.            android:layout_width="fill_parent"
28.            android:layout_height="wrap_content"
29.            android:minLines="3"
30.            android:id="@+id/content"
31.        />
32.
33.        <LinearLayout
34.            android:orientation="horizontal"
35.            android:layout_width="fill_parent"
36.            android:layout_height="fill_parent">
37.
38.            <Button
39.                android:layout_width="wrap_content"
40.                android:layout_height="wrap_content"
41.                android:id="@+id/button"
42.                android:text="@string/save"
43.            />
44.
45.            <Button
46.                android:layout_width="wrap_content"
47.                android:layout_height="wrap_content"
48.                android:id="@+id/read"
49.                android:text="@string/read"
50.            />
51.        </LinearLayout>
52.    </LinearLayout>
```

(3) 在 src 目录 com.bcpl.activity 包下，创建 FileUtil.java 文件，代码如下。

```java
1.   package com.bcpl.activity;
2.   import java.io.ByteArrayOutputStream;
3.   import java.io.FileInputStream;
4.   import java.io.FileOutputStream;
5.
6.   import android.content.Context;
7.   import android.util.Log;
8.
9.   /**
10.   * 文件保存与读取功能实现类
11.   * @author Administrator
12.   */
13.  public class FileUtil{
14.
15.      public static final String TAG = "FileService";
16.      private Context context;
17.
18.      //得到传入的上下文对象的引用
19.      public FileUtil(Context context) {
20.          this.context = context;
21.      }
22.      public FileUtil(){
23.
24.      }
25.
26.      /**
27.       * 保存文件
28.       *
29.       * @param fileName 文件名
30.       * @param content  文件内容
31.       * @throws Exception
32.       */
33.      public void save(String fileName, String content) throws Exception {
34.
35.          //由于页面输入的都是文本信息，所以当文件名不是以.txt 后缀名结尾时，自
             //动加上.txt 后缀
36.          if (!fileName.endsWith(".txt")) {
37.              fileName = fileName + ".txt";
38.          }
39.
40.          byte[] buf = fileName.getBytes("iso8859-1");
41.
```

```java
42.         Log.e(TAG, new String(buf,"utf-8"));
43.
44.         fileName = new String(buf,"utf-8");
45.
46.         Log.e(TAG, fileName);
47.
48.         //如果希望文件被其他应用读和写，可以传入：
49.         //openFileOutput("output.txt", Context.MODE_WORLD_READABLE
            //+ Context.MODE_WORLD_WRITEABLE);
50.
51.         FileOutputStream fos = context.openFileOutput(fileName,
            context.MODE_PRIVATE);
52.         fos.write(content.getBytes());
53.         fos.close();
54.     }
55.
56.     /**
57.      * 读取文件内容
58.      *
59.      * @param fileName 文件名
60.      * @return 文件内容
61.      * @throws Exception
62.      */
63.     public String read(String fileName) throws Exception {
64.
65.         //由于页面输入的都是文本信息，所以当文件名不是以.txt后缀名结尾时，自动加上.txt后缀
66.         if (!fileName.endsWith(".txt")) {
67.             fileName = fileName + ".txt";
68.         }
69.
70.         FileInputStream fis = context.openFileInput(fileName);
71.         ByteArrayOutputStream baos = new ByteArrayOutputStream();
72.
73.         byte[] buf = new byte[1024];
74.         int len = 0;
75.
76.         //将读取后的数据放置在内存中——ByteArrayOutputStream
77.         while ((len = fis.read(buf)) != -1) {
78.             baos.write(buf, 0, len);
79.         }
80.
81.         fis.close();
82.         baos.close();
83.
```

```
84.         //返回内存中存储的数据
85.         return baos.toString();
86.     }
87.    }
88.
89. }
```

（4）修改 src 目录下包 com.bcpl.activity 中的 MainActivity.java 文件，代码如下。

```
1.  package com.bcpl.activity;
2.  import android.app.Activity;
3.  import android.os.Bundle;
4.  import android.util.Log;
5.  import android.view.View;
6.  import android.widget.Button;
7.  import android.widget.EditText;
8.  import android.widget.Toast;
9.  public class MainActivity extends Activity {
10.     /** Called when the activity is first created. */
11.
12.     //得到 FileUtil 对象
13.     private FileUtil fileService = new FileUtil(this);
14.     //定义视图中的 filename 输入框对象
15.     private EditText fileNameText;
16.     //定义视图中的 contentText 输入框对象
17.     private EditText contentText;
18.     //定义一个 Toast 提示对象
19.     private Toast toast;
20.     @Override
21.     public void onCreate(Bundle savedInstanceState) {
22.         super.onCreate(savedInstanceState);
23.         setContentView(R.layout.main);
24.
25.         //得到视图中的两个输入框和两个按钮的对象引用
26.         Button button = (Button)this.findViewById(R.id.button);
27.         Button read = (Button)this.findViewById(R.id.read);
28.         fileNameText = (EditText) this.findViewById(R.id.filename);
29.         contentText = (EditText) this.findViewById(R.id.content);
30.
31.         //为"保存文件"按钮添加保存事件
32.         button.setOnClickListener(new View.OnClickListener() {
33.             @Override
34.             public void onClick(View v) {
35.
36.                 String fileName = fileNameText.getText().toString();
```

```
37.            String content = contentText.getText().toString();
38.
39.            //当文件名为空的时候，提示用户文件名为空，并记录日志
40.            if(isEmpty(fileName)) {
41.                toast = Toast.makeText(MainActivity.this, R.
                   string.empty_filename, Toast.LENGTH_LONG);
42.                toast.setMargin(RESULT_CANCELED, 0.345f);
43.                toast.show();
44.                Log.w(fileService.TAG, "The file name is empty");
45.                return;
46.            }
47.
48.            //当文件内容为空的时候，提示用户文件内容为空，并记录日志
49.            if(isEmpty(content)) {
50.                toast = Toast.makeText(MainActivity.this, R.
                   string.empty_content, Toast.LENGTH_LONG);
51.                toast.setMargin(RESULT_CANCELED, 0.345f);
52.                toast.show();
53.                Log.w(fileService.TAG, "The file content is empty");
54.                return;
55.            }
56.
57.            //当文件名和内容都不为空的时候，调用 fileService 的 save 方法
58.            //当成功执行的时候，提示用户保存成功，并记录日志
59.            //当出现异常的时候，提示用户保存失败，并记录日志
60.            try {
61.                fileService.save(fileName, content);
62.                toast = Toast.makeText(MainActivity.this, R.
                   string.success, Toast.LENGTH_LONG);
63.                toast.setMargin(RESULT_CANCELED, 0.345f);
64.                toast.show();
65.                Log.i(fileService.TAG, "The file save successful");
66.            } catch (Exception e) {
67.                toast = Toast.makeText(MainActivity.this, R.
                   string.fail, Toast.LENGTH_LONG);
68.                toast.setMargin(RESULT_CANCELED, 0.345f);
69.                toast.show();
70.                Log.e(fileService.TAG, "The file save failed");
71.            }
72.
73.        }
74.    });
75.
76.
```

```
77.         //为"读文件"按钮添加读取事件
78.         read.setOnClickListener(new View.OnClickListener() {
79.             @Override
80.             public void onClick(View v) {
81.
82.                 //得到"文件名"输入框中的值
83.                 String fileName = fileNameText.getText().toString();
84.
85.                 //如果文件名为空,则提示用户输入文件名,并记录日志
86.                 if(isEmpty(fileName)) {
87.                     toast = Toast.makeText(MainActivity.this, R.
                            string.empty_filename, Toast.LENGTH_LONG);
88.                     toast.setMargin(RESULT_CANCELED, 0.345f);
89.                     toast.show();
90.                     Log.w(fileService.TAG, "The file name is empty");
91.                     return;
92.                 }
93.
94.                 //调用fileService的read方法,并将读取出来的内容放入到文本内容输入框里面
95.                 //如果成功执行,提示用户读取成功,并记录日志
96.                 //如果出现异常信息(例如文件不存在),提示用户读取失败,并记录日志
97.                 try {
98.                     contentText.setText(fileService.read(fileName));
99.                     toast = Toast.makeText(MainActivity.this, R.
                            string.read_success, Toast.LENGTH_LONG);
100.                    toast.setMargin(RESULT_CANCELED, 0.345f);
101.                    toast.show();
102.                    Log.i(fileService.TAG, "The file read successful");
103.                } catch (Exception e) {
104.                    toast = Toast.makeText(MainActivity.this, R.
                            string.read_fail, Toast.LENGTH_LONG);
105.                    toast.setMargin(RESULT_CANCELED, 0.345f);
106.                    toast.show();
107.                    Log.e(fileService.TAG, "The file read failed");
108.                }
109.            }
110.        });
111.
112.
113.    }
114.
115.    // isEmpty方法,判断字符串是否为空
116.    private boolean isEmpty(String s) {
117.        if(s == null || "".equals(s.trim())) {
```

```
118.            return true;
119.       }
120.       return false;
121.    }
122.
123. }
```

（5）修改 res 目录下 values 文件下的 strings.xml 文件，代码如下。

```
1.  <?xml version="1.0" encoding="utf-8"?>
2.  <resources>
3.      <string name="hello">Hello World, MainActivity!</string>
4.      <string name="app_name">FileWriteAndReadDemo</string>
5.      <string name="filename">文件名</string>
6.      <string name="read">读文件</string>
7.      <string name="save">保存文件</string>
8.      <string name="content">文件内容</string>
9.      <string name="success">保存成功</string>
10.     <string name="fail">保存失败</string>
11.     <string name="empty_filename">空文件名</string>
12.     <string name="read_success">读取成功</string>
13.     <string name="read_fail">读取失败</string>
14.     <string name="empty_content">空文件内容</string>
15. </resources>
```

（6）如需存入 SD 卡，需要在 AndroidManifest.xml 文件中的<manifest>节点中添加读写文件的权限，具体代码见后续案例。

（7）部署工程 FileWriteAndReadDemo，程序运行效果如图 8-20 所示。

输入存储的文件名及文件内容，单击"保存文件"按钮，如果保存成功，Toast 会显示保存成功提示信息，如图 8-21 和图 8-22 所示，文件存储到目录/data/data/com.bcpl.activity/files/下（打开 DDMS 视图下的 File Explorer 面板进行查看），如图 8-23 所示。

图 8-20　FileWriteAndReadDemo 运行效果

图 8-21　保存文件成功

```
E  02-03 14:51:27.721     827    827    com.bcpl.activity    FileService    test.txt
I  02-03 14:51:27.751     827    827    com.bcpl.activity    FileService    The file save successful
```

图 8-22　保存成功日志

```
▲ 📁 com.bcpl.activity                         2015-02-03  14:51  drwxr-x--x
    ▷ 📁 cache                                 2015-02-03  14:49  drwxrwx--x
    ▲ 📁 files                                 2015-02-03  14:51  drwxrwx--x
        📄 test.txt                         24 2015-02-03  14:51  -rw-rw----
    📁 lib                                     2015-02-03  14:49  lrwxrwxrwx  -> /data/a...
```

图 8-23　文件保存路径

下次启动后，在"文件名"编辑框中输入文件名，单击"读文件"按钮，文件内容会自动读取出来。然后 Toast 会显示提示信息，logcat 显示读取成功，如图 8-24 和图 8-25 所示。如果没有填写文件名，单击"读文件"按钮，Toast 信息会提示"空文件名"，LogCat 会显示"The file name is empty"。

图 8-24　文件读取成功

```
E  02-03 14:51:45.780     827    827    com.bcpl.activity    FileService    test.txt
I  02-03 14:51:45.800     827    827    com.bcpl.activity    FileService    The file save successful
I  02-03 14:57:26.911     827    827    com.bcpl.activity    FileService    The file read successful
```

图 8-25　LogCat 显示文件读取成功

8.3.2　SD 卡存储及应用

SD 卡（Secure Digital Memory Card）是 Android 的外部存储设备，广泛使用于数码设备上，Android 系统提供了对 SD 卡的便捷的访问方法。

前面讲述了使用 Activity 的 openFileOutput()方法保存文件，文件是存放在手机自身空间内的，一般手机的自身存储空间不大，如果要存放像视频这样的大文件，人们通常把它放置在外部的存储设备 SD 卡中。

SD 卡适用于保存大尺寸的文件或者是一些无须设置访问权限的文件，可以保存录制的大容量的视频文件和音频文件等。

SD 卡使用的是 FAT（File Allocation Table）的文件系统，不支持访问模式和权限控制，但可以通过 Linux 文件系统的文件访问权限的控制保证文件的私密性。

1．SD 卡创建方式

Android 模拟器支持 SD 卡，但模拟器中没有默认的 SD 卡，应用程序开发人员必须在模拟器中手工添加 SD 卡的映像文件。

创建 SD 卡有两种方式：一种是在 Eclipse 创建模拟器时创建 SD 卡；另一种是使用 <Android SDK>/tools 目录下的 mksdcard 工具创建 SD 卡映像文件。

在控制台窗口中进入 Android SDK 安装路径的 tools 目录下，使用 mksdcard 工具，命令如下：

```
mksdcard -l SDCa 1024M d:\android\sdcard_file
```

第一个参数-l 表示后面的字符串是 SD 卡的标签，这个新建立的 SD 卡的标签是 SDCa。

第二个参数 1024M 表示 SD 卡的容量是 1GB。

第三个参数表示 SD 卡映像文件的保存位置，上面的命令将映像保存在 D:\android 目录下 sdcard_file 文件中。在 CMD 中执行该命令后，则可在所指定的目录中找到生产的 SD 卡映像文件。

2．访问 SD 卡

在编程访问 SD 卡，往 SD 卡中存放文件之前，首先程序需要先判断手机是否装有 SD 卡（检测系统的/sdcard 目录是否可用），并且可以进行读写。如果不可用，则说明设备中的 SD 卡已经被移除（如用在 Android 模拟器中则表明 SD 卡映像没有被正确加载）；如果可用，则直接通过使用标准的 Java.io.File 类进行访问，使用代码如下。

```
1.   if(Environment.getExternalStorageState().
2.      equals(Environment.MEDIA_MOUNTED)){
3.         //获取 SD 卡目录
4.         File sdCardDir = Environment.getExternalStorageDirectory();
5.         File saveFile = new File(sdCardDir,"test.txt");
6.         FileOutputStream outStream = new FileOutputStream(saveFile);
7.         outStream.write("How are you! ".getBytes());
8.         outStream.close();
9.   }
```

上述代码中 Environment.getExternalStorageState()方法用于获取 SD 卡的状态，如果手机中装有 SD 卡，并且可以进行读写，那么方法返回的状态等于 Environment.MEDIA_MOUNTED。

第 4 行 Environment.getExternalStorageDirectory()方法用于获取 SD 卡的目录。

或使用下面的代码完成。

```
1.  File saveFile = new File("/sdcard/test.txt");
2.  FileOutputStream outStream = new FileOutputStream(saveFile);
3.  outStream.write("How are you! ".getBytes());
4.  outStream.close();
```

注意： 在处理中文字符时需要注意编码问题，发送和接收、保存和读取都采用相同的字符编码，一般采用 utf-8 编码，以防出现乱码。

上述介绍了 SD 卡的创建方式及访问方法，下面通过一个案例，详细介绍访问 SD 卡的应用。

（1）创建一个新的 Android 工程，工程名为 SDCardDemo，目标 API 选择 17（即 Android 4.2 版本），应用程序名为 SDCardDemo，包名为 com.bcpl.activity，创建的 Activity 的名字为 MainActivity，最小 SDK 版本根据选择的目标 API 会自动添加为 8，创建项目工程。

（2）修改 res 目录下 layout 文件夹中的 activity_main.xml 文件，设置线性布局，添加两个 EditText 控件和两个 Button 控件描述，并设置相关属性，代码如下所示。

```
1.  <?xml version="1.0" encoding="utf-8"?>
2.  <LinearLayout xmlns:android="http://schemas.android.com/apk/res/android"
3.      android:orientation="vertical"
4.      android:layout_width="fill_parent"
5.      android:layout_height="fill_parent">
6.  <EditText android:id="@+id/edit1"
7.      android:layout_width="fill_parent"
8.      android:layout_height="wrap_content"
9.      android:lines="4"/>
10. <Button android:id="@+id/write"
11.     android:layout_width="wrap_content"
12.     android:layout_height="wrap_content"
13.     android:text="@string/write"/>
14. <EditText android:id="@+id/edit2"
15.     android:layout_width="fill_parent"
16.     android:layout_height="wrap_content"
17.     android:editable="false"
18.     android:cursorVisible="false"
19.     android:lines="4"/>
20. <Button android:id="@+id/read"
21.     android:layout_width="wrap_content"
22.     android:layout_height="wrap_content"
23.     android:text="@string/read"/>
24. </LinearLayout>
```

（3）修改 res 目录下 values 文件夹中的 strings.xml 文件，代码如下所示。

```xml
1. <?xml version="1.0" encoding="utf-8"?>
2. <resources>
3.     <string name="hello">Hello World, MainActivity!</string>
4.     <string name="app_name">SDCardDemo</string>
5.     <string name="read">从 SD 卡读取</string>
6.     <string name="write">写入 SD 卡</string>
7. </resources>
```

(4) 修改 src 目录中 com.bcpl.activity 包下的 MainActivity.java 文件，代码如下。

```java
1.  package com.bcpl.activity;
2.  import java.io.BufferedReader;
3.  import java.io.File;
4.  import java.io.FileInputStream;
5.  import java.io.FileOutputStream;
6.  import java.io.FileWriter;
7.  import java.io.IOException;
8.  import java.io.InputStreamReader;
9.  import java.io.OutputStreamWriter;
10. import java.io.PrintWriter;
11. import android.app.Activity;
12. import android.os.Bundle;
13. import android.os.Environment;
14. import android.view.View;
15. import android.view.View.OnClickListener;
16. import android.widget.Button;
17. import android.widget.EditText;
18.
19. public class MainActivity extends Activity {
20.
21.     private Button btn_read, btn_write;
22.     final String FILE_NAME = "/myfile.txt";
23.
24.     @Override
25.     public void onCreate(Bundle savedInstanceState)
26.     {
27.         super.onCreate(savedInstanceState);
28.         setContentView(R.layout.main);
29.         // 获取两个按钮
30.         this.btn_read = (Button) findViewById(R.id.read);
31.         this.btn_write = (Button) findViewById(R.id.write);
32.         // 获取两个文本框
33.         final EditText edit1 = (EditText) findViewById(R.id.edit1);
34.         final EditText edit2 = (EditText) findViewById(R.id.edit2);
35.         // 为 write 按钮绑定事件监听器
```

```
36.        this.btn_write.setOnClickListener(new OnClickListener()
37.        {
38.            @Override
39.            public void onClick(View source)
40.            {
41.                // 将edit1中的内容写入文件中
42.                write(edit1.getText().toString());
43.                edit1.setText("");
44.            }
45.        });
46.
47.        this.btn_read.setOnClickListener(new OnClickListener()
48.        {
49.            @Override
50.            public void onClick(View v)
51.            {
52.                // 读取指定文件中的内容，并显示出来
53.                edit2.setText(read());
54.            }
55.        });
56.    }
57.
58.    private String read()
59.    {
60.        BufferedReader br = null;
61.        try
62.        {
63.            //如果手机插入了SD卡，而且应用程序具有访问SD卡的权限
64.            if (Environment.getExternalStorageState()
65.                .equals(Environment.MEDIA_MOUNTED))
66.            {
67.                //获取SD卡对应的存储目录
68.                File sdCardDir = Environment.getExternalStorageDirectory();
69.                //获取指定文件对应的输入流
70.                FileInputStream fis = new FileInputStream(sdCardDir
71.                    .getCanonicalPath() + FILE_NAME);
72.                //构造BufferedReader从文件中读取
73.                br = new BufferedReader(new
74.                    InputStreamReader(fis));
75.                StringBuilder sb = new StringBuilder("");
76.                String line = null;
77.                while((line = br.readLine()) != null)
78.                {
79.                    sb.append(line);
```

```
80.                }
81.                return sb.toString();
82.            }
83.        }
84.        catch (Exception e)
85.        {
86.            e.printStackTrace();
87.        }
88.        finally{
89.            if(br != null){
90.                try {
91.                    br.close();
92.                } catch (IOException e) {
93.                    // TODO Auto-generated catch block
94.                    e.printStackTrace();
95.                }
96.                br = null;
97.            }
98.        }
99.        return null;
100.    }
101.
102.    private void write(String content)
103.    {
104.        PrintWriter pw = null;
105.        try
106.        {
107.            //如果手机插入了SD卡,而且应用程序具有访问SD卡的权限
108.            if (Environment.getExternalStorageState()
109.                .equals(Environment.MEDIA_MOUNTED))
110.            {
111.                //获取SD卡的目录
112.                File sdCardDir = Environment.getExternalStorageDirectory();
113.                File targetFile = new File(sdCardDir.getCanonicalPath()
114.                    + FILE_NAME);
115.                //构造PrintWriter对象向文件中写入
116.                FileOutputStream fos = new FileOutputStream(targetFile);
117.                pw = new PrintWriter(new OutputStreamWriter(fos));
118.                pw.write(content);
119.                pw.flush();
120.
121.            }
```

```
122.            }
123.            catch (Exception e)
124.            {
125.                e.printStackTrace();
126.            }
127.            finally{
128.                if(pw != null){
129.                    pw.close();
130.                    pw = null;
131.                }
132.            }
133.        }
134.    }
```

（5）在 AndroidManifest.xml 文件中，在<manifest>根节点下添加在 SD 卡中创建、删除、写入数据的权限，代码如下。

```
1.  <!-- 在 SD 卡中创建与删除文件权限 -->
2.  <uses-permission
    android:name="android.permission.MOUNT_UNMOUNT_FILESYSTEMS"/>
3.  <!-- 向 SD 卡写入数据权限 -->
4.  <uses-permission
    android:name="android.permission.WRITE_EXTERNAL_STORAGE"/>
```

此外，在 AndroidManifest.xml 中<uses-sdk android:minSdkVersion="11" />的最小 SDK 版本需为 11 以上。

（6）部署运行 SDCardDemo 工程，然后在"写入 SD 卡"按钮上方的编辑框中输入 "this is a sdcard data app"，单击"写入 SD 卡"按钮，如图 8-26 所示。数据写入到 mnt/sdcard/myfile.txt 文件中，如图 8-27 所示，在 myfile.txt 文件中存储刚才写入的内容。具体导出文件步骤及操作详见后续案例，此处不再赘述。

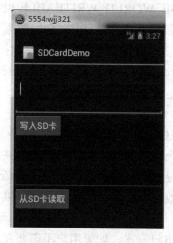

图 8-26　写入 SD 卡内容

sdcard 2015-02-03 15:26 d---rwxr-x
Android 2015-01-31 16:16 d---rwxr-x
LOST.DIR 2015-01-31 15:14 d---rwxr-x
myfile.txt 18 2015-02-03 15:26 ----rwxr-x

图 8-27 存储路径

单击“从 SD 卡读取”按钮，运行结果显示如图 8-28 所示。

5554:wjj321
3:29
SDCardDemo
写入SD卡
this is a card app
从SD卡读取

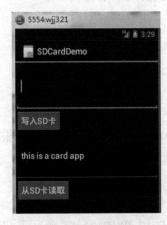

图 8-28 读取 SD 卡内容效果

8.4 数据共享访问

8.4.1 ContentProvider 简介

ContentProvider 类位于 android.content 包下，ContentProvider（数据提供者）是在应用程序间共享数据的一种接口机制。

虽然在前面章节的讲述中，通过指定文件的操作模式为 Context.MODE_WORLD_READABLE 或 Context.MODE_WORLD_WRITEABLE 也可以对外共享数据，但如果采用文件操作模式对外共享数据，数据的访问方式会因数据存储的方式而不同，导致数据的访问方式无法统一。例如，采用 XML 文件对外共享数据，需要进行 XML 解析才能读取数据；采用 sharedpreferences 共享数据，需要使用 sharedpreferences API 读取数据。

1. ContentProvider 的作用

在 Android 系统中，ContentProvider 的作用是对外共享数据，也就是说 ContentProvider 提供了在多个应用程序之间统一的数据共享方法，将需要共享的数据封装起来，提供了一组供其他应用程序调用的接口，通过 ContentResolver 来操作数据。应用程序可以指定需要共享的数据，而其他应用程序则可以在不知道数据来源、路径的情况下，对共享数据进行查询、添加、删除和更新等操作，使用 ContentProvider 对外共享数据的好处是统一了数据的访问方式。如果用户不需要在多个应用程序之间共享数据，可以通过前面讲述的 SQLiteDatabase 创建数据库的方式，实现数据内部共享。

2．ContentProvider 调用原理

ContentProvider 在创建和使用前，需要先通过数据库、文件系统或网络实现底层数据存储，然后自定义类继承 ContentProvider 类，并在其中实现基本数据操作的接口函数，包括添加、删除、查找和更新等功能。

ContentProvider 的接口函数不能直接使用，需要使用 ContentResolver 对象，通过 URI 间接调用 ContentProvider。

用户使用 ContentResolver 对象与 ContentProvider 进行交互，而 ContentResolver 则通过 URI 确定需要访问的 ContentProvider 的数据集。ContentResolver 对象与 ContentProvider 的调用关系如图 8-29 所示。

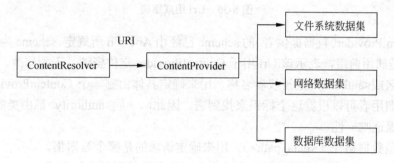

图 8-29　ContentResolver 与 ContentProvider 调用关系

其中，ContentProvider 负责组织应用程序的数据；向其他应用程序提供数据。

ContentResolver 则负责获取 ContentProvider 提供的数据；修改/添加/删除更新数据等。

8.4.2　Uri、UriMatcher 和 ContentUris 简介

1．Uri 简介

Uri 代表了要操作的数据 Uri 的信息，主要有以下两部分。

（1）需要操作的 ContentProvider；

（2）对 ContentProvider 中的什么数据进行操作，通过 Uri 来确定。

下面分别就上述 Uri 包含的两部分进行介绍。

1）ContentProvider 数据模式

ContentProvider 的数据模式类似于数据库的数据表，每行是一条记录，每列具有相同的数据类型，每条记录都包含一个长型的字段_ID，用来唯一标识每条记录。

ContentProvider 可以提供多个数据集，调用者使用 URI 对不同的数据集的数据进行操作。

ContentProvider 数据模式如表 8-5 所示。

表 8-5 ContentProvider 数据模式

_ID	NAME
1	John
2	Sam

2) Uri

Uri 用来定位任何远程或本地的可用资源,在 ContentProvider 中使用的 Uri 通常由以下几部分组成,如图 8-30 所示。

图 8-30 Uri 组成结构

ContentProvider(数据提供者)的 scheme 已经由 Android 所规定,scheme 为 content://;content://是通用前缀,表示该 Uri 用于 ContentProvider 定位资源,无须修改。

主机名或<authority>是授权者名称,用来确定具体由哪一个 ContentProvider 提供资源,外部调用者可以根据这个标识来找到它。因此,一般<authority>都由类的小写全称组成,以保证唯一性。

路径是数据路径(<data_path>),用来确定请求的是哪个数据集。

如果 ContentProvider 仅提供一个数据集,数据路径则是可以省略的。

如果 ContentProvider 提供多个数据集,数据路径则必须指明具体是哪一个数据集。

数据集的数据路径可以写成多段格式,例如/wjj /house 和/wjj /tea。<id>是数据编号,用来唯一确定数据集中的一条记录,用来匹配数据集中_ID 字段的值。

如果请求的数据并不只限于一条数据,则<id>可以省略。

Android SDK 推荐的方法是:在提供数据表字段中包含一个 ID,在创建表时 INTEGER PRIMARY KEY AUTOINCREMENT 标识此 ID 字段。

例如:

wjj/1:表示要操作 wjj 表中 id 为 1 的记录。

wjj/1/name:表示要操作 wjj 表中 id 为 1 的记录的 name 字段。

/wjj:表示要操作 wjj 表中的所有记录。

注意:如上述调用关系中所述,要操作的数据不一定来自数据库,也可以是文件系统、XML 或网络等其他存储方式。例如,要操作 XML 文件中 wjj 节点下的 name 节点,构建的路径为/wjj/name。

如果要把一个字符串转换成 Uri,可以使用 Uri 类中的 parse()方法,如下:

```
Uri uri = Uri.parse("content://com.hisoft. provider.helloprovider/wjj ")
```

2. UriMatcher 类简介

上述 Uri 代表了要操作的数据,需要解析 Uri 并从 Uri 中获取数据。

UriMatcher 类是 Android 系统提供了的用于操作 Uri 的工具类。它用于匹配 Uri，用法如下。

（1）注册需要匹配 Uri 路径，如下：

```
UriMatcher  sMatcher = new UriMatcher(UriMatcher.NO_MATCH);
//常量UriMatcher.NO_MATCH 表示不匹配任何路径的返回码
//如果 match()方法匹配 content:// com.hisoft. provider.helloprovider/wjj
//路径，返回匹配码为1
sMatcher.addURI("com.hisoft. provider.helloprovider ", "wjj", 1);//添加
//需要匹配 uri，如果匹配就会返回匹配码
//如果 match()方法匹配 content:// com.hisoft. provider.helloprovider/wjj/1
//路径，返回匹配码为 2
```

上述代码中 addURI()方法的声明语法：

```
public void  addURI(String authority, String path, int code)
```

其中，authority 表示匹配的授权者名称；
path 表示数据路径；
code 表示返回代码。

（2）使用 sMatcher.match(uri)方法对输入的 Uri 进行匹配。

如果匹配就返回匹配码，匹配码是调用 addURI()方法传入的第三个参数，假设匹配 content:// com.hisoft. provider.helloprovider/wjj 路径，返回的匹配码为 1，代码如下。

```
sMatcher.addURI("com.hisoft. provider.helloprovider ", "wjj /#", 2);
//#号为通配符
switch (sMatcher.match(Uri.parse("content:// com.hisoft. provider.
helloprovider/wjj /1"))) {
   case 1
     break;
   case 2
     break;
   default://不匹配
     break;
}
```

3．ContentUris 类简介

ContentUris 类也是 Android 系统提供的用于操作 Uri 的工具类，用于操作 Uri 路径后面的 ID 部分，它有两个比较常用的方法：withAppendedId(uri, id)和 parseId(uri)方法。withAppendedId(uri, id)用于为路径加上 ID 部分，代码如下：

```
Uri uri = Uri.parse("content:// com.hisoft. provider.helloprovider/wjj")
Uri resultUri = ContentUris.withAppendedId(uri, 1);
//生成后的 Uri 为: content:// com.hisoft. provider.helloprovider/wjj/1
```

parseId(uri)方法用于从路径中获取 ID 部分：

```
Uri uri = Uri.parse("content:// com.hisoft. provider.helloprovider/wjj/1")
long personid = ContentUris.parseId(uri);//获取的结果为:1
```

8.4.3 创建 ContentProvider

ContentProvider 的创建分为以下三步。

（1）自定义类继承 ContentProvider，并重载 ContentProvider 的 6 个方法。

新创建的自定义类继承 ContentProvider 后，需要重载 6 个方法，代码如下。

```
1.  public class ContentProviderDemo extends ContentProvider{
2.     public boolean onCreate();//初始化底层数据集和建立数据连接等工作
3.     public Uri insert(Uri uri, ContentValues values) ;// 添加数据集
4.     public int delete(Uri uri, String selection, String[] selectionArgs);
5.      //删除数据集
6.     public int update(Uri uri, ContentValues values, String selection,
       String[] selectionArgs);// 更新数据集
7.     public Cursor query(Uri uri, String[] projection,String selection,
       String[] selectionArgs, String sortOrder); //查询数据集
8.     public String getType(Uri uri)// 返回指定 Uri 的 MIME 数据类型
9.  }
```

注意：

如果 Uri 是单条数据，则返回的 MIME 数据类型应以 vnd.android.cursor.item 开头。

如果 Uri 是多条数据，则返回的 MIME 数据类型应以 vnd.android.cursor.dir/开头。

（2）实现 UriMatcher

在新创建的 ContentProvider 类中，通过创建一个 UriMatcher，用于判断 URI 是单条数据还是多条数据。通常为了便于判断和使用 Uri，一般将 Uri 的授权者名称和数据路径等内容声明为静态常量，并声明 CONTENT_URI。

```
1.  public static final String AUTHORITY = " com.hisoft.helloprovider ";
2.  public static final String PATH_SINGLE = "wjj /#";
3.  public static final String PATH_MULTIPLE = "wjj";
4.  public static final String CONTENT_URI_STRING = "content://" +
    AUTHORITY + "/" + PATH_MULTIPLE;
5.  public static final Uri  CONTENT_URI = Uri.parse(CONTENT_URI_STRING);
6.  private static final int MULTIPLE_WJJ = 1;
7.  private static final int SINGLE_WJJ = 2;
8.
9.  private static final UriMatcher uriMatcher;
10. static {
11.    uriMatcher = new UriMatcher(UriMatcher.NO_MATCH);
12.    uriMatcher.addURI(AUTHORITY, PATH_SINGLE, MULTIPLE_WJJ);
```

```
13.        uriMatcher.addURI(AUTHORITY, PATH_MULTIPLE, SINGLE_WJJ);
14. }
```

然后，在使用 UriMatcher 时，则可以直接调用 match()函数，对指定的 URI 进行判断，代码如下。

```
switch(uriMatcher.match(uri)){
    case MULTIPLE_WJJ:
        //多条数据的处理过程
        break;
    case SINGLE_WJJ:
        //单条数据的处理过程
        break;
    default:
        throw new IllegalArgumentException("非法的 URI:" + uri);
}
```

（3）在 AndroidManifest.xml 文件中注册 ContentProvider。

实现完成上述 ContentProvider 类的代码后，需要在 AndroidManifest.xml 文件中进行注册，在<application>根节点下，添加<provider>标签，并设置属性，代码如下。

```
<provider android:name = ".HelloProvider"
          android:authorities = "com.hisoft.helloprovider"/>
<!--注册了一个授权者名称为 com.hisoft.helloprovider 的 ContentProvider，其实
现类是 HelloProvider-->
```

8.4.4 ContentResolver 操作数据

使用 ContentResolver 类可以完成外部应用对 ContentProvider 中的数据进行添加、删除、修改和查询操作。ContentResolver 对象的创建，可以使用 Activity 提供的 getContentResolver()方法。ContentResolver 类具有以下方法。

public Uri insert(Uri uri, ContentValues values)：该方法用于往 ContentProvider 中添加数据。

public int delete(Uri uri, String selection, String[] selectionArgs)：该方法用于从 ContentProvider 中删除数据。

public int update(Uri uri, ContentValues values, String selection, String[] selectionArgs)：该方法用于更新 ContentProvider 中的数据。

public Cursor query(Uri uri, String[] projection, String selection, String[] selectionArgs, String sortOrder)：该方法用于从 ContentProvider 中获取数据。

这些方法的第一个参数为 Uri，代表要操作的 ContentProvider 和对其中的什么数据进行操作。示例代码如下。

```
1.    ContentResolver resolver = getContentResolver();
```

```
2.  Uri uri = Uri.parse("content:// com.hisoft.helloprovider/wjj ");
3.  //添加一条记录
4.  ContentValues values = new ContentValues();
5.  values.put("name", "John");
6.  values.put("age", 20);
7.  resolver.insert(uri, values);
8.  //获取 wjj 表中所有记录
9.  Cursor cursor = resolver.query(uri, null, null, null, "usrid desc");
10. while(cursor.moveToNext()){
11.   Log.i("ContentTest", "usrid="+ cursor.getInt(0)+ ",name="+ cursor.
      getString(1));
12. }
13. //把 id 为 1 的记录的 name 字段值更新为 lisi
14. ContentValues updateValues = new ContentValues();
15. updateValues.put("name", "lisi");
16. Uri updateIdUri = ContentUris.withAppendedId(uri, 2);
17. resolver.update(updateIdUri, updateValues, null, null);
18. //删除 id 为 2 的记录
19. Uri deleteIdUri = ContentUris.withAppendedId(uri, 2);
20. resolver.delete(deleteIdUri, null, null);
```

8.4.5 ContentProvider 应用

前面介绍了 ContentProvider 的调用关系、创建 ContentProvider 的步骤，以及 ContentResolver 操作数据的方法。下面通过一个读取 SQLite 数据库数据案例，详细介绍 ContentProvider 的应用。

（1）创建一个新的 Android 工程，工程名为 ContentProviderDemo，目标 API 选择 17（即 Android 4.2 版本），应用程序名为 ContentProviderDemo，包名为 com.bcpl.activity，创建的 Activity 的名字为 MainActivity，最小 SDK 版本根据选择的目标 API 会自动添加为 8，创建项目工程。

（2）修改 res 目录下 layout 文件夹中的 activity_main.xml 文件，设置线性布局中的嵌套线性布局，添加两个 EditText 控件、一个 Button 控件、一个 TextView 和一个 ScrollView 控件描述，并设置相关属性，代码如下所示。

```
1.  <?xml version="1.0" encoding="utf-8"?>
2.  <LinearLayout xmlns:android="http://schemas.android.com/apk/res/android"
3.      android:orientation="vertical"
4.      android:layout_width="fill_parent"
5.      android:layout_height="fill_parent"
6.      >
7.      <LinearLayout
8.          android:orientation="horizontal"
9.          android:layout_width="fill_parent"
```

```
10.        android:layout_height="wrap_content"
11.    >
12.      <TextView
13.        android:layout_width="wrap_content"
14.        android:layout_height="wrap_content"
15.        android:text="请输入姓名："
16.        android:textColor="@android:color/white"
17.        android:textSize="18dip"
18.        android:paddingRight="3dip"
19.      />
20.      <EditText
21.        android:text=""
22.        android:id="@+id/EditText01"
23.        android:layout_width="150dip"
24.        android:layout_height="wrap_content">
25.      </EditText>
26.      <Button
27.        android:text="查询"
28.        android:id="@+id/Button01"
29.        android:layout_width="wrap_content"
30.        android:layout_height="wrap_content">
31.      </Button>
32.    </LinearLayout>
33.    <ScrollView
34.      android:id="@+id/ScrollView01"
35.      android:layout_width="fill_parent"
36.      android:layout_height="wrap_content">
37.      <EditText
38.        android:id="@+id/EditText02"
39.        android:layout_width="fill_parent"
40.        android:layout_height="wrap_content">
41.      </EditText>
42.    </ScrollView>
43. </LinearLayout>
```

（3）修改 src 目录中 com.bcpl.activity 包下的 MainActivity.java 文件，读取上一个案例创建的数据库数据，代码如下。

```
1.  package com.bcpl.activity;
2.  import android.app.Activity;
3.  import android.content.ContentResolver;
4.  import android.database.Cursor;
5.  import android.net.Uri;
6.  import android.os.Bundle;
```

```
7.   import android.view.View;
8.   import android.view.View.OnClickListener;
9.   import android.widget.Button;
10.  import android.widget.EditText;
11.
12.  public class MainActivity extends Activity {
13.
14.      private ContentResolver cr;
15.
16.      @Override
17.      public void onCreate(Bundle savedInstanceState) {
18.          super.onCreate(savedInstanceState);
19.          setContentView(R.layout.main);
20.
21.          cr=this.getContentResolver();
22.
23.          //初始化"查询"按钮
24.          Button b=(Button)this.findViewById(R.id.Button01);
25.          b.setOnClickListener(
26.            new OnClickListener()
27.            {
28.               @Override
29.               public void onClick(View v) {
30.                   EditText et=(EditText)findViewById(R.id.EditText01);
31.                   String stuname=et.getText().toString().trim();
32.
33.                   Cursor cur=cr.query
34.                   (
35.            Uri.parse("content://com.hisoft.activity.mycontentprovider/stu"),
36.                   new String[]{"stuno","stuname","stuage","stuclass"},
37.                   "stuname=?",
38.                   new String[]{stuname},
39.                   "stuage ASC"
40.                   );
41.
42.                   appendMessage("学号\t\t姓名\t\t年龄\t班级");
43.                   while(cur.moveToNext())
44.                   {
45.                       String stuno=cur.getString(0);
46.                       String sname=cur.getString(1);
47.                       int stuage=cur.getInt(2);
48.                       String stuclass=cur.getString(3);
49.                       appendMessage(stuno+"\t"+sname+"\t\t"+stuage+"\t"+
                           stuclass);
```

```
50.             }
51.                 cur.close();
52.             }
53.         }
54.     );
55. }
56.
57. //向文本区中添加文本
58. public void appendMessage(String msg)
59. {
60.     EditText et=(EditText)this.findViewById(R.id.EditText02);
61.     et.append(msg+"\n");
62. }
63. }
```

(4)部署 ContentProviderDemo 工程,程序运行后,如图 8-31 所示。在编辑框中输入查询的姓名,然后单击"查询"按钮,程序从 SQLite 数据库中读取数据,结果如图 8-32 所示。

图 8-31　ContentProviderDemo 运行效果

图 8-32　查询结果

8.5　网络存储应用

前面介绍的 4 种存储都将数据存储在本地设备上,本节介绍的是另外一种存储(获取)数据的方式,通过网络来实现数据的存储和获取。通过网络来获取和保存数据资源,这个方法需要设备保持网络连接状态,所以相对存在一些限制。经常用于相关操作的两个类,分别是 java.net.*和 android.net.*。

下面通过一个使用模拟器给 Gmail 邮箱发邮件的案例,详细介绍网络存储的应用。

(1)创建一个新的 Android 工程,工程名为 NetWorkDemo,目标 API 选择 17(即 Android 4.2 版本),应用程序名为 NetWorkDemo,包名为 com.bcpl.activity,创建的 Activity 的名字为 MainActivity,最小 SDK 版本根据选择的目标 API 会自动添加为 8,创建项目工程。

（2）修改 res 目录下 layout 文件夹中的 activity_main.xml 文件，设置线性布局中的嵌套线性布局，添加一个 EditText 控件、一个 TextView 描述，并设置相关属性，代码如下所示。

```
1.  <?xml version="1.0" encoding="utf-8"?>
2.  <LinearLayout xmlns:android="http://schemas.android.com/apk/res/android"
3.      android:orientation="vertical"
4.      android:layout_width="fill_parent"
5.      android:layout_height="fill_parent"
6.      >
7.  <TextView
8.      android:layout_width="fill_parent"
9.      android:layout_height="wrap_content"
10.     android:text="@string/hello"
11.     />
12.  <EditText
13.      android:id="@+id/EditText01"
14.      android:layout_width="fill_parent"
15.      android:layout_height="wrap_content">
16.  </EditText>
17.  </LinearLayout>
```

（3）修改 src 目录中 com.bcpl.activity 包下的 MainActivity.java 文件，实现在文本框中输入邮件内容，邮件主题为"网络存储"，按返回键，则调用邮件系统，代码如下。

```
1.  package com.bcpl.activity;
2.
3.  import android.app.Activity;
4.  import android.content.Intent;
5.  import android.net.Uri;
6.  import android.os.Bundle;
7.  import android.view.KeyEvent;
8.  import android.widget.EditText;
9.
10.
11.
12. public class NetWorkDemoActivity extends Activity {
13.     private EditText mEditText;
14.
15.     /** Called when the activity is first created. */
16.     public void onCreate(Bundle savedInstanceState) {
17.         super.onCreate(savedInstanceState);
18.         setContentView(R.layout.main);
19.         mEditText = (EditText) findViewById(R.id.EditText01);
```

```
20.     }
21.
22.     @Override
23.     public boolean onKeyDown(int keyCode, KeyEvent event) {
24.         // TODO Auto-generated method stub
25.         if (keyCode == KeyEvent.KEYCODE_BACK) {
26.             final Intent intent = new Intent(android.content.Intent.ACTION_SEND);
27.             intent.putExtra(android.content.Intent.EXTRA_EMAIL, new String[]{"wjjyu@gmail.com"});
28.             intent.setType("plain/text");
29.             intent.putExtra(android.content.Intent.EXTRA_SUBJECT,"网络存储");
30.             intent.putExtra(android.content.Intent.EXTRA_TEXT,
31.                 String.valueOf(mEditText.getText()));
32.             startActivity(Intent.createChooser(intent, "Send mail..."));
33.             this.finish();
34.             return true;
35.         }
36.         return super.onKeyDown(keyCode, event);
37.     }
38. }
```

（4）设置模拟器邮件系统配置，选择 E-mail 图标，设置 Gmail 账户的用户名和密码，连接 Gmail 服务器，测试连通。

（5）部署运行 NetWorkDemo 工程，运行后在编辑框中输入邮件内容"测试邮件"，如图 8-33 所示。

然后按返回键，自动调用邮件系统，如图 8-34 所示，然后单击 Send 按钮，即可发送邮件到设定的 Gmail 邮箱。

图 8-33　测试邮件

图 8-34　发送邮件

8.6 数据存储项目案例

学习目标：学习、掌握 SQLite 数据库创建、操作、管理及应用。

案例描述：通过"妈咪宝贝"案例中的妈妈日记本，实现妈妈每天日记的添加、数据库保存。

案例要点：SQLite 数据库常用操作方法。

案例步骤：

（1）创建工程 Android_Contacts，选择 Android 4.2 作为目标平台。

（2）在 res 目录 layout 文件下创建 main.xml 文件，代码如下。

```xml
1.  <?xml version="1.0" encoding="utf-8"?>
2.  <LinearLayout xmlns:android="http://schemas.android.com/apk/res/android"
3.      android:orientation="vertical" android:layout_width="fill_parent"
4.      android:layout_height="fill_parent">
5.      <TextView android:layout_height="wrap_content" android:id="@+id/tw"
6.          android:text="妈妈日记本" android:layout_width="fill_parent"
7.          android:gravity="center" android:textColor="#FFFFFF"></TextView>
8.
9.      <TextView android:id="@+id/android:empty"
10.         android:layout_width="wrap_content" android:layout_height=
            "wrap_content"
11.         android:text="点击菜单添加日记" android:textColor="#000000">
            </TextView>
12.     <ListView android:id="@+id/android:list"
13.         android:layout_width="fill_parent" android:layout_height=
            "fill_parent"></ListView>
14.
15. </LinearLayout>
```

（3）在 src 目录下 com.bcpl.baby.diary 包中，创建 DiaryDao.java，代码如下。

```java
1.  package com.bcpl.baby.diary;
2.
3.  import android.content.ContentValues;
4.  import android.content.Context;
5.  import android.database.Cursor;
6.  import android.database.sqlite.SQLiteDatabase;
7.  import android.database.sqlite.SQLiteOpenHelper;
8.  import android.util.Log;
9.  public class DiaryDao {
10.     //上下文
11.     private Context context = null;
```

```
12.
13.     private static final String DB_NAME = "diary.db";
14.     private static final String TB_NAME = "TB_DIARY";
15.     private static final String F_ID = "_id";
16.     public static final String F_TITLE = "title";
17.     public static final String F_CONTENT = "cc";
18.     public static final String F_CREATED = "date";
19.     //初始化成员并打开、关闭数据库
20.     private SQLiteDatabase db;
21.     private DiaryHelper dbHelper;
22.     public DiaryDao(Context context) {
23.         // TODO Auto-generated constructor stub
24.         this.context = context;
25.     }
26.     /**
27.     * 构建helper创建数据库和表并获取可用的SQLite Datebase
28.     */
29.     public void open(){
30.         //构建helper对象，创建数据库与表
31.         dbHelper = new DiaryHelper(context);
32.         //并获取可用的SQLite Database
33.         db = dbHelper.getWritableDatabase();
34.     }
35.     /**
36.     * 关闭数据库
37.     */
38.     public void close(){
39.         db.close();
40.     }
41.     /**
42.     * 增加记录
43.     * @param title
44.     * @param context
45.     * @return
46.     */
47.     public long addDiary(String title, String content){
48.         ContentValues initValues = new ContentValues();
49.         initValues.put(F_TITLE, title);
50.         initValues.put(F_CONTENT, content);
51.         initValues.put(F_CREATED, "2011-04-21");
52.         return db.insert(TB_NAME, null, initValues);
53.     }
54.     /**
55.     * 删除记录
```

```java
56.     * @param rowId
57.     * @return
58.     */
59.
60.
61.    public int delete(long rowId){
62.        return db.delete(TB_NAME, F_ID+"="+rowId, null);
63.    }
64.
65.
66.    /**
67.     * 查询记录
68.     * @return
69.     */
70.    public Cursor getAll(){
71.        String col[] = new String[]{F_ID,F_TITLE,F_CONTENT,F_CREATED};
72.        return db.rawQuery("select * from tb_diary",null);
73.
74.    }
75.
76.    public Cursor getDiary(long rowId){
77.        String col[] = new String[]{F_TITLE,F_CONTENT};
78.        Log.d("GET Diary","row id " + rowId);
79.        return db.rawQuery("select * from tb_diary where " + F_ID +"=
              " + rowId , null);
80.
81.    }
82.    /**
83.     * 更新记录
84.     * @param rowId
85.     * @param title
86.     * @param context
87.     * @return
88.     */
89.    public int update(long rowId,String title,String content){
90.        ContentValues initValues = new ContentValues();
91.        initValues.put(F_TITLE, title);
92.        initValues.put(F_CONTENT, content);
93.        initValues.put(F_CREATED, "2011-01-02");
94.        return db.update(TB_NAME,initValues,F_ID+" = "+rowId,null);
95.    }
96.    /**
97.     * 创建内部类，实现初始化数据库与表
98.     * @author Administrator
```

```
99.     *
100.    */
101.   class DiaryHelper extends SQLiteOpenHelper {
102.
103.       /**
104.        * 创建数据库
105.        * @param context
106.        */
107.       public DiaryHelper(Context context) {
108.           // 创建数据库
109.           super(context, DB_NAME, null, 1);
110.       }
111.   /**
112.    * 创建表
113.    */
114.       @Override
115.       public void onCreate(SQLiteDatabase db) {
116.           // TODO Auto-generated method stub
117.           db.execSQL("create table " + TB_NAME +
118.                   " ("
119.                   + F_ID + " integer primary key AUTOINCREMENT,"
120.                   + F_TITLE + " text, "
121.                   + F_CONTENT + " text, "
122.                   + F_CREATED + " text ) ");
123.       }
124.
125.       /**
126.        * 更新
127.        */
128.       @Override
129.       public void onUpgrade(SQLiteDatabase db, int arg1, int arg2) {
130.           // TODO Auto-generated method stub
131.           db.execSQL("drop table if exists "+TB_NAME);
132.           onCreate(db);
133.       }
134.
135.       }
136.   }
```

（4）在 src 目录下 com.bcpl.baby.diary 包中，创建 Diary.java，代码如下。

```
1.  package com.bcpl.baby.diary;
2.
3.
```

```java
4.  import com.bcpl.baby.R;
5.  import android.app.ListActivity;
6.  import android.content.Intent;
7.  import android.database.Cursor;
8.  import android.os.Bundle;
9.  import android.view.ContextMenu;
10. import android.view.Menu;
11. import android.view.MenuItem;
12. import android.view.View;
13. import android.view.ContextMenu.ContextMenuInfo;
14. import android.widget.ListView;
15. import android.widget.SimpleCursorAdapter;
16. import android.widget.AdapterView.AdapterContextMenuInfo;
17.
18. public class Diary extends  ListActivity {
19.     /** 声明组件*/
20.     private ListView listView = null;
21.
22.     /** 菜单按钮 id*/
23.     private static final int M_ADD = Menu.FIRST;
24.     private static final int M_UPDATE = Menu.FIRST + 1;
25.     private static final int M_DEL = Menu.FIRST + 2;
26.     private static final int M_SELECT = Menu.FIRST + 3;
27.     int idGroup = 0;
28.     /** menuItem id*/
29.     int orderMenuItem1 = Menu.NONE;
30.     int orderMenuItem2 = Menu.NONE + 1;
31.     int orderMenuItem3 = Menu.NONE + 2;
32.     /** 声明数据库访问对象 */
33.     private DiaryDao dao = null;
34.     private Cursor cursor = null;
35.     private long listId = 0;
36.     private SimpleCursorAdapter simpleCursorAdapter;
37.     boolean isCreate = true;
38.     /** Called when the activity is first created. */
39.     @Override
40.     public void onCreate(Bundle savedInstanceState) {
41.         super.onCreate(savedInstanceState);
42.         setContentView(R.layout.main);
43.         /** 创建数据库访问对象*/
44.         dao = new DiaryDao(Diary.this);
45.         /** 通过 dao 调用 open()方法打开数据库，表*/
46.         dao.open();
47.         readerListView();
```

```
48.         registerForContextMenu(getListView());
49.     }
50.
51.     private void readerListView() {
52.         /** 通过查询表中所有数据获得结果集*/
53.         cursor = dao.getAll();
54.         /** 管理结果集，当activity停止时，它将自动调用deativate()方法，当
                activity重新启动时，它将自动调用requery()方法。当activity被销毁
55.          * 时，结果将会自动关闭*/
56.         startManagingCursor(cursor);
57.         /** 简单结果集适配器：它是将模板和数据结合在一起*/
58.         simpleCursorAdapter = new SimpleCursorAdapter(Diary.this,
                android.R.layout.simple_expandable_list_item_1,cursor,new
59.                 String[]{DiaryDao.F_TITLE},new int[]{android.R.id.text1});
60.         /** 将适配器设置给ListView：把模板和数据放到ListView里*/
61.         setListAdapter(simpleCursorAdapter);
62.         /**给ListView注册长按菜单,通过getListView()获得ListView组件**/
63.
64.     }
65.
66.     /**
67.      * 创建menu菜单
68.      * @param menu
69.      */
70.     @Override
71.     public boolean onCreateOptionsMenu(Menu menu) {
72.         // TODO Auto-generated method stub
73.
74.         /** 建立menu*/
75.         menu.add(idGroup,M_ADD,orderMenuItem1,"添加");
76.
77.         menu.setGroupCheckable(idGroup, true, true);
78.
79.         return super.onCreateOptionsMenu(menu);
80.     }
81.     /**
82.      * 动态创建menu菜单
83.      */
84.     @Override
85.     public boolean onPrepareOptionsMenu(Menu menu) {
86.         // TODO Auto-generated method stub
87.         boolean isHave = getListAdapter().getCount()>0?true:false;
88.         if(isHave){
89.             if(isCreate){
```

```
90.              menu.add(idGroup,M_UPDATE,orderMenuItem2,"修改");
91.              menu.add(idGroup,M_DEL,orderMenuItem3,"删除");
92.              isCreate = false;
93.          }
94.      }else{
95.          menu.removeItem(M_UPDATE);
96.          menu.removeItem(M_DEL);
97.          isCreate = true;
98.      }
99.      return super.onPrepareOptionsMenu(menu);
100.
101. }
102.
103. /**
104.  * menu菜单的选择事件
105.  * @param menu
106.  */
107. @Override
108. public boolean onOptionsItemSelected(MenuItem item) {
109.     // TODO Auto-generated method stub
110.     long currentItem = listId ;
111.     switch(item.getItemId())
112.     {
113.     case M_ADD:
114.         Intent intent = new Intent(Diary.this,EditDiary.class);
115.         intent.putExtra("option", 0);
116.         startActivity(intent);
117.         Diary.this.finish();
118.         dao.close();
119.         break;
120.     case M_UPDATE:
121.         intent = new Intent(Diary.this,EditDiary.class);
122.         intent.putExtra("option", 2);
123.         intent.putExtra("itemID", currentItem);
124.         startActivity(intent);
125.         dao.close();
126.         Diary.this.finish();
127.         break;
128.     case M_DEL:
129.
130.         //Toast.makeText(Diary.this, " currentItem " + currentItem,
                     //Toast.LENGTH_LONG).show();
131.         dao.delete(currentItem);
132.         readerListView();
```

```
133.                break;
134.            }
135.            return super.onOptionsItemSelected(item);
136.    }
137.
138.    /**
139.     * 创建长按菜单（上下文菜单）
140.     * @param ContextMenu menu
141.     * @param View v
142.     */
143.    @Override
144.    public void onCreateContextMenu(ContextMenu menu, View v,
145.            ContextMenuInfo menuInfo) {
146.        // TODO Auto-generated method stub
147.        menu.add(0,M_UPDATE,0,"修改");
148.        menu.add(0,M_DEL,0,"删除");
149.        menu.add(0,M_SELECT,0,"查看");
150.        super.onCreateContextMenu(menu, v, menuInfo);
151.    }
152.    /**
153.     * 长按菜单（上下文菜单）事件
154.     * @param MenuItem item
155.     */
156.    @Override
157.    public boolean onContextItemSelected(MenuItem item) {
158.        // TODO Auto-generated method stub
159.        AdapterContextMenuInfo info = (AdapterContextMenuInfo)item.getMenuInfo();
160.        long itemID = info.id;
161.        Intent intent;
162.        switch(item.getItemId())
163.        {
164.        case M_UPDATE:
165.            intent = new Intent(Diary.this,EditDiary.class);
166.            intent.putExtra("option", 2);
167.            intent.putExtra("itemID", itemID);
168.            startActivity(intent);
169.            dao.close();
170.            Diary.this.finish();
171.            break;
172.        case M_DEL:
173.            dao.delete(itemID);
174.            readerListView();
175.            break;
```

```
176.            case M_SELECT:
177.                //Toast.makeText(Diary.this, "" + itemID, Toast.
                    //LENGTH_SHORT).show();
178.                intent = new Intent(Diary.this,EditDiary.class);
179.                intent.putExtra("option", 1);
180.                intent.putExtra("itemID", itemID);
181.                startActivity(intent);
182.                dao.close();
183.                Diary.this.finish();
184.
185.                break;
186.        }
187.        return super.onContextItemSelected(item);
188.    }
189.    @Override
190.    protected void onListItemClick(ListView l, View v, int position, long id) {
191.        // TODO Auto-generated method stub
192.        super.onListItemClick(l, v, position, id);
193.        //数据库中的id号赋给listId
194.        listId = id;
195.    }
196. }
```

(5) 在 src 目录下 com.bcpl.baby.diary 包中，创建 EditDiary.java，代码如下。

```
1.  package com.bcpl.baby.diary;
2.  import com.bcpl.baby.R;
3.  import android.app.Activity;
4.  import android.content.Intent;
5.  import android.database.Cursor;
6.  import android.os.Bundle;
7.  import android.view.KeyEvent;
8.  import android.view.View;
9.  import android.view.View.OnClickListener;
10. import android.widget.Button;
11. import android.widget.EditText;
12. import android.widget.Toast;
13.
14. public class EditDiary extends Activity{
15.     /**声明组件*/
16.     private Button saveButton = null;
17.     private EditText titleEditText = null;
18.     private EditText conTextEditText = null;
19.
```

```
20.    private   DiaryDao dao = null;
21.    private Bundle bundle = null;
22.    private long itemId = 0;
23.    private int controlId = 0;
24.
25.      @Override
26.      protected void onCreate(Bundle savedInstanceState) {
27.          // TODO Auto-generated method stub
28.          super.onCreate(savedInstanceState);
29.          setContentView(R.layout.editdiary);
30.          bundle = getIntent().getExtras();
31.          controlId = bundle.getInt("option");
32.          saveButton =(Button)findViewById(R.id.bsave);
33.          titleEditText =(EditText)findViewById(R.id.etitle);
34.          conTextEditText =(EditText)findViewById(R.id.econtent);
35.       dao = new DiaryDao(EditDiary.this);
36.       dao.open();
37.       switch(controlId)
38.       {
39.       case Const.M_ADD:
40.
41.           saveButton.setOnClickListener(new OnClickListener() {
42.
43.               @Override
44.               public void onClick(View v) {
45.                   // TODO Auto-generated method stub
46.
47.                   String title = titleEditText.getText().toString();
48.                   String context = conTextEditText.getText().toString();
49. //
50.                   long id = dao.addDiary(title, context);
51. //                 Toast.makeText(EditDiary.this, "   " +id ,Toast.
  //LENGTH_SHORT).show();
52.                   if(id>0)
53.                      Toast.makeText(EditDiary.this, "保存成功",
                         Toast.LENGTH_SHORT).show();
54.                   else
55.                      Toast.makeText(EditDiary.this, "保存失败",
                         Toast.LENGTH_SHORT).show();
56.
57.                   Intent intent = new Intent(EditDiary.this,
                      Diary.class);
58.                   startActivity(intent);
59.                   EditDiary.this.finish();
```

```
60.                    dao.close();
61.                }
62.            });
63.            break;
64.        case Const.M_SELECT:
65.            itemId = bundle.getLong("itemID");
66. //         Toast.makeText(EditDiary.this, "" + itemId, Toast.
    //LENGTH_SHORT).show();
67.            Cursor cursor = dao.getDiary(itemId);
68.            startManagingCursor(cursor);
69.            int  string =cursor.getColumnIndex(DiaryDao.F_TITLE);
70.            cursor.moveToNext();
71. //         Toast.makeText(EditDiary.this, "" + string + "    "+c,
    //Toast.LENGTH_SHORT).show();
72.            titleEditText.setText(cursor.getString(string));
73.            conTextEditText.setText(cursor.getString(cursor.getCol
               umnIndex(DiaryDao.F_CONTENT)));
74.            titleEditText.setEnabled(false);
75.            conTextEditText.setEnabled(false);
76.            break;
77.        case Const.M_UPDATE:
78.            itemId = bundle.getLong("itemID");
79. //         Toast.makeText(EditDiary.this, "" + itemId, Toast.
    //LENGTH_SHORT).show();
80.            cursor = dao.getDiary(itemId);
81.            startManagingCursor(cursor);
82.             string =cursor.getColumnIndex(DiaryDao.F_TITLE);
83.            cursor.moveToNext();
84. //         Toast.makeText(EditDiary.this, "" + string + "    "+c,
    //Toast.LENGTH_SHORT).show();
85.            titleEditText.setText(cursor.getString(string));
86.            conTextEditText.setText(cursor.getString(cursor.
               getColumnIndex(DiaryDao.F_CONTENT)));
87.          saveButton.setOnClickListener(new OnClickListener() {
88.
89.              @Override
90.              public void onClick(View v) {
91.                  // TODO Auto-generated method stub
92.                  String title = titleEditText.getText().toString();
93.                  String context = conTextEditText.getText().toString();
94.                  long id = dao.update(itemId, title, context);
95. //               Toast.makeText(EditDiary.this, "   " +id ,Toast.
    //LENGTH_SHORT).show();
96.                  if(id>0)
```

```
97.                    Toast.makeText(EditDiary.this, "保存成功",
                           Toast.LENGTH_SHORT).show();
98.                 else
99.                    Toast.makeText(EditDiary.this, "保存失败",
                           Toast.LENGTH_SHORT).show();
100.
101.                 Intent intent = new Intent(EditDiary.this,
                        Diary.class);
102.                 startActivity(intent);
103.                 dao.close();
104.                 EditDiary.this.finish();
105.              }
106.          });
107.          break;
108.       }
109.
110.
111.    }
112.
113.    @Override
114.    public boolean onKeyDown(int keyCode, KeyEvent event) {
115.       // TODO Auto-generated method stub
116.       if(keyCode == KeyEvent.KEYCODE_BACK)
117.       {
118.          switch(controlId)
119.          {
120.          case Const.M_SELECT:
121.             Intent intent = new Intent(EditDiary.this,Diary.
                    class);
122.             startActivity(intent);
123.             EditDiary.this.finish();
124.             dao.close();
125.             break;
126.          }
127.       }
128.       return super.onKeyDown(keyCode, event);
129.    }
130.
131. }
```

（6）部署项目工程，程序运行后，单击"妈妈篇"图标，可以进入妈妈篇功能：营养食谱、妈妈讲故事、妈妈日记本、宝宝教育，如图8-35所示。

单击"妈妈日记本"图标，可以进入写日记状态，单击"添加"按钮可以进入添加日记界面，通过添加标题、内容，然后单击"保存"按钮保存，如图8-36～图8-38所示。

图 8-35 妈妈篇

图 8-36 妈妈日记本添加界面

图 8-37 妈妈日记本添加内容界面

图 8-38 妈妈日记本编辑界面

保存后自动返回妈妈日记本的首页，在这里可以通过"添加"、"修改"、"删除"按钮来对每一篇日记进行添加、修改、删除，如图 8-39 所示。

图 8-39 妈妈日记本修改界面

习 题

一、简答题

1. 常用的 SharedPreferences 访问模式有哪些？

2. 在 Android 系统中，SharedPreferences 读取本地应用程序和其他应用程序的方法有什么不同？

3. Android 系统支持的文件操作模式主要有哪些？它们之间的区别是什么？

4. 在 AndroidManifest.xml 文件中如何添加注册 SD 卡访问应用权限？

二、实训

要求：

在本章项目案例的基础上，完成妈妈日记模糊搜索功能。

第 9 章

Google 位置应用服务开发

学习目标

本章介绍了地理位置定位服务、Google Map 应用等。读者通过本章的学习,能够深入熟悉 Android 系统通信服务,掌握以下知识要点。

(1) SmsManager 管理短信息操作及应用。

(2) TelephonyManager、onCallStateChanged、onServiceStateChanged、onSignalStrengthChanged 等类及其常用方法。

9.1 地理位置定位服务

地理位置定位服务(Location-Based Service,LBS),又称移动定位服务,是指通过移动运营商提供的无线电通信网络(如 GSM 网、CDMA 网)或外部定位方式(如 GPS)获取移动终端用户的位置信息(地理坐标)。在 GIS(Geographic Information System,地理信息系统)平台的支持下,为用户提供相应服务的一种增值业务。它通常包含两方面服务:一是确定移动设备或用户所在的地理位置;二是提供与位置相关的各类信息服务。

LBS 在移动服务方面的基本原理是:当移动用户需要信息服务或监控管理中心需要对某移动终端进行移动计算时,首先移动终端通过内嵌的定位设备如 GNSS(GPS/GLONASS)获得终端本身当前的空间位置数据,并实时地通过无线通信把数据上传送到服务中心,然后服务中心 GIS 服务器根据终端的地理位置、服务要求进行空间分析、决策,进而再下传至移动终端或中心的计算机。

目前,LBS 开发应用的版本有 Web 版、移动客户端版、平板电脑版,如人人网的人人报到、Firefox 地理位置定位、Google 的 Google Maps Coordinate 等,提供地图定位服务的有 Google 地图、百度地图、搜狗地图等。

基于位置定位的服务通常使用经度和纬度来定位地理位置,Google Maps 库提供了地理编码器,实现经纬度和真实地址(如街道门牌号等)的映射互换。本书主要介绍 Android 平台下对 Google 位置定位服务和 Map 的应用开发。

9.1.1 Android Location API 简介

Google 对于 Android 平台提供了 Location 包和 Maps 扩展库,以开发基于移动客户端的 Google Map 地图应用。在 Android 中,android.location 包提供了访问设备位置信息的服务,在 location 包中,其主要是通过核心组件 LocationManager 类来实现设备的定位、跟踪和趋近提示,提供上述系统服务。LocationManager 类不能被直接用来创建对象实例化,而是通过调用 Context 对象的 getSystemService(Context.LOCATION_SERVICE)方法来获取对象句柄,创建 LocationManager 对象实例。下面就上述的 LocationManager 类以及相关类的常用属性和方法进行介绍。

1. LocationManager 类

通过 LocationManager 可以实现设备的定位、跟踪和趋近提示。上面已讲解过它可以通过 Context.getSystemService(Context.LOCATION_SERVICE)方法来获得 LocationManager 类对象实例。其常用的属性、方法如表 9-1 所示。

表 9-1 LocationManager 类对象常用的属性、方法

属性和方法	描述
GPS_PROVIDER	静态字符串常量,表明 LocationProvider 是 GPS
NETWORK_PROVIDER	静态字符串常量,表明 LocationProvider 是网络
addGpsStatusListener(GpsStatus.Listener listener)	添加一个 GPS 状态监听器
addProximityAlert(double latitude, double longitude, float radius, long expiration, PendingIntent intent)	添加一个趋近警告
getAllProviders()	获得所有的 LocationProvider 列表
getBestProvider(Criteria criteria, boolean enabledOnly)	根据 Criteria 返回最适合的 LocationProvider
getLastKnownLocation(String provider)	根据 Provider 获得位置信息
getProvider(String name)	获得指定名称的 LocationProvider
getProvider(boolean enableOnly)	获得可利用的 LocationProvider 列表
removeProximityAlert(PendingIntent intent)	删除趋近警告
requestLocationUpdates(String provider, long minTime, float minDistance, PendingIntent intent)	通过给定的 Provider 名称,周期性地通知当前 Activity
requestLocationUpdates(String provider, long minTime, float minDistance, LocationListener listener)	通过给定的 Provider 名称,并将其绑定指定的 LocationListener 监听器

2. LocationProvider 类

LocationProvider 位于 android.location 包下,是位置提供者的抽象超类,用来描述位置提供者,设置位置提供者的一些属性,周期性报告设备的地理位置信息。可以通过 Criteria 类来为 LocationProvider 对象设置条件,获得用户定义的 LocationProvider 对象。其常用的方法和属性如表 9-2 所示。

表 9-2 LocationProvider 对象常用的方法和属性

属性或方法名称	描述
AVAILABLE	静态整型常量，标示是否可利用
OUT_OF_SERVICE	静态整型常量，不在服务区
TEMPORAILY_UNAVAILABLE	静态整型常量，临时不可利用
getAccuracy()	获得精度
getName()	获得位置提供者的名称
getPowerRequirement()	获得电源需求
hasMonetaryCost()	如果 provider 收费返回 true，免费返回 false
requiresCell()	是否需要访问基站网络
requiresNetWork()	是否需要 Intent 网络数据
requiresSatelite()	是否需要访问卫星
supportsAltitude()	是否能够提供高度信息
supportsBearing()	是否能够提供方向信息
supportsSpeed()	是否能够提供速度信息

3. Criteria 类

Criteria 类也位于 android.location 包下，它封装了用于获得 LocationProvider 的条件，可以根据指定的 Criteria 条件来过滤获得 LocationProvider。其常用属性和方法如表 9-3 所示。

表 9-3 Criteria 类常用属性和方法

属性或访求名称	描述
ACCERACY_COARSE	粗略精确度
ACCURACY_FINE	较高精确度
POWER_HIGH	用电消耗高
POWER_LOW	用电消耗低
isAlititudeRequried()	返回 Provider 是否需要高度信息
isBearingRequired()	返回 Provider 是否需要方位信息
isSpeedRequried()	返回 Provider 是否需要速度信息
isCostAllowed()	是否允许产生费用
setAccuracy(int accuracy)	设置 Provider 的精确度
setAltitudeRequired (boolean altitudeRequired)	设置 Provider 是否需要高度信息
setBearingRequired (boolean bearingRequired)	设置 Provider 是否需要方位信息
setCostAllowed (boolean costAllowed)	设置 Provider 是否产生费用
setSpeedAccuracy (int accuracy)	设置 Provider 是否需要速度信息
getAccuracy()	获得精度

说明：通常也可以在 AndroidManifest.xml 文件中添加地理位置访问精度的权限设置，如下：

```
<uses-permission android:name="android.permission.ACCESS_FINE_LOCATION"/>
<uses-permission android:name="android.permission.ACCESS_COARSE_LOCATION"/>
```

4. Location 类

Location 类也位于 android.location 包下，用于描述当前设备的地理位置信息，包括经纬度、方向、高度和速度等。开发者可以通过 LocationManager.getLastKnownLocation（String provider）方法获得 Location 实例。其常用的方法、属性如表 9-4 所示。

表 9-4 Location 类常用的方法、属性

方 法	描 述
public float getAccuracy ()	获得精确度
public double getAltitude ()	获得高度
public float getBearing ()	获得方向
public double getLatitude ()	获取经度
public double getLongitude ()	获得纬度
public float getSpeed ()	获得速度

从上述得知，开发者借助 LocationManager 类的对象生成的系统服务，来调用或使用 Android 平台下的 GPS 服务，通过 Context.getSystemService(Context.LOCATION_SERVIER)方法来获取 LocationManager 对象实例，然后可以通过 LocationProvider 来描述位置提供者。此外，可以用上述的 Criteria 类来设置自定义满足用户的最佳要求。最后，通过 LocationManager.getLastKnownLocation(String provider)方法可以获得 Location 实例，然后用 Location 类获取自己所在的位置信息，如经纬度等信息。

5. LocationListener 类

LocationListener 类位于 android.location 包下，用于接收从 LocationManager 的位置发生改变时的通知。如果 LocationListener 被注册添加到 LocationManager 对象，并且此 LocationManager 对象调用了 requestLocationUpdates(String, long, float, LocationListener) 方法，那么接口中的相关方法将会被调用。

LocationListener 类常用的方法如表 9-5 所示。

表 9-5 LocationListener 类常用的方法

方 法 名	描 述
onLocationChanged (Location location)	此方法在当位置发生改变后被调用。这里可以没有限制地使用 Location 对象。 参数为位置发生变化后的新位置
onProviderDisabled(String provider)	此方法在 provider 被用户关闭后被调用，如果基于一个已经关闭了的 provider 调用 requestLocationUpdates 方法被调用，那么这个方法理解被调用。 参数为与之关联的 Location Provider 名称
onPorviderEnabled (Location location)	此方法在 provider 被用户开启后调用。 参数为 provider 与之关联的 Location Provider 名称

续表

方 法 名	描 述
onStatusChanged (String provider, int Status, Bundle extras)	此方法在 Provider 的状态在可用、暂时不可用和无服务三个状态直接切换时被调用。 参数：provider，与变化相关的 Location Provider 名称。 Status，如果服务已停止，并且在短时间内不会改变，状态码为 OUT_OF_SERVICE；如果服务暂时停止，并且在短时间内会恢复，状态码为 TEMPORARILY_UNAVAILABLE；如果服务正常有效，状态码为 AVAILABLE。 extras，一组可选参数，其包含 provider 的特定状态

9.1.2 获取位置定位

上面介绍了在 Android 中应用地理位置定位服务常用的包及类和方法。下面通过一个获取地理位置定位服务的应用案例，详细介绍 Location API 的应用。

（1）创建一个新的 Android 工程，工程名为 LocationDemo，目标 API 选择 17（即 Android 4.2 版本），应用程序名为 LocationDemo1，包名为 com.bcpl，创建的 Activity 的名字为 MainActivity，最小 SDK 版本根据选择的目标 API 会自动添加为 8。

（2）修改 res 目录下 layout 文件夹中的 activity_main.xml 文件，设置线性布局，添加一个 EditText 控件、描述，并设置相关属性，代码如下所示。

```
1.  <?xml version="1.0" encoding="utf-8"?>
2.  <LinearLayout xmlns:android="http://schemas.android.com/apk/res/android"
3.      android:orientation="vertical"
4.      android:layout_width="fill_parent"
5.      android:layout_height="fill_parent"
6.      >
7.  <TextView
8.      android:layout_width="fill_parent"
9.      android:layout_height="wrap_content"
10.     android:text="@string/hello"
11.      android:id="@+id/textView1"
12.
13.     />
14. </LinearLayout>
```

（3）修改 src 目录中 com.bcpl.activity 包下的 MainActivity.java 文件，代码如下。

```
1.  package com.bcpl.activity;
2.  import android.app.Activity;
3.  import android.content.Context;
4.  import android.location.Location;
```

```
5.    import android.location.LocationListener;
6.    import android.location.LocationManager;
7.    import android.os.Bundle;
8.    import android.widget.EditText;
9.
10.   public class MainActivity extends Activity {
11.       // 定义LocationManager对象
12.       private LocationManager locManager;
13.       // 定义程序界面中的EditText组件
14.       private EditText show;
15.
16.       @Override
17.       public void onCreate(Bundle savedInstanceState)
18.       {
19.           super.onCreate(savedInstanceState);
20.           setContentView(R.layout.main);
21.           // 获取程序界面上的EditText组件
22.           show = (EditText) findViewById(R.id.show);
23.           // 创建LocationManager对象
24.           locManager = (LocationManager) getSystemService
              (Context.LOCATION_SERVICE);
25.           // 从GPS获取最近的定位信息
26.           Location location = locManager.getLastKnownLocation(
27.               LocationManager.GPS_PROVIDER);
28.           // 使用location根据EditText的显示
29.           updateView(location);
30.           // 设置每3秒获取一次GPS的定位信息
31.           locManager.requestLocationUpdates(LocationManager.GPS_
              PROVIDER
32.               , 3000, 8, new LocationListener()
33.           {
34.               @Override
35.               public void onLocationChanged(Location location)
36.               {
37.                   // 当GPS定位信息发生改变时,更新位置
38.                   updateView(location);
39.               }
40.
41.               @Override
42.               public void onProviderDisabled(String provider)
43.               {
44.                   updateView(null);
45.               }
46.
```

```
47.            @Override
48.            public void onProviderEnabled(String provider)
49.            {
50.                // 当GPS LocationProvider可用时，更新位置
51.                updateView(locManager
52.                    .getLastKnownLocation(provider));
53.            }
54.
55.            @Override
56.            public void onStatusChanged(String provider, int status,
57.                Bundle extras)
58.            {
59.            }
60.        });
61.    }
62.
63.    // 更新EditText中显示的内容
64.    public void updateView(Location newLocation)
65.    {
66.        if (newLocation != null)
67.        {
68.            StringBuilder sb = new StringBuilder();
69.            sb.append("实时的位置信息：\n");
70.            sb.append("经度：");
71.            sb.append(newLocation.getLongitude());
72.            sb.append("\n纬度：");
73.            sb.append(newLocation.getLatitude());
74.            sb.append("\n高度：");
75.            sb.append(newLocation.getAltitude());
76.            sb.append("\n速度：");
77.            sb.append(newLocation.getSpeed());
78.            sb.append("\n方向：");
79.            sb.append(newLocation.getBearing());
80.            show.setText(sb.toString());
81.        }
82.        else
83.        {
84.            // 如果传入的Location对象为空则清空EditText
85.            show.setText("");
86.        }
87.    }
88. }
```

（4）修改AndroidManifest.xml文件，在<manifest>根目录下添加获取定位信息权限，

代码如下:

1.   <!-- 授权获取定位信息 -->
2.   <uses-permission android:name="android.permission.ACCESS_FINE_LOCATION"/>
3.   <uses-permission android:name="android.permission.ACCESS_COARSE_LOCATION"/>

（5）部署 LocationDemo 应用项目工程，如果在模拟器中运行程序，调试经纬度，需要打开 DDMS，在 Emulator Control 面板中，选择 Location Controls→Manual→Decimal 单选按钮，然后在 Longitude 和 Latitude 中输入经纬度，单击 Send 按钮，如图 9-1 所示。由于使用模拟器的缘故，高度、速度、方向为 0，如果在真机上测试，则会显示真值。运行如图 9-2 所示。

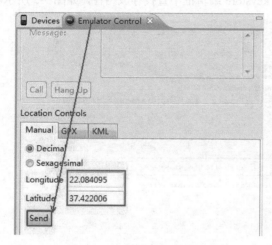

图 9-1　设置经纬度值并发送　　　　图 9-2　读取经纬度、高度、速度、方向数据

9.2　Google Map 应用

9.2.1　Google Map API 简介

Google 提供的 Maps 扩展库位于 com.google.android.maps 包中，它不是 Android SDK 标准库。扩展包 com.google.android.maps 中包含一系列用于在 Google Map 上显示、控制和层叠信息的功能类。在 Maps 库中关键的类是 MapView，在 Android 标准库中它是 ViewGroup 的子类。MapView 显示的地图和数据来自于 Google Maps 服务器，当 MapView 成为焦点时，它能捕获按键、触摸来自动拉伸缩放地图，它能够提供基本的 UI 元素供用户控制、操作地图。

在应用程序中使用 Maps 扩展库，需要安装关于 Google APIs 的 add-on 文件夹，如果使用 Android SDK，则不需要安装 add-on，因为它已经被预装在 Android SDK 包中。

通常 MapView 类提供封装了 Google Maps API 的适配器，供开发者在应用程序中通过类的方法操纵 Google Maps 的数据。MapView 对象以并列块的形式显示从 Google 地图服务器上下载的地图,但在应用 Google 地图服务器数据之前,必须先进行注册并申请 Maps API Key,下面就对申请 Maps API Key 进行 Google Map 应用进行讲解。

9.2.2 申请 Map API KEY 和创建 AVD

1. 申请 Map API KEY

在开发应用 Google Map 之前,必须先申请 Google Map API Key,其过程如下。

(1) 在使用 JDK 内置的 keytool 工具之前,必须先确定默认的 keystore 存储位置,在 Eclipse 中,选择 Windows→Preferences 命令打开对话框,然后在对话框中选择 Android→Build,再在右侧的 Default debug keystore 编辑框中查找存放路径,如图 9-3 所示。

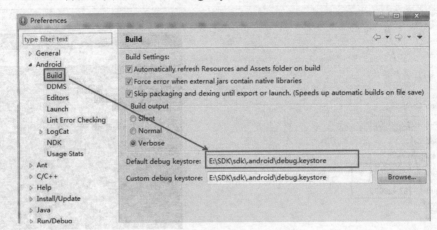

图 9-3 编辑 keystore 存放路径

(2) 打开命令窗口,输入命令使用 keytool 工具生成 SHA1 指纹,命令如下:

```
C:\>keytool -list -alias androiddebugkey -keystore "E:\SDK\sdk\.android\debug.keystore" -storepass android -keypass android
```

执行上述命令后,生成的 SHA1 指纹(与 V1 版本不同,V1 生成的是 MD5)如图 9-4 所示。

图 9-4 命令生成 SHA1 指纹

(3) 生成 SHA1 指纹后,在浏览器地址栏中输入 "https://code.google.com/apis/console/?noredirect",打开 Map API Key 申请页面,输入上述生成的 SHA1 指纹,在申请 Map API Key 之前,必须先创建 Gmail 账户登录,如果是第一次,需要创建项目,默认情况会创建 API Project 的项目,如图 9-5 所示。

第 9 章 Google 位置应用服务开发

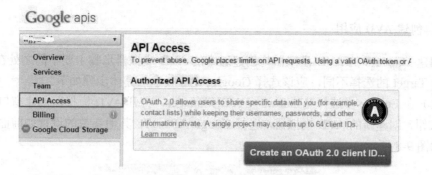

图 9-5　申请 Map API Key

（4）单击 Create new Android Key 按钮，弹出 Google API Key 生成页面，输入上述生成的 SHA1 指纹和应用程序包名，如图 9-6 所示。

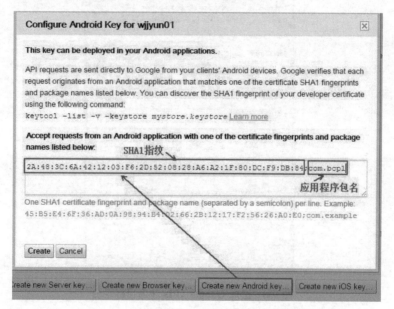

图 9-6　生成 Map API Key 页面

（5）单击 Create 按钮，生成绑定应用程序包名的 API Key，如图 9-7 所示。

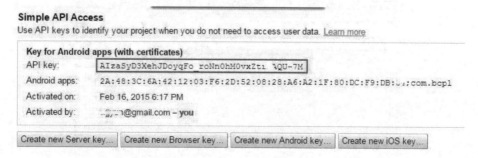

图 9-7　API Key 生成

2. 创建 AVD 应用

创建 Map API 的 AVD 模拟器与创建通常的 AVD 模拟器步骤主要不同之处在于，目标平台 Target 的选择不同，应该选择 Google APIs 版本，具体步骤如下。

（1）在 Eclipse 中选择 Windows→AVD Manager，打开 AVD 模拟器创建窗口，单击 New 按钮，创建新的 AVD 模拟器，选择 Google APIs 17，其他参数的设置如前面章节所述，如图 9-8 所示。

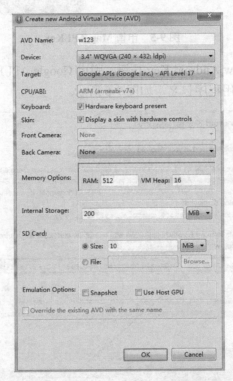

图 9-8　选择 Google APIs 作为目标平台

（2）单击 Create AVD 按钮，生成 AVD 模拟器，选中新创建的 AVD 模拟器，然后单击 Start 按钮，启动新创建的 AVD 模拟器，如图 9-9 所示。

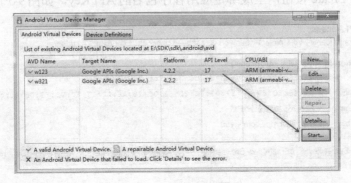

图 9-9　AVD 创建成功

9.3 项目案例

学习目标：学习 Google Map 的基本方法、属性的设置等应用。

案例描述：申请 Google Map KEY，使用 AbsoluteLayout 布局和 MapView 控件、Button 按钮，调用 Map 应用实现公司位置的准确定位和地图显示。

案例要点：MapView 控件描述、Google Map KEY 申请及添加、Map 相关方法。

案例实施：

（1）创建工程 Project_Chapter_9，选择目标平台为 Google APIs（API Level 17），右击新创建的工程名，选择 Properties 命令打开对话框，然后选中 Android，添加 google_play_services_lib 包，如图 9-10 所示。

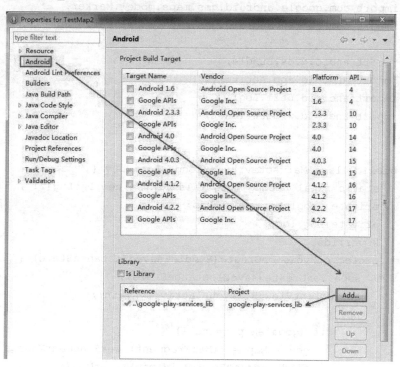

图 9-10　添加 google-play-services_lib 包

（2）修改 res/layout 目录下的 activity_main.xml，添加 Map Fragment，代码如下：

```
1.  <fragment
2.      android:id="@+id/map"
3.      android:name="com.google.android.gms.maps.MapFragment"
4.      android:layout_width="match_parent"
5.      android:layout_height="match_parent"/>
```

（3）在上述的 Google Console 创建的 API 工程中，单击左侧的 Services 按钮，选择

Google Maps Android API V2 状态为 on,如图 9-11 所示。

图 9-11 打开 Google Maps API V2 应用

(4) 在 src 目录包 com.bcpl 下,添加 GoogleMap 和 MapFragment 引用,创建 MainActivity.java 文件,代码如下。

```
1.   package com.bcpl;
2.   import com.google.android.gms.maps.GoogleMap;
3.   import com.google.android.gms.maps.MapFragment;
4.   import com.google.android.gms.maps.model.LatLng;
5.   import com.google.android.gms.maps.model.Marker;
6.   import com.google.android.gms.maps.model.MarkerOptions;
7.
8.   import android.os.Bundle;
9.   import android.app.Activity;
10.  import android.util.Log;
11.  import android.widget.Toast;
12.
13.
14.  public class MainActivity extends Activity {
15.     static final LatLng TutorialsPoint = new LatLng(21 , 57);
16.     private GoogleMap googleMap;
17.     private Log log;
18.     @Override
19.     protected void onCreate(Bundle savedInstanceState) {
20.        super.onCreate(savedInstanceState);
21.        setContentView(R.layout.activity_main);
22.        try {
23.           if (googleMap == null) {
24.              googleMap = ((MapFragment) getFragmentManager().
25.              findFragmentById(R.id.map)).getMap();
26.              log.i("tag","this is a test");
27.
28.           }
29.           googleMap.setMapType(GoogleMap.MAP_TYPE_HYBRID);
30.           Marker TP = googleMap.addMarker(new MarkerOptions().
31.           position(TutorialsPoint).title("TutorialsPoint"));
32.        } catch (Exception e) {
33.           e.printStackTrace();
34.        }
35.
```

```
36.    }
37.
38. }
```

（5）修改 AndroidManifest.xml 文件，在代码中添加 Google Map 的使用权限和申请的 Map API KEY 及访问网络的权限，代码如下。

```
1.  <?xml version="1.0" encoding="utf-8"?>
2.  <manifest xmlns:android="http://schemas.android.com/apk/res/android"
3.    package="com.bcpl"
4.    android:versionCode="1"
5.    android:versionName="1.0" >
6.
7.
8.    <uses-permission android:name="com.bcpl.googlemaps.permission.MAPS_RECEIVE" />
9.
10.   <uses-sdk
11.     android:minSdkVersion="12"
12.     android:targetSdkVersion="17" />
13.   <permission
14.     android:name="com.bcpl.googlemaps.permission.MAPS_RECEIVE"
15.     android:protectionLevel="signature" />
16.
17.
18.   <uses-permission android:name="android.permission.ACCESS_NETWORK_STATE" />
19.   <uses-permission android:name="android.permission.INTERNET" />
20.   <uses-permission android:name="com.google.android.providers.gsf.permission.READ_GSERVICES" />
21.   <uses-permission android:name="android.permission.WRITE_EXTERNAL_STORAGE" />
22.
23.   <uses-permission android:name="android.permission.ACCESS_COARSE_LOCATION" />
24.   <uses-permission android:name="android.permission.ACCESS_FINE_LOCATION" />
25.
26.   <uses-feature
27.     android:glEsVersion="0x00020000"
28.     android:required="true" />
29.
30.   <application
31.     android:allowBackup="true"
```

```
32.        android:icon="@drawable/ic_launcher"
33.        android:label="@string/app_name"
34.        android:theme="@style/AppTheme" >
35.        <activity
36.            android:name="com.bcpl.MainActivity"
37.            android:label="@string/app_name" >
38.            <intent-filter>
39.                <action android:name="android.intent.action.MAIN" />
40.
41.                <category android:name="android.intent.category.LAUNCHER"/>
42.            </intent-filter>
43.        </activity>
44.     <meta-data android:name="com.google.android.maps.v2.API_KEY"
45.                android:value="AIzaSyD3XehJDoyqFo_roNn0hM0vxZtrTAQU-7M"/>
46.    </application>
47.
48. </manifest>
```

习 题

一、简答题

1. 简述 LBS 中地理编码器（Grocoder）所使用的编码函数（前向和反向）分别如何应用。它的作用主要是什么？

2. LocationManager 类主要作用是什么？如何进行实例化？

3. LocationManager、LocationProvider、Criteria、Location 4 个类之间的应用关系是什么？

4. MapView 类和 MapFragment 有什么区别？

5. 在 Android 系统中，生成 MD5 和 SHA1 指纹有什么区别？

二、实训

要求：

在本章项目案例的基础上，要求完成基于 Google Map 的打车软件应用，用户输入起始位置和终点位置后，单击"查询"按钮，给用户显示地图中周围的车辆。

第 10 章

Android 物联网应用开发基础

学习目标

本章主要介绍了物联网的发展历史、物联网的体系框架（感知层、网络层和应用层）及应用协议、物联网的关键技术、当前物联网应用的主要操作系统、物联网的终端设备、Android 硬件传感器、物联网终端数据采集、数据传输、服务器端与客户端的通信模式及数据传输格式、物联网传感数据的图形应用等。使读者通过本章的学习，能够深入熟悉物联网系统通信模式、Android 与物联网结合数据应用开发等，掌握以下知识要点：

（1）物联网的体系框架构成及分层模式。
（2）物联网的关键技术、物联网操作系统、物联网操作系统与移动互联网的关系。
（3）物联网设备的分类、终端与网关。
（4）Android 硬件传感器的类别及应用。
（5）物联网终端数据采集的方式、数据传输的格式及应用协议。
（6）物联网传感数据的移动终端图形展示及绘制。

10.1 物联网概述

10.1.1 物联网简介

物联网（Internet of Things）的概念最早是由 MIT Auto-ID Center 在 1999 年提出的，它最初的定义为：把所有物品通过射频识别等信息传感设备与互联网连接起来，实现智能化管理。通过它可以获取无处不在的信息，实现物与物、物与人之间的信息交互、智能化识别和管理，实现信息基础设施与物理基础设施的全面融合，并最终形成统一的智能基础设施。

近年来，物联网在互联网的浪潮中，经历了提出、发展、快速发展的过程。2003 年，UID Center（Ubiquitous ID Center，泛在识别中心）在日本成立，负责研究和推广自动识别的核心技术，即在所有物品上植入微型芯片，组建网络进行通信。2004 年，日本提出 u-Japan 构想，希望到 2010 年将日本建设成一个"Anytime, Anywhere, Anything, Anyone"

都可以上网的环境。同年，韩国政府制定了 u-Korea 战略，韩国信通部发布《数字时代的人本主义：IT839 战略》以具体呼应 u-Korea。

2005 年，突尼斯信息社会世界峰会上国际电信联盟指出，物联网时代即将来临，智能嵌入式技术将得到广泛应用。

2008 年，IBM 提出"智慧地球"概念，以"互联网+物联网=智慧地球"为其发展战略。同年 10 月，欧洲物联网大会在法国召开，就 EPCglobal 网络结构进行了交流并就建立公平、分布式管理的唯一标识符达成了共识。

2009 年 6 月，欧盟执委会发表了题为"Internet of Things – An action plan for Europe"的物联网行动方案，描绘了物联网技术应用的前景。

2013 年欧盟通过了"地平线 2020"科研计划，就物联网领域的传感器、架构、标识、安全和隐私、语义互操作等方面研究进行重点支持。

在国内，物联网的发展大致可以分为以下两个阶段。

起步发展阶段（2003—2008 年）：2003 年 12 月，国家标准化管理委员会与科技部在北京召开了"物流信息新技术——物联网及产品电子代码（EPC）研讨会暨第一次物流信息新技术联席会议"。2004 年，EPCglobal China（全球产品电子代码中国）成立。2005 年，上海电子标签与物联网产学研联盟成立。2008 年，中国科学院无锡微纳米传感网工程技术研发中心成立，成为物联网研究的核心单位。

快速发展阶段（2009 年之后）：2009 年 8 月，温家宝总理在无锡考察传感网产业发展时明确指示要早一点谋划未来，早一点攻破核心技术，并且明确要求尽快建立中国的传感信息中心，或者叫"感知中国"中心。同年 9 月，工业信息部传感器网络标准化工作小组成立，标志着我国将加快制定符合我国发展需求的传感网技术标准。2010 年，上海物联网中心成立，致力引领中国物联网技术标准的制定和研究。同年，中国物联网标准联合工作组成立，2013 年 2 月，国务院发布《关于推进物联网有序健康发展的指导意见》（国发〔2013〕7 号），针对物联网发展面临的突出问题，以及长远发展的需要，从全局性和顶层设计的角度进行了系统考虑，确立了发展目标，明确了下一阶段的发展思路。物联网在电力、医疗、交通等各个领域得到了广泛深入的应用。

物联网早期应用的方面有 3M 公司在英国图书馆的 RFID 自助结算系统、温哥华酒店的 RFID 追踪洗涤物品等，其主要关注于三个方面：全面改制、可靠传输、智能处理。目前，全球物联网应用仍处于发展初期，物联网在行业领域的应用逐步广泛深入，在公共市场的应用已经得到推广，M2M（机器与机器通信）、车联网、智能电网是近两年全球发展较快的重点应用领域。

10.1.2 物联网体系框架及应用协议

2009 年，欧洲物联网研究项目工作组（CERP-IoT）制定的《物联网战略研究路线图》将物联网研究分为 10 个层面，分别为：①感知，ID 发布机制与识别；②物联网宏观架构；③通信（OSI 物理与链路层）；④组网（OSI 网络层）；⑤软件平台、中间件；⑥硬件；⑦情报提炼；⑧搜索引擎；⑨能源管理；⑩安全。

物联网体系结构分为三个层次，分别为感知层、网络层和应用层。其框架如图 10-1 所示。

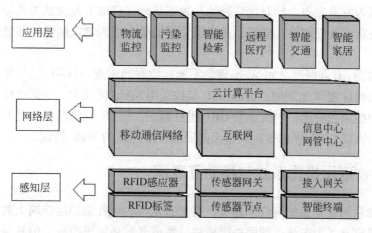

图 10-1　物联网体系框架

感知层，通常由 WSN、传感器和控制器组成，用于数据采集及最终控制，即利用 RFID、传感器、二维码等随时随地获取物体的信息；网络层，网络层为原有的互联网、电信网或者电视网，主要完成信息的远距离传输等功能；应用层，把感知层得到的信息进行处理，实现智能化识别、定位、跟踪、监控和管理等实际应用，主要完成服务发现和服务呈现的工作。

另外，也有把物联网的体系框架泛化为 5 层，分别依次为感知层、接入层、网络层、支撑层、应用层。其中，接入层主要完成各类设备的网络接入，该层重点强调各类接入方式，比如 3G/4G、Mesh 网络、Wi-Fi、有线或者卫星等方式；支撑层又称中间件或者业务层，主要完成信息的表达与处理，最终达到语义互操作和信息共享的目的；对上提供统一的接口与虚拟化支撑，虚拟化包括计算虚拟化和存储虚拟化等内容，较为典型的技术是上述的云计算平台。

由于物联网是互联网的延伸，其网络层的通信同样也基于 TCP/IP，但其接入层协议比较多，主要有内网协议（RFID、ZigBee、蓝牙）、外网协议（Wi-Fi、2G、3G、LTRE）、IPv4 及 IPv6。关于协议技术的详细介绍已经超出本书的讨论范围，感兴趣的读者可参考相关书籍。

10.1.3　物联网关键技术

物联网关键技术涉及许多方面，从广义的方面来看，其包含了物联网体系接结构设计（基于表征状态转移风格（RESTful）的体系结构）、短距离通信技术（1GHz 以下频段，802.11ah；802.15.4q 等）、无线传感网 IP 技术、物联网语义技术（从传感网本体定义向网络/服务/资源本体延伸）、通信技术（蓝牙、Wi-Fi 等）以及安全和隐私技术、标准、软件服务与算法、硬件等。

从网络层面，3GPP 正在研究机器类通信（Machine Type Communications，MTC）

和智能终端对现有网络架构的影响。

平台层面，通过 RESTful、MQTT、Socket 等协议向第三方开发者开放数据读取或控制能力成为发展重点，针对物联网的开发工具包（SDK）及调试工具等逐渐出现，与应用程序商店、社交网络、微博、搜索引擎等的融合成为物联网感知信息分发共享的未来方向。

架构层面，互联网理念和 Web 理念不断向物联网渗透，ITU-T 已经发布了基于 Web 的物联网架构标准即 Y.2063；oneM2M 架构采用 Web 化理念，一切可访问的数据、对象、实体均抽象为资源，由统一的 URI 进行标识；IETF 制定的资源受限物体应用层协议（CoAP 协议），即是 REST 风格面向资源受限网络的 Web 协议。

10.1.4 物联网操作系统与移动互联网

关于物联网操作系统有许多定义，有把它定义为运行感知层终端上的最重要的系统软件，也有定义其为内核、辅助外围模块、集成开发环境等组成。但其最基本的实现思想是：实现了应用软件和硬件的分离；硬件驱动程序与操作系统内核的分离。目前，物联网操作系统并没有统一版本，其主要的有 Android、HelloX（国内）、mbed OS（ARM）、Windows Compact、Windows Embedded OS 等，HelloX 和 mbed OS 提出的物联网操作系统架构如图 10-2 和图 10-3 所示，其他的操作系统架构基本类似，但由于 Android 在移动互联网方面的优势，而且它是完全免费开放的操作系统（Android NDK，Android Native 开放包供开发者使用 C 语言或者 C++语言开发），围绕它已经有许多应用开发，它实际上已经逐渐成为物联网标准操作系统。

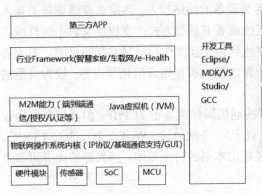

图 10-2　HelloX 物联网操作系统架构　　图 10-3　mbed OS 物联网操作系统架构

上述 HelloX 物联网操作系统架构与 Android 移动操作系统的相同之处如下。

（1）通过定义标准的硬件驱动程序接口，实现操作系统内核与硬件驱动程序的分离。

（2）通过引入 Java 虚拟机，并定义基于 C 语言的标准 API 接口，实现应用程序与硬件的分离。

不同之处如下。

（1）物联网操作系统的整体映像尺寸，必须是能够高度伸缩的，以适应硬件资源受限的应用场景。既能够适应手环等硬件资源相对丰富的应用场景，也能够适应环境监测器等不是非常智能的应用。

（2）除提供 Java 接口外，物联网操作系统还应该提供标准的 C 语言接口，以应对高效率、高实时性的应用。如物联网终端的生产厂商可以使用 C 接口，开发针对该硬件的高效应用，第三方企业则可以使用 Java 接口，开发可广泛移植到同类设备上的应用。

（3）物联网操作系统的驱动程序框架，应该设计得足够灵活和有足够兼容性，并能够动态加载和卸载设备驱动程序。

（4）物联网操作系统的版本分支或者变种数量会非常大。针对每个行业，甚至每种硬件（如汽车、冰箱等），都会有一个对应的版本，这需要编译开发工具进行良好的支持。

事实上，由于 Android 系统开源，目前结合物联网的 Android 操作系统已经成形并应用。

物联网应用过程中涉及物联网终端、操作系统、远程服务器，其应用模式主要有以下几种。

（1）物联网终端、操作系统、应用 APP 交互模式：物联网终端（汽车、冰箱、门锁、追踪卡等）上运行物联网操作系统，应用（APP）运行在物联网操作系统上。

（2）物联网终端、应用 APP、智能手机交互模式：智能手机通过本地连接通道（蓝牙、Wi-Fi、ZigBee 等）连接到物联网终端，控制终端上的 APP 的安装和卸载，以及 M2M 终端的相关配置（安全信息等）。

（3）APP 应用程序运行于物联网终端，是基于 Client/Server 模式（如智能手机的微信 APP），则物联网终端需要与 APP 的"应用程序后台"进行交互，实现业务逻辑。

（4）物联网终端与"终端管理后台"（通常物联网终端制造厂商建立并维护）建立持久的通信连接，用于实时更新物联网操作系统内核版本、实时更新物联网终端的硬件驱动程序等。

（5）物联网终端与终端直接通信，物联网终端之间通过本地通信连接通道（蓝牙、Wi-Fi、ZigBee 等）直接进行通信，无须借助后台。如汽车到达路口后，可以与信号灯通信，向信号灯注册，信号灯能够掌握各个方向的排队汽车数量，然后根据数量来决定信号的变换，达到优化交通的目的。

物联网终端之间的直接通信（端端通信）是物联网关键能力之一，也是物联网区别于移动互联网的关键。

10.1.5　物联网未来发展

随着物联网在行业领域的应用逐步广泛深入，M2M（机器与机器通信）、车联网、智能电网成为其近两年全球发展较快的重点应用领域。目前，M2M 已经形成完整产业链和内在驱动力的应用，全球已有四百多家移动运营商提供 M2M 服务，应用在安防、汽车、工业检测、自动化、医疗和智慧能源管理等领域。物联网与移动互联网在终端、网络、平台及架构各个层面融合发展。基于开源硬件和开放平台的物联网设备开发新模式，已经得到广泛的应用。

借鉴互联网的开放理念设计具有可扩展性、泛接入性、服务保证性、松耦合特性、自主性、泛在共享协同 6 个方面基本特性的物联网架构是未来研究的重点，也是物联网发展的关键。

当前，移动互联网正进入高速普及期，成功的产品和服务模式不断向其他产业领域延伸渗透，而处于起步阶段的物联网，也开始融入移动互联网元素，移动互联网与物联网的结合成为物联网发展最有市场潜力和创新空间的方向，涵盖移动智能终端集成传感器和新型人机交互等技术支撑融合类应用、基于移动智能终端的融合应用等方面。这也是本书引入基于 Android 物联网开发的最重要的原因。

物联网产生的海量数据以及对采集到的海量信息进行大规模甄别和筛选、数据存储、数据挖掘、数据处理、决策分析等，将进一步推动物联网应用的规模化发展和产生新的价值空间。同时，智慧城市的建设也为物联网应用提供巨大的市场需求，也是当前和未来物联网推动城市治理、民生服务以及产业发展的强大驱动力。

10.2 物联网设备

10.2.1 物联网终端

物联网终端是物联网中连接感知层和传输网络层，实现数据采集以及数据传输的设备。它担负着数据采集、初步处理、加密、传输等多种功能。物联网终端通常由外围感知（传感）接口、中央处理模块和外部通信接口三部分组成，通过外围感知接口与传感设备连接，如 RFID 读卡器，红外感应器，环境传感器等，读取传感设备的数据并经中央处理模块处理后，按照网络协议，通过外部通信接口（GPRS 模块、以太网接口、Wi-Fi 等）发送到以太网中指定数据中心处理平台，进行数据处理和分发。

根据不同的划分标准，物联网终端设备划分不同。

（1）从行业应用上可以分为工业设备检测终端、设施农业检测终端、物流 RFID 识别终端、电力系统检测终端、安防视频监测终端等。

工业设备检测终端通常有采集位移传感器、位置传感器（GPS）、震动传感器、液位传感器、压力传感器、温度传感器、协调器、多普勒等，如图 10-4 所示。

设施农业检测终端通常有空气温湿度传感器、土壤温度传感器、土壤水分传感器、光照传感器、气体含量传感器等用来采集数据。

物流 RFID 识别终端分为固定式、车载式和手持式。固定式一般安装在仓库门口或其他货物通道，车载式安装在物流运输车中，手持式则由使用者手持使用。固定式一般只有识别功能，用于跟踪

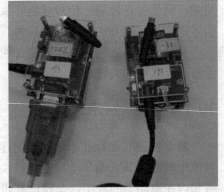

图 10-4　协调器、多普勒

货物的入库和出库，车载式和手持式中一般具有 GPS 定位功能和基本的 RFID 标签扫描功能，用来识别货物的状态、位置、性能等参数。

(2) 从使用场所可以分为固定终端、移动终端和手持终端。

(3) 从传输方式可以分为以太网终端、Wi-Fi 终端、2G 终端、3G 终端等。

(4) 从使用扩展性分为单一功能终端和通用智能终端两种，单一终端一般外部接口较少，设计简单，仅满足单一应用或单一应用的部分扩展。如汽车监控用的图像传输服务终端、电力监测用的终端、物流用的 RFID 终端等。通用智能终端外部接口较多，设计复杂，能满足两种或更多场合的应用。它可以通过内部软件的设置、修改应用参数，或通过硬件模块的拆卸来满足不同的应用需求。

(5) 从传输通路可以分为数据透传终端和非数据透传终端。数据透传终端在输入口与应用软件之间建立起数据传输通路，使数据可以通过模块的输入口输入，通过软件原封不动地输出，表现给外界的方式相当于一个透明的通道。非数据透传终端将外部多接口的采集数据通过终端内的处理器合并后传输，具有多路同时传输的优点，同时减少了终端数量。

10.2.2 物联网网关

目前，关于物联网网关设备还没有统一的定义及标准，通常是指将多种接入手段整合起来，统一互联到接入网络的关键设备，网关可满足局部区域短距离通信的接入需求，实现与公共网络的连接，同时完成转发、控制、信令交换和编解码等功能，而终端管理、安全认证等功能保证了物联网业务的质量和安全。

从广义的角度来看，物联网网关设备是感知接入层、数据处理层和传输应用层三层涉及的所有硬件设备和软件的总称，如图 10-5 所示。从狭义的角度来看，其仅仅是基于 ZigBee 技术的物联网数据采集模块或数据处理模块。

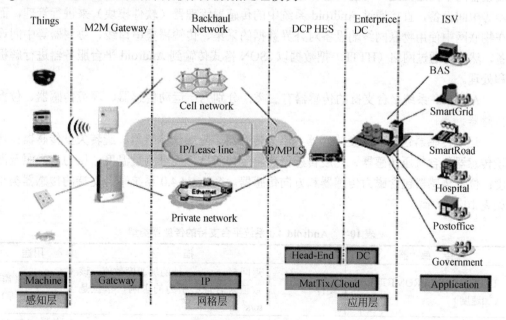

图 10-5 物联网网关

按照不同的标准可以将网关分成不同的类型。

按连接网络划分分为以下几种。

（1）局域网/主机网关。局域网/主机网关主要在大型计算机系统和个人计算机之间提供连接服务。

（2）局域网/局域网网关。这种类型的网关与局域网/主机网关很类似，不同的是这种网关主要用于连接多个使用不同通信协议或数据传输格式的局域网。目前大多数网关都是属于这类网关。

（3）因特网/局域网网关。这种网关主要用于局域网和因特网间的访问和连接控制。

按产品功能划分分为以下几种。

（1）数据网关。数据网关通常在多个使用不同协议及数据格式的网络间提供数据转换功能。

（2）应用网关。应用网关是在使用不同数据格式的环境中，进行数据翻译功能的专用系统。

（3）安全网关。安全网关是各种提供系统（或者网络）安全保障的硬件设备或软件的统称，它是各种技术的有机结合，保护范围从低层次的协议数据包到高层次的具体应用。

10.3　Android 硬件传感器

目前安装 Android 系统的设备大多已经内置了传感器，它们与上述讲解的物联网终端设备的不同之处是：内置传感器的设备对外界数据的采集、传输不需要串口线、协调器等中间设备，直接通过 Android 系统中的传感器管理器（软件模块）来进行管理，而在物联网中使用物联网终端设备对外界数据的采集、传输需要串口线、协调器等中间设备，然后再通过网络（HTTP）把数据以 JSON 格式传输到 Android 平台服务器进行解析和处理。

Android 系统平台支持的传感器有三类，分别为：运动传感器、环境传感器、位置传感器。

运动传感器包含加速度传感器、重力传感器、陀螺仪传感器、旋转矢量传感器；环境传感器包含气压传感器、光度传感器和温度传感器，用来测量温度、压力、光照及湿度；位置传感器包含磁力传感器和方向传感器。Android 4.0 系统平台支持的传感器类型如表 10-1 所示。

表 10-1　Android 4.0 系统平台支持的传感器类型

传　感　器	类型	描　　　述	用途
TYPE_ACCELEROMETER（加速度）	硬件	返回 x、y、z 三轴的加速度数值，该数值包含地心引力的影响，单位是 m/s^2	运动探测（震动、倾斜等）
TYPE_AMBIENT_TEMPERATURE（温度）	硬件	温度传感器返回当前周围的温度(°C)	监测空气的温度

续表

传 感 器	类型	描 述	用途
TYPE_GRAVITY（重力）	软件或硬件	返回 x、y、z 三轴的加速度数值单位是 m/s^2	运动探测（震动、倾斜等）
TYPE_GYROSCOPE（陀螺仪）	硬件	返回 x、y、z 三轴的角加速度数据。角加速度的单位是 radians/second	旋转探测（旋转、翻转等）
TYPE_LIGHT（光线）	硬件	光线感应传感器检测实时的光线强度，光强单位是 lx	控制屏幕的亮度
TYPE_LINEAR_ACCELERATION（线性）	软件或硬件	线性加速度传感器是加速度传感器减去重力影响获取的数据。返回 x、y、z 三轴的值	监控单轴的加速度
TYPE_MAGNETIC_FIELD（磁力）	硬件	返回 x、y、z 三轴的环境磁场数据	磁力数据由电子罗盘传感器提供
TYPE_ORIENTATION（方向）	软件	返回三轴的角度数据，方向数据的单位是角度	提供设备的位置
TYPE_PRESSURE（压力）	硬件	压力传感器返回当前的压强，单位是百帕斯卡（hPa）	监测空气压力变化
TYPE_PROXIMITY（近距离）	硬件	检测物体与手机的距离，单位是 cm，通常用于决定是否电话耳机	近距离传感器可用于接听电话时自动关闭 LCD 屏幕以节省电量
TYPE_RELATIVE_HUMIDITY（湿度）	硬件	监测周围湿度(%)	监测绝对、相对湿度
TYPE_ROTATION_VECTOR（旋转矢量）	软件或硬件	旋转矢量代表设备的方向，是一个将坐标轴和角度混合计算得到的数据	监测旋转
TYPE_TEMPERATURE（温度）	硬件	监测温度(℃)。这个传感器在 API Level 14 之后被 TYPE_AMBIENT_TEMPERATURE 传感器代替	监测温度

说明：表 10-1 中 TYPE_RELATIVE_HUMIDITY 和 TYPE_AMBIENT_TEMPERATURE 在 Android 2.3 系统平台中不支持。

在 Android 系统中访问和控制传感器，获取传感器的数据需要通过 Android 系统传感器框架，Android 传感器框架在 android.hardware 包下，包含的类和接口有：SensorManager 类、Sensor 类、SensorEvent 类、SensorEventListener 接口。

Android 内置传感器应用及配置步骤。

（1）通过 SensorManager 类的 getSystemService()方法获取设备上传感器服务器的引用。

```
private SensorManager mSensorManager;
mSensorManager = (SensorManager) getSystemService(Context.SENSOR_SERVICE);
```

（2）使用 getSensorList()方法，获取设备上所有的传感器。

```
List<Sensor> deviceSensors = mSensorManager.getSensorList(Sensor.TYPE_ALL);
```

或获取指定的重力传感器。

```
List<Sensor> deviceSensors = mSensorManager.getSensorList(Sensor.TYPE_GRAVITY);
```

（3）通过调用 getDefaultSensor()方法，判断设备上是否存在某一类型的传感器，如磁力传感器。

```
private SensorManager mSensorManager;
 mSensorManager = (SensorManager) getSystemService(Context.SENSOR_SERVICE);
if (mSensorManager.getDefaultSensor(Sensor.TYPE_MAGNETIC_FIELD) != null){
   //存在磁力传感器
   }
else {
   //  不存在磁力传感器
   }
```

注意：并非所有的传感器在每个设备上都可以用，所以在使用前需要检查传感器是否可用。

（4）通过 SensorEventListener 接口的 onAccuracyChanged()和 onSensorChanged()方法监听传感器事件。下面通过光度传感器来示例，首先是在 activity_main.xml 中创建 TextView 控件，然后把获取的光线传感器的数据在 TextView 控件中进行显示。

```
1.  public class SensorActivity extends Activity implements SensorEventListener {
2.    private SensorManager mSensorManager;
3.    private Sensor mLight;
4.    Private TextView tv;
5.    @Override
6.    public final void onCreate(Bundle savedInstanceState) {
7.      super.onCreate(savedInstanceState);
8.      setContentView(R.layout.activity_main);
9.      tv=(TextView)findViewById(R.id.tv);
10.     mSensorManager = (SensorManager) getSystemService(Context.
        SENSOR_SERVICE);
11.     mLight = mSensorManager.getDefaultSensor(Sensor.TYPE_LIGHT);
12.   }
13.
14.   @Override
15.   public final void onAccuracyChanged(Sensor sensor, int accuracy) {
16.     // 如果传感器精确度发生变化，在这里进行处理
17.   }
18.
```

```
19.    @Override
20.    public final void onSensorChanged(SensorEvent event) {
21.      // 光线传感器返回单个值
22.      // 许多传感器返回三个值，分别对应x,y,z轴
23.      float lux = event.values[0];
24.     tv.setText(" "+lux);
25.    }
26.
27.    @Override
28.    protected void onResume() {
29.      super.onResume();
30.      mSensorManager.registerListener(this, mLight, SensorManager.
          SENSOR_DELAY_NORMAL);
31.    }
32.
33.    @Override
34.    protected void onPause() {
35.      super.onPause();
36.      mSensorManager.unregisterListener(this);
37.    }
38. }
```

为了保证传感器应用的兼容性，在发布的 APP 应用中可以使用 getDefaultSensor() 方法来判断传感器存在与不存在的情况下的处理机制。此外，Android 还提供了在 AndroidManifest.xml 中使用标签<uses-feature>来过滤传感器的应用（在设备中不存在指定传感器的情况下）。如在 AndroidManifest.xml 中加速度传感器应用的配置：

```
<uses-feature    android:name="android.hardware.sensor.accelerometer"
android:required="true" />
```

注意：传感器的代码并不能在模拟器中应用，因模拟器中没有模拟器传感器，必须应用在真实的设备上。

10.4 物联网终端数据采集应用开发

基于物联网终端的数据采集应用可以分为两种模式。一种模式是物联网终端（采集数据）通过 ZigBee 协议发送到协调器，然后协调器通过串口线把数据传输到 PC（或服务器），最后运行在 PC（或服务器）上的基于 Android 系统平台的 Android 终端界面显示处理后的数据信息，如图 10-6 所示。

另外一种模式是物联网终端（采集数据）通过 ZigBee 协议发送到协调器，然后协调器通过串口线把数据传输到 Web 服务器，Web 服务器把数据存储到数据库，然后 Android 移动客户终端通过 HTTP 发送请求给服务器，服务器接到请求后进行验证。如果是合法

的身份的请求，则从数据库提取数据，并进行响应回复到移动客户终端，在此请求和响应过程中，数据的传输格式通常以 JSON 格式封装。这种模式也是通用的物联网软件设计架构，具体模式如图 10-7 所示。

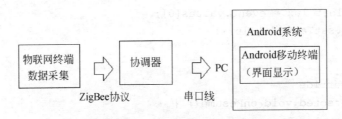

图 10-6　物联网采集终端与移动显示终端工作模式

图 10-7　基于物联网的软件设计架构

上述讲解了 Android 系统设备内置传感器的应用，下面就采用第一种应用模式的物联网终端设备采集外部环境的光照应用开发进行介绍。

物联网温湿度光感传感器（如图 10-8 所示）应用主要涉及服务器端、客户端（Android）、终端设备三部分，传感器终端设备（终端节点）通过无线（ZigBee 协议）将状态和数据发送到协调器（如图 10-9 所示），协调器通过串口线（如图 10-10 所示）将数据传输到 PC 端，然后 Android 客户端读取串口数据，其主要步骤如下。

图 10-8　光敏传感器

图 10-9　协调器

图 10-10　串口线

（1）创建串行接口类，实现对 COM 端口的操作。

```
1.  package android.serialport.sam;
2.  import java.io.File;
3.  import java.io.IOException;
4.  import java.security.InvalidParameterException;
5.  import android.content.SharedPreferences;
6.  import android.serialport.api.SerialPort;
7.  import android.serialport.api.SerialPortFinder;
8.  import android.util.Log;
9.  public class Application extends android.app.Application {
10.     public SerialPortFinder mSerialPortFinder = new SerialPortFinder();
11.     private SerialPort mSerialPort = null;
12.     private String path;
13.     private int baudrate;
14.
15.     public SerialPort getSerialPort(File device,int baud) throws
        SecurityException, IOException, InvalidParameterException {
16.
17.         closeSerialPort();
18.         mSerialPort = null;
19.
20.         if (mSerialPort == null) {
21.         /* Read serial port parameters */
22.             SharedPreferences sp = getSharedPreferences("android.
                zigbee_preferences", MODE_PRIVATE);
23.             path = device.getAbsolutePath();
24.             baudrate = baud;
25.         /* Check parameters */
26.             if ( (path.length() == 0) || (baudrate == -1)) {
27.                 throw new InvalidParameterException();
28.             }
29.
30.         /* Open the serial port */
31.             mSerialPort = new SerialPort(new File(path), baudrate,
0);
32.             Log.v("debug", device.getAbsolutePath());
33.         }
34.         return mSerialPort;
35.     }
36.
37.     public String getPathString(){
38.         return path;
39.     }
```

```
40.
41.    public int getbaudrate(){
42.        return baudrate;
43.    }
44.
45.    public void closeSerialPort() {
46.        if (mSerialPort != null) {
47.            Log.v("debug", "closeSerialPort");
48.            mSerialPort.close();
49.            mSerialPort = null;
50.        }
51.    }
52. }
```

（2）创建抽象类实现初始化串口，并读取数据，代码如下。

```
1.  package android.serialport.sam;
2.
3.  import java.io.File;
4.  import java.io.IOException;
5.  import java.io.InputStream;
6.  import java.io.OutputStream;
7.  import java.security.InvalidParameterException;
8.  import com.bcpl.lightSensor.R;
9.  import android.app.Activity;
10. import android.app.AlertDialog;
11. import android.content.DialogInterface;
12. import android.content.DialogInterface.OnClickListener;
13. import android.os.Bundle;
14. import android.serialport.api.SerialPort;
15. import android.util.Log;
16. import android.view.MotionEvent;
17.
18. public abstract class SerialPortActivity extends Activity {
19.
20.     protected Application mApplication;
21.     protected SerialPort mSerialPort;
22.     protected OutputStream mOutputStream;
23.     private InputStream mInputStream;
24.     private ReadThread mReadThread;
25.     private class ReadThread extends Thread {
26.
27.         @Override
28.         public void run() {
29.             super.run();
```

```
30.            while(!isInterrupted()) {
31.                int size;
32.                try {
33.                    byte[] buffer = new byte[34];
34.                    if (mInputStream == null) return;
35.                    int readCount = 0;
36.                    readCount += mInputStream.read(buffer, readCount,
                           34 - readCount);
37.                                onDataReceived(buffer, readCount);
38.                } catch (IOException e) {
39.                    e.printStackTrace();
40.                    return;
41.                }
42.            }
43.        }
44.    }
45.
46.    private void DisplayError(int resourceId) {
47.        AlertDialog.Builder b = new AlertDialog.Builder(this);
48.        b.setTitle("Error");
49.        b.setMessage(resourceId);
50.        b.setPositiveButton("OK", new OnClickListener() {
51.            public void onClick(DialogInterface dialog, int which) {
52.                SerialPortActivity.this.finish();
53.            }
54.        });
55.        b.show();
56.    }
57.
58.    @Override
59.    protected void onCreate(Bundle savedInstanceState) {
60.        super.onCreate(savedInstanceState);
61.    }
62.
63.    public void initSerial(File device,int baudate){
64.        mApplication = (Application) getApplication();
65.        try {
66.            mSerialPort = mApplication.getSerialPort(device.
                   getAbsoluteFile(),baudate);
67.            mOutputStream = mSerialPort.getOutputStream();
68.            mInputStream = mSerialPort.getInputStream();
69.
70.            /* Create a receiving thread */
71.            mReadThread = new ReadThread();
```

```
72.            mReadThread.start();
73.        } catch (SecurityException e) {
74.            DisplayError(R.string.error_security);
75.        } catch (IOException e) {
76.            DisplayError(R.string.error_unknown);
77.        } catch (InvalidParameterException e) {
78.            DisplayError(R.string.error_configuration);
79.        }
80.    }
81.
82.    public String getDevPath(){
83.        return mApplication.getPathString();
84.    }
85.
86.    public int getBaudate(){
87.        return mApplication.getbaudrate();
88.    }
89.
90.    public void closeRead(){
91.        if (mReadThread != null) {
92.            mReadThread.interrupt();
93.        }
94.    }
95.    protected abstract void onDataReceived(final byte[] buffer, final int size);
96.    @Override
97.    protected void onDestroy() {
98.        if (mReadThread != null)
99.            mReadThread.interrupt();
100.       try {
101.           mInputStream.close();
102.       } catch (IOException e) {
103.           // TODO Auto-generated catch block
104.           e.printStackTrace();
105.       }
106.       mApplication.closeSerialPort();
107.       mSerialPort = null;
108.       super.onDestroy();
109.    }
110. }
```

（3）在 res 目录下 layout 文件夹中修改 activity_main.xml 文件，在线性布局 LinearLayout 中添加线性布局 LinearLayout，并添加 TextView 控件，具体如下。

```
1.  <?xml version="1.0" encoding="utf-8"?>
2.  <LinearLayout xmlns:android="http://schemas.android.com/apk/res/android"
3.      android:id="@+id/btnInfoLayout"
4.      android:layout_width="fill_parent"
5.      android:layout_height="wrap_content"
6.      android:orientation="horizontal" >
7.      <LinearLayout
8.          android:layout_width="fill_parent"
9.          android:layout_height="wrap_content"
10.         android:orientation="vertical" >
11.         <TextView
12.             android:id="@+id/lightTitle"
13.             android:layout_width="fill_parent"
14.             android:layout_height="wrap_content"
15.             android:layout_margin="3dip"
16.             android:background="#bb000000"
17.             android:gravity="center"
18.             android:textColor="#ffffffff"
19.             android:textSize="15pt" />
20.         <LinearLayout
21.             android:layout_width="fill_parent"
22.             android:layout_height="wrap_content"
23.             android:orientation="horizontal" >
24.             <TextView
25.                 android:id="@+id/info"
26.                 android:layout_width="fill_parent"
27.                 android:layout_height="wrap_content"
28.                 android:layout_margin="3dip"
29.                 android:background="#ffffff"
30.                 android:gravity="left"
31.                 android:textColor="#000000"
32.                 android:textSize="15pt"></TextView>
33.         </LinearLayout>
34.     </LinearLayout>
```

（4）创建 MainActivity 类，读取串口数据，并在界面 TextView 中显示。

```
1.  package com.bcpl.lightSensor;
2.
3.  import java.io.File;
4.  import java.text.SimpleDateFormat;
5.  import java.util.Date;
6.  import java.util.Timer;
7.  import java.util.TimerTask;
8.  import com.zigbee.pro.receDataPro;
```

```java
9.  import android.content.Context;
10. import android.os.Bundle;
11. import android.os.Handler;
12. import android.os.Message;
13. import android.serialport.sam.SerialPortActivity;
14. import android.view.KeyEvent;
15. import android.view.WindowManager;
16. import android.widget.TextView;
17.
18. public class MainActivity extends SerialPortActivity {
19.     private static int loc = 0;
20.     private static byte[] RcvBuf;
21.     public static Context mContext = null;
22.     // ZigBee 数据命令字
23.     final static int FRAME_START_UP =0xA5;
24.     final static int COMMAND_HANDSHAKE = 0xAA;
25.     final static int COMMAND_DATABRIDGE = 0xDA;
26.     final static int COMMAND_DEVICE_REPORT = 0xFD;
27.     final static int COMMAND_SUBDEVICE_INFO = 0xF4;
28.     final static int COMMAND_HANDSHAKE_ANSWER=0xAD;
29.     private byte[] actAddr;
30.
31.     private String sensorString = "lightSensorDemo";
32.     private String sAddrString = "null";
33.     private String sensorDataString = "null";
34.     private final String comDev= "com9";
35.     private final int comDaud = 115200;
36.
37.     private TextView lightTitle;
38.     private TextView info;
39.     private receDataPro dataPro;
40.
41.     @Override
42.     protected void onCreate(Bundle savedInstanceState) {
43.         super.onCreate(savedInstanceState);
44.         setContentView(R.layout.activity_main.xml);
45.         //打开串口"com9", Windows 环境下
46.         initSerial(new File(comDev), comDaud);
47.         //设定 activity 在屏幕中的位置
48.         WindowManager.LayoutParams params = getWindow().getAttributes();
49.         params.height = 240;
50.         params.width = 400;
51.         getWindow().setAttributes(params);
52.
```

```java
53.         RcvBuf = new byte[1024];
54.     //RcvBuf存储从串口中收到的数据，作为大的缓冲区存在
55.         actAddr = new byte[2];           //存储当前激活终端的地址
56.
57.         dataPro = new receDataPro();
58.     //启动后，协调器发送握手指令
59.         Message message = new Message();
60.         message.what = 1;
61.         handler.sendMessage(message);
62.
63.         lightTitle = (TextView) findViewById(R.id.lightTitle);
64.         info = (TextView) findViewById(R.id.info);
65.         lightTitle.setText("终端信息");
66.         showLightSensorInfo();
67.     }
68.
69.     //消息处理
70.     final Handler handler = new Handler() {
71.         public void handleMessage(Message msg) {
72.             switch (msg.what) {
73.             case 1: //发送握手指令
74.                 dataPro.sendHandShake(mOutputStream);
75.                 break;
76.             case 2://获取终端设备的数据
77.                 dataPro.RequestDeviceData(actAddr, mOutputStream);
78.                 break;
79.             default:
80.                 break;
81.             }
82.         }
83.     };
84.
85.     //接收串口数据
86.     @Override
87.     protected void onDataReceived(final byte[] buffer, final int size) {
88.         // TODO Auto-generated method stub
89.         runOnUiThread(new Runnable() {
90.             public void run() {
91.                 // 收到数据后，利用串口状态机，读取串口数据
92.                 if (size > 0) {
93.                     if (loc + size > 1024) {
94.                         loc = 0;
95.                     }
96.                     System.arraycopy(buffer, 0, RcvBuf, loc, size);
```

```java
97.                 loc += size;
98.                 controllerRecieveData(RcvBuf, loc);
99.             }
100.         }
101.     });
102. }
103.
104. //利用串口数据状态机,循环处理接收到的数据
105. private void controllerRecieveData(byte[] buffer, int size) {
106.     byte[] Buf = new byte[2048];
107.     System.arraycopy(buffer, 0, Buf, 0, size);
108.     int machineState = 0; // 状态机标示位
109.     int cmdBegin = 0;
110.     int datalen = size;
111.     for (int i = 0; i < datalen;) {
112.         if (machineState == 0) {
113.             if ((Buf[i] & 0xFF) != FRAME_START_UP) {
                                                         //协议开头标志位
114.                 i++;
115.                 continue;
116.             }
117.             cmdBegin = i;
118.             machineState = 1;
119.             i++;
120.             continue;
121.         } else if (machineState == 1) {
122.             if ((Buf[i] & 0xFF) == COMMAND_HANDSHAKE_ANSWER) {
123. //握手回答指令
124.                 machineState = 0;
125.                 i++;
126.             }
127.             if ((Buf[i] & 0xFF) == COMMAND_HANDSHAKE) {
128.                 //收到终端设备发上来的握手指令,协调器回复握手指令
129.                 Message message = new Message();
130.                 message.what = 1;
131.                 handler.sendMessage(message);
132.                 // 握手成功
133.                 machineState = 0;
134.                 i++;
135.             } else if ((Buf[i] & 0xFF) == COMMAND_DATABRIDGE) {
136.                 //收到终端设备的信息,包含设备信息和数据信息两种
137.                 i++;
138.                 int dataleft = datalen - i;
139.                 int cmdLen;
```

```
140.                    if (dataleft < 3) {
141.                        break;
142.                    }
143.                    cmdLen = 3 + (Buf[i + 2] & 0xFF);
144.                    if (dataleft < cmdLen) {
145.                        break; // 数据不完整，退出
146.                    }
147.                    //数据完整，取指令数据
148.                    byte[] tempBuf = new byte[cmdLen + 1];
149.                    System.arraycopy(Buf, cmdBegin, tempBuf, 0, cmdLen + 1);
150.
151.                    //处理具体的数据
152.                    if (DeviceDataReceived(tempBuf) == 0)
153.                        break;
154.
155.                    i += cmdLen;
156.                    machineState = 0;
157.                } else {
158.                    i++;
159.                }
160.            } else {
161.                i++;
162.            }
163.        }
164.        if (machineState == 1) {//如果没有收到完整数据,则继续等待
165.            System.arraycopy(RcvBuf, cmdBegin, RcvBuf, 0, datalen - cmdBegin);
166.            loc = datalen - cmdBegin;
167.        } else {
168.            loc = 0;
169.        }
170.    }
171.
172.    //解析具体的指令,数据
173.    public int DeviceDataReceived(byte[] data) {
174.        byte[] sAddr = new byte[2];
175.        byte[] nSub = new byte[1];
176.        int start = 5;
177.        if (data.length <= 5)
178.            return 0;
179.        if ((data[start] & 0xFF) == COMMAND_DEVICE_REPORT) {
180.            //设备报告指令,报告地址
181.            sAddr = dataPro.getRepAddr(data, start);
```

```
182.            //获取到设备地址
183.                dataPro.reportSubInfo(sAddr, mOutputStream);
184.            //发送报告指令的回复指令
185.                nSub[0] = 0x00;
186.                dataPro.RequestDeviceInfo(sAddr, nSub, mOutputStream);
187.            } else if ((data[start] & 0xFF) == COMMAND_SUBDEVICE_INFO){
188.            //设备信息报告指令,报告地址,终端设备名
189.            String sensorName = dataPro.getRepSubInfo(data, start);
190.            //判断收到终端设备名称是否为当前需要的
191.                if (sensorName.equals(sensorString)) {
192.                    System.arraycopy(data, 2, actAddr, 0, 2);
193.            //保存当前终端的地址到激活终端的地址
194.                    sAddrString = receDataPro.Bytes2HexString(actAddr);
195.                    TimerTask task = new TimerTask() {
196.                        @Override
197.                        public void run() {
198.                            // TODO Auto-generated method stub
199.                            Message message = new Message();
200.                            message.what = 2;
201.                            handler.sendMessage(message);
202.                        }
203.                    };
204.
205.                    Timer timer = new Timer();
206.                    timer.schedule(task, 1000, 1000);
207.            }
208.            } else if ((data[1] & 0xFF) == COMMAND_DATABRIDGE) {
209.            sensorDataString = dataPro.getRepSubData(data, start);
210.                showLightSensorInfo();
211.            }
212.            return 1;
213.        }
214.
215.        //显示节点的信息与数据
216.        void showLightSensorInfo() {
217.            //获取当前接收到数据的时间
218.            SimpleDateFormat formatter = new SimpleDateFormat("HH:mm:ss");
219.            Date curDate = new Date(System.currentTimeMillis());
              // 获取当前时间
220.            String timeString = formatter.format(curDate);
221.            //在Android界面上显示终端设备数据
222.            String devInfo = "终端设备名称:" + sensorString + "\n" + "
                地址:" + sAddrString
223.                    + "\n" + "数   据:" + sensorDataString + "\n" + "
```

```
                    采集时间:"
224.                    + timeString;
225.            info.setText(devInfo);
226.        }
227.    }
```

10.5 物联网传感数据图形应用

上述介绍了物联网数据采集软件应用设计的第一种模式，下面就第二种模式进行介绍，因第一种模式已经详细介绍了物联网采集终端和服务器之间的数据通信过程，本部分只就 Android 移动客户端与服务器之间的通信进行介绍，并就通过 HTTP 获取的采集数据处理及如何以图形的方式进行展示进行详述，考虑到阅读本书的读者和学生，并不一定有物联网终端采集硬件设备，本应用单独创建了采集数据服务类（提供随机的采集数据模拟真实采集数据），同时，也提供了真实的服务器与协调器约定的通信方式（实践中，一般公司都有自己单独约定方式，并没有统一的方式）。

（1）创建传感数据图形工程，修改 res 目录中 layout 文件夹下的 activity_main.xml 文件，添加线性布局并进行嵌套，在线性布局中添加 TextView 和自己定义的画布控件类。此处由于篇幅有限就不再写其他控件在布局文件中的定义代码，只写出自定义画布控件类引用的代码如下。

```
1.    <LinearLayout
2.        android:layout_width="match_parent"
3.        android:layout_height="match_parent"
4.        android:orientation="vertical" >
5.        <com.sean.pm25app.PmView
6.            android:id="@+id/pmView"
7.            android:layout_width="match_parent"
8.            android:layout_height="match_parent"/>
9.    </LinearLayout>
```

（2）创建采集数据服务类 PM25Service.java（提供 PM2.5 数据加随机数），把自动生成的数据写入 SQLite 数据库，并把数据广播发送出去。代码如下。

```
1.    public class PM25Service extends Service {
2.
3.        NetManager netManager;
4.        SubThread subThread;
5.        IotDAO iotDao;
6.        @Override
7.        public void onCreate() {
8.            // TODO Auto-generated method stub
9.            super.onCreate();
```

```
10.        Log.d(DataConst.TAG, "service onCreate ...:");
11.        netManager = new NetManager(this);
12.        iotDao = new IotDAO(this);
13.        subThread = new SubThread();
14.        subThread.start();
15.    }
16.
17.    @Override
18.    @Deprecated
19.    public void onStart(Intent intent, int startId) {
20.        // TODO Auto-generated method stub
21.        super.onStart(intent, startId);
22.    }
23.
24.    @Override
25.    public void onDestroy() {
26.        // TODO Auto-generated method stub
27.        subThread.isRun = false;
28.        subThread.interrupt();
29.        super.onDestroy();
30.
31.    }
32.
33.    @Override
34.    public IBinder onBind(Intent arg0) {
35.        // TODO Auto-generated method stub
36.        return null;
37.    }
38.    class SubThread extends Thread{
39.        boolean isRun = true;
40.        public void run(){
41.            Log.d(DataConst.TAG, "subThread start...:");
42.            while(isRun){
43.                int newValuePM25 = netManager.connServerGetPM25();
44.                newValuePM25 =(int)(Math.random()*100)+newValuePM25;
45.                // 写数据库
46.                iotDao.insert(newValuePM25, DataConst.IOT_TYPE_PM);
47.                // 发送广播
48.                Intent intent = new Intent(DataConst.ACT_PM25);
49.                intent.putExtra("value", newValuePM25);
50.                sendBroadcast(intent);
51.                Log.d(DataConst.TAG, "subThread pm2.5 :"+newValuePM25);
                  int newValueCO2 = netManager.connServerGetCO2();
52.            newValueCO2 =(int)(Math.random()*100)+newValueCO2;
```

```
53.                    // 写数据库
54.                    iotDao.insert(newValueCO2, DataConst.IOT_TYPE_CO2);
55.                    // 发送广播
56.                    Intent intent2 = new Intent(DataConst.ACT_PM25);
57.                    intent2.putExtra("value", newValueCO2);
58.                    sendBroadcast(intent2);
59.                    Log.d(DataConst.TAG, "subThread pm2.5 :"+newValueCO2);
60. //
61.                    try {
62.                        Thread.sleep(1000*5);
63.                    } catch (InterruptedException e) {
64.                        // TODO Auto-generated catch block
65.                        e.printStackTrace();
66.                    }
67.                }
68.                Log.d(DataConst.TAG, "subThread end...:");
69.            }
70.        }
71. }
```

（3）创建画图类，分别设置了画布的底色、画线的颜色，并自己计算画布中的坐标，在数据变化的点画圆圈。代码如下。

```
1.
2.  public class PmView extends View {
3.      // 坐标集合
4.      ArrayList<Integer> pts = new ArrayList<Integer>(0);
5.      public ArrayList<Integer> getPts() {
6.          return pts;
7.      }
8.      public void setPts(ArrayList<Integer> pts) {
9.          this.pts = pts;
10.     }
11.     public PmView(Context context) {
12.         super(context);
13.
14.         // TODO Auto-generated constructor stub
15.     }
16.     public PmView(Context context, AttributeSet attrs, int defStyleAttr) {
17.         super(context, attrs, defStyleAttr);
18.         // TODO Auto-generated constructor stub
19.     }
20.
21.     public PmView(Context context, AttributeSet attrs) {
```

```
22.        super(context, attrs);
23.        // TODO Auto-generated constructor stub
24.    }
25.
26.    @Override
27.    protected void onDraw(Canvas canvas) {
28.        // TODO Auto-generated method stub
29.
30.        super.onDraw(canvas);
31.        canvas.drawColor(Color.WHITE);
32.        Paint paint = new Paint();
33.        paint.setAntiAlias(true);
34.        paint.setColor(0x8800ffff);
35.        paint.setStrokeWidth(2);
36.        // 画线
37.        float[] fPtx = new float[pts.size()*2];
38.
39.        int x = 20;
40.        Log.d(DataConst.TAG,"pts:"+pts.toString() );
41.        int j = 0;
42.        for (int i = 0; i < pts.size(); i++) {
43.            //x 坐标,自己计算
44.            fPtx[j] = x;
45.            fPtx[j+1] = pts.get(i);
46.            x=x+60;
47.            j=j+2;
48.        }
49.
50.        Log.d(DataConst.TAG,"fPts x"+Arrays.toString(fPtx) );
51.        Bitmap pBitmap =((BitmapDrawable)getResources().getDrawable
               (R.drawable.pm2_point_)).getBitmap();
52.        // 画圆
53.        for (int i = 0; i < fPtx.length; i+=2) {
54.            //canvas.drawCircle(fPtx[i],fPtx[i+1] ,5, paint);
55.            canvas.drawBitmap(pBitmap, fPtx[i], fPtx[i+1], paint);
56.        }
57.        // 画折线
58.        if(fPtx.length>4){
59.            //canvas.drawLines(fPtx, 0, fPtx.length, paint);
60.            //canvas.drawLines(fPtx, 2, fPtx.length-2, paint);
61.            Path linePath = new Path();
62.            paint.setStyle(Paint.Style.STROKE);
63.            linePath.moveTo(fPtx[0],fPtx[1]);
64.            for (int i = 2; i < fPtx.length; i+=2) {
```

```
65.                linePath.lineTo(fPtx[i],fPtx[i+1]);
66.            }
67.            //linePath.close();
68.            canvas.drawPath(linePath, paint);
69.        }
70.        //canvas.drawBitmap(bitmap, 10, 220, paint);
71.        //canvas.drawText("", 10, 180, paint);
72.
73.    }
74. }
```

(4) 创建 MainActivity，启动数据采集服务类，接收广播的数据，调用画布类，在界面上进行显示，代码如下。

```
1.  public class MainActivity extends Activity {
2.
3.      String value="";
4.      private TextView viewPM,viewPMNow,week,day,time;
5.      PM25BroadcastReceiver pm25Receiver;
6.      private PmView pmView;
7.      IotDAO iotDao ;
8.      private Button btn_open,btn_close;
9.      Handler handler=new Handler();
10.     Handler handler2=new Handler();
11.
12.     @Override
13.     protected void onCreate(Bundle savedInstanceState) {
14.         super.onCreate(savedInstanceState);
15.         setContentView(R.layout.activity_main);
16.         viewPM = (TextView)findViewById(R.id.viewPM);
17.         week = (TextView)findViewById(R.id.week);
18.         day = (TextView)findViewById(R.id.day);
19.         time = (TextView)findViewById(R.id.time);
20.         pmView = (PmView)findViewById(R.id.pmView);
21.         pm25Receiver = new PM25BroadcastReceiver();
22.         iotDao = new IotDAO(this);
23.         IntentFilter filter = new IntentFilter();
24.         filter.addAction(DataConst.ACT_PM25);
25.         super.registerReceiver(pm25Receiver, filter);
26.         new Thread(task).start();
27.     }
28.     boolean run = true;
29.       Runnable task = new Runnable()
30.       {
```

```
31.       public void run()
32.       {
33.         if (MainActivity.this.run) {
34.           MainActivity.this.handler.postDelayed(this, 1000L);
35.         }
36.         MainActivity.this.time.setText(TimeUtil.getTime());
37.         MainActivity.this.day.setText(TimeUtil.getDay());
38.         Log.d(DataConst.TAG, "getweek");
39.         MainActivity.this.week.setText(TimeUtil.getWeek());
40.
41.       }
42.     };
43.     @Override
44.     protected void onDestroy() {
45.         // TODO Auto-generated method stub
46.         super.onDestroy();
47.         super.unregisterReceiver(pm25Receiver);
48.     }
49.
50.     public void btnStart(View v){
51.         Log.d(DataConst.TAG,"btnStart");
52.         Intent intent = new Intent(this,PM25Service.class);
53.         startService(intent);
54.
55.     }
56.
57.     public void btnClose(View v){
58.         Log.d(DataConst.TAG,"btnClose");
59.         Intent intent = new Intent(this,PM25Service.class);
60.         stopService(intent);
61.     }
62.     public void btnShow(View v){
63.         Log.d(DataConst.TAG,"btnbtnShow");
64.
65.         ArrayList<Integer> data =
66.                 iotDao.getDataByType(DataConst.IOT_TYPE_PM);
67.         pmView.setPts(data);
68.         pmView.invalidate();
69.         Log.d(DataConst.ACT_PM25,"data:"+data.toString());
70.
71.     }
72.
73.     @Override
74.     public boolean onCreateOptionsMenu(Menu menu) {
```

```
75.          //Inflate the menu; this adds items to the action bar if it is present.
76.          //getMenuInflater().inflate(R.menu.main, menu);
77.
78.          return true;
79.      }
80.
81.      class PM25BroadcastReceiver extends BroadcastReceiver{
82.
83.          @Override
84.          public void onReceive(Context arg0, Intent intent) {
85.              // TODO Auto-generated method stub
86.              // 从广播中提取数据，并展示在页面上
87.              String action = intent.getAction();
88.              if(DataConst.ACT_PM25.equals(action)){
89.                  value = intent.getIntExtra("value", 0)+"";
90.                  viewPM.setText(value);
91.                  ArrayList<Integer> data =
92.                          iotDao.getDataByType(DataConst.IOT_TYPE_PM);
93.                  pmView.setPts(data);
94.                  pmView.invalidate();
95.              }
96.          }
97.
98.      }
99.
100.
101.
102.
103.  }
```

（5）程序运行结果如图 10-11 所示。

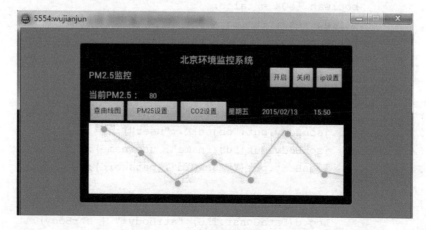

图 10-11　物联网终端数据图形显示

10.6 项目案例

学习目标：学习物联网网关服务器与移动客户端数据传输、移动客户端绘制图形的基本方法及应用。

案例描述：移动客户端以 JSONBody 封装用户名、密码，向服务器发起 post 请求，服务器接受请求，并进行服务器数据库验证（用户验证），验证通过后，以服务器与客户端约定的数据格式 JSONObject 返回，用 key-value 的形式提取 PM2.5 的值，然后在客户端调用画图类再使用图形的方式绘制 PM2.5 数据曲线图。

案例要点：HttpPost、HttpResponse、JSONObject、Canvas 相关方法的应用。

案例实施：

（1）创建工程 Project_Chapter_10，选择目标平台为 API Level 17，包名为 com.sean.pm25app，在包下创建连接服务器的类 NetManager.java，代码如下。

```
1.  public class NetManager {
2.      public static final String TAG = "pm25app";
3.      private String url = "";
4.      private String ip="";
5.      public NetManager(Context ctx) {
6.      ServerConfig conf = new ServerConfig(ctx);
7.      conf.initConfig();
8.      url = "http://" + conf.getIp() + ":" + conf.getPort() +
        "/index.cgi";
9.      ip="http://"+conf.getIp1()+"."+conf.getIp2()+"."+conf.getIp3()+".
        "+conf.getIp4()+":" + conf.getPort() + "/index.cgi";
10.     }
11.
12.     public boolean connServerReg(String username, String password) {
13.         boolean isOk = false;
14.         String resultString = "";
15.         HttpPost post = new HttpPost(ip);
16. //      HttpPost post = new HttpPost(url);
17.         JSONObject jsonBody = new JSONObject();
18.         try {
19.             jsonBody.put("action", "register");
20.             jsonBody.put("object", "user");
21.             jsonBody.put("username", username);
22.             jsonBody.put("password", password);
23.
24.             String strBody = jsonBody.toString();
25.             Log.d(DataConst.TAG, "strBody:" + strBody);
26.             post.setEntity(new ByteArrayEntity(strBody.getBytes("UTF-8")));
```

```java
27.
28.         HttpResponse resp = new DefaultHttpClient().execute(post);
29.         String result = EntityUtils.toString(resp.getEntity());
30.         Log.d(TAG, "conn server...reg result:" + result);
31.         JSONObject resultJson = new JSONObject(result);
32.
33.         resultString = resultJson.getString("result");
34.         if ("ok".equals(resultString)) {
35.             isOk = true;
36.         }
37.
38.         Log.d(TAG, "conn server..reg resultString:" + resultString);
39.     } catch (JSONException e) {
40.         // TODO Auto-generated catch block
41.         e.printStackTrace();
42.     } catch (UnsupportedEncodingException e) {
43.         // TODO Auto-generated catch block
44.         e.printStackTrace();
45.     } catch (ClientProtocolException e) {
46.         // TODO Auto-generated catch block
47.         e.printStackTrace();
48.     } catch (IOException e) {
49.         // TODO Auto-generated catch block
50.         e.printStackTrace();
51.     }
52.     return isOk;
53. }
54.
55.
56. public boolean connServerLogin(String username, String password) {
57.     boolean isOk = false;
58.     String resultString = "";
59.     HttpPost post = new HttpPost(ip);
60.     JSONObject jsonBody = new JSONObject();
61.     try {
62.         jsonBody.put("action", "login");
63.         jsonBody.put("object", "user");
64.         jsonBody.put("username", username);
65.         jsonBody.put("password", password);
66.         String strBody = jsonBody.toString();
67.         post.setEntity(new ByteArrayEntity(strBody.getBytes("UTF-8")));
68.         Log.d(TAG, "conn server...login send:" + strBody);
69.         HttpResponse resp = new DefaultHttpClient().execute(post);
70.         String result = EntityUtils.toString(resp.getEntity());
```

```
71.            Log.d(TAG, "conn server...login result:" + result);
72.            JSONObject resultJson = new JSONObject(result);
73.
74.            resultString = resultJson.getString("result");
75.            if ("ok".equals(resultString)) {
76.                isOk = true;
77.            }
78.
79.            Log.d(TAG, "conn server..login resultString:" + resultString);
80.        } catch (JSONException e) {
81.            // TODO Auto-generated catch block
82.            e.printStackTrace();
83.        } catch (UnsupportedEncodingException e) {
84.            // TODO Auto-generated catch block
85.            e.printStackTrace();
86.        } catch (ClientProtocolException e) {
87.            // TODO Auto-generated catch block
88.            e.printStackTrace();
89.        } catch (IOException e) {
90.            // TODO Auto-generated catch block
91.            e.printStackTrace();
92.        }
93.        return isOk;
94.    }
95.
96.    public int connServerGetPM25() {
97.        int value = 0;
98.        HttpPost post = new HttpPost(ip);
99.        JSONObject jsonBody = new JSONObject();
100.       try {
101.           jsonBody.put("action", "get");
102.           jsonBody.put("object", "light");
103.           String strBody = jsonBody.toString();
104.           post.setEntity(new ByteArrayEntity(strBody.getBytes("UTF-8")));
105.           HttpResponse resp = new DefaultHttpClient().execute(post);
106.           String result = EntityUtils.toString(resp.getEntity());
107.           Log.d(TAG, "conn server...pm2.5 result:" + result);
108.           JSONObject resultJson = new JSONObject(result);
109.           value = resultJson.getInt("pm2.5");
110.           Log.d(TAG, "conn server...pm2.5 value:" + value);
111.       } catch (JSONException e) {
112.           // TODO Auto-generated catch block
```

```
113.                    e.printStackTrace();
114.                } catch (UnsupportedEncodingException e) {
115.                    // TODO Auto-generated catch block
116.                    e.printStackTrace();
117.                } catch (ClientProtocolException e) {
118.                    // TODO Auto-generated catch block
119.                    e.printStackTrace();
120.                } catch (IOException e) {
121.                    // TODO Auto-generated catch block
122.                    e.printStackTrace();
123.                }
124.                return value;
125.            }
126.
127.        }
```

（2）创建登录类 LoginAct.java，在单击"登录"按钮时，调用连接服务器类 NetManager.java，如果验证成功则调用 MainActivity.java；如果验证失败，则提示失败信息，代码如下。

```
1.  public class LoginAct extends Activity {
2.      ImageButton imgBtnLoginAuto;//声明 imgBtnLoginAuto 按钮
3.      EditText editUser, editPassword;//声明 editUser,editPassword 按钮
4.      boolean isAuto;// 声明 isAuto
5.      UserConfig userConfig;// 声明 userConfig
6.      Handler handler = new Handler() {// 声明并初始化一个 handler 对象
7.
8.          @Override
9.          public void handleMessage(Message msg) {//创建 handleMessage 方法
10.                                                 //在发送消息时该方法会自动调用
11.             // TODO Auto-generated method stub
12.             super.handleMessage(msg);//
13.             switch (msg.what) {// 通过得到 msg.what 的属性的值做判断
14.             case DataConst.UP_EDIT://  如果是等于
15.                 editUser.setEnabled(true);// 设置按钮可用
16.                 editPassword.setEnabled(true);// 设置按钮可用
17.                 imgBtnLoginAuto.setEnabled(true);// 设置按钮可用
18.                 break;
19.             case DataConst.TOAST:// 否则就提示 msg.obj 的 toast
20.                 Tools.toast(LoginAct.this, (String) msg.obj);
                                  //调用 Tools 里的 toast 显示消息
21.                 break;
22.             default:
23.                 break;
```

```
24.            }
25.         }
26.
27.      };
28.
29.      @Override
30.      protected void onCreate(Bundle savedInstanceState) {//
31.          super.onCreate(savedInstanceState);
32.          setContentView(R.layout.act_login);//设置注册显示界面是 act_login
33.          userConfig = new UserConfig(this);//实例化对象
34.          userConfig.initUserConfig();
                         //调用 userConfig 里的 initUserConfig 方法
35.          initView();// 调用 initView 方法
36.      }
37.
38.      public void btnRegDialog(View v) {//
39.          Log.d(DataConst.TAG, "btnRegDialog......");//打印正在登录没有问题
40.          RegDialog regDialog = new RegDialog(this, R.style.MyDialog);
                         //声明并实例化 regDialog 对象
41.          regDialog.setContentView(R.layout.act_reg);
                         //设置注册显示界面是 act_reg
42.          regDialog.initView(handler);
                         //调用 regDialog 里 initview（初始化视图）的 handler
43.          regDialog.show();//让 regDialog 显示出来
44.      }
45.
46.      public void initView() {
47.          imgBtnLoginAuto = (ImageButton) findViewById(R.id.imgBtn
             LoginAuto);//获取 imgBtnLoginAuto 图片按钮
48.          editUser = (EditText) findViewById(R.id.editUser);
             //获取 editUser 编辑框
49.          editPassword = (EditText) findViewById(R.id.editPassword);
             //获取 editPassword 编辑框
50.          //读配置文件
51.          isAuto = userConfig.isAutoLogin();//初始化 isAuto 对象
52.          if (isAuto) {//
53.              editUser.setText(userConfig.getUsername());//从 userConfig
                 （用户配置）里得到用户名 editUser 并显示在 editUser 里
54.              editPassword.setText(userConfig.getPassword());//从 userConfig
                 （用户配置）里得到用户名 editUser 并显示在 editPassword 里
55.              imgBtnLoginAuto.setBackgroundResource(R.drawable.
                 login_auto_);//从 R.drawable 里得到 login_auto_图片
56.              editUser.setEnabled(false);// 设置按钮不可用
57.              editPassword.setEnabled(false);// 设置按钮不可用
```

```
58.            imgBtnLoginAuto.setEnabled(false);// 设置按钮不可用
59.            connServerLogin(userConfig.getUsername(), userConfig.
               getPassword());// connServerLogin 里传入 Username、Password
60.
61.        } else {
62.            imgBtnLoginAuto.setBackgroundResource(R.drawable.
               login_auto);// 否则从 R.drawable 里得到 login_auto 图片
63.        }
64.
65.    }
66.
67.
68.    public void btnLoginStart(View v) {// 开始登录按钮
69.        String username = editUser.getText().toString();
           //username 获取 editUser 的编辑框的内容
70.        String password = editPassword.getText().toString();
           //password 获取 edit
71.                                                //password 的编辑框的内容
72.        connServerLogin(username, password);//调用 connServerLogin 方法
73.    }
74.
75.    public void connServerLogin(final String n, final String p) {
           //声明了两个
76.                                                  //String 类型的参数
77.        new Thread() {// 创建一个线程
78.            public void run() {
79.                NetManager netManager = new NetManager(LoginAct.this);
                   //声明并实例化 NetManager 对象
80.                boolean bl = netManager.connServerLogin(n, p);//定义
                   //一个 boolean 的 bl 并初始化为 netManager 里 connServerLogin
81.                Message message = new Message();//声明并实例化 Message
82.                message.what = DataConst.TOAST;//获取 message 属性的值
83.                message.obj = "正在登录中....";//
84.                handler.sendMessage(message);// 发送消息
85.                try {
86.                    Thread.sleep(3000);// 线程休眠 3 秒钟
87.                } catch (InterruptedException e) {//
88.                    // TODO Auto-generated catch block
89.                    e.printStackTrace();
90.                }
91.                String hint = "登录成功...";//字符串的显示是登录成功
92.                if (bl) {// 如果登录成功
93.                    Intent intent = new Intent(LoginAct.this,
94.                        MainActivity.class);//从 LoginAct 跳转到
```

```
                        //MainActivity
95.             startActivity(intent);//开始 intent 活动
96.         } else {
97.             hint = "登录失败...";//否则登录失败
98.             Message msg = new Message();//声明并实例化 Message
99.             message.what = DataConst.TOAST;
                //Message 的属性的值赋予确定
100.            message.obj = hint;//
101.            handler.sendEmptyMessage(DataConst.TOAST);
                //发送空消息
102.            handler.sendEmptyMessageDelayed(DataConst.
                UP_EDIT, 1000);//延迟 1 秒发送消息
103.        }
104.
105.      }
106.   }.start();// 开始线程
107.   }
108. }
```

（3）创建 MainActivity.java 和采集数据服务（获取类）PM25Service.java，具体代码与 10.5 节案例一样，此处不再重复。

（4）部署工程到移动客户端，启动应用 APP，输入用户名和密码，单击"登录"按钮后，如图 10-12 所示。移动客户端发送的数据格式如图 10-13 所示。单击"开启"按钮，开始从服务器端获取 PM2.5 的值，然后在移动客户端绘制。单击"查曲线图"按钮，开始绘制 PM2.5 曲线图，如图 10-14 所示。

图 10-12 输入用户名和密码进行登录

```
02-14 14:37:59.822    793    825    com.sean.pm25app    pm25app    conn server...login send:{"action":"login","password":"123","username":"root"
                                                                    ,"object":"user"}
```

图 10-13 单击"登录"按钮后连接服务器的数据 JSON 格式

第 10 章 Android 物联网应用开发基础   489

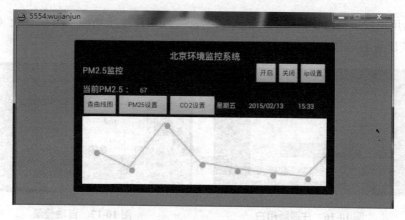

图 10-14　PM2.5 数据曲线图

单击"PM25 设置"按钮，弹出设置预警值界面，如图 10-15 所示。关于 PM2.5 预警值设置界面及数据保存，具体见代码，此处不再列出进行介绍。

图 10-15　设置 PM2.5 预警值

习　　题

一、简答题

1．简述物联网体系框架结构及其对应的应用协议。

2．物联网的关键技术主要有哪些？它们与移动互联网关键技术之间有何联系及区别？

3．JSON 格式数据如何封装和解析？

4．如何应用 HttpPost、HttpResponse 等类，在客户端和服务器之间进行数据的传输？

5．物联网终端、协调器及服务器网关之间数据流工作模式是什么？

二、实训

要求：

在本章项目案例的基础上，要求完成基于 PM2.5 数据超过预警后，自动响起手机铃声，并给用户发送提示短消息的功能。同时，完成用户注册功能及自动登录（保存到数据库）和 IP 设置功能，如图 10-16～10-18 所示。

图 10-16　注册新用户

图 10-17　自动登录

图 10-18　设置 IP

参 考 文 献

[1] Juhani Lehtimaki. Smashing Android UI: Responsive User Interfaces and Design Patterns for Android Phones and Tablets. Wiley, 2012.
[2] 罗升阳. Android 系统源代码情景分析. 北京：电子工业出版社，2012.
[3] Reto Meier. Professional Android 4 Application Development. Wrox; 3 edition，2012.
[4] Wei-Meng Lee. Beginning Android 4 Application Development. Wrox; 1 edition, 2012.
[5] Jason Ostrander. Android UI Fundamentals: Develop & Design. Peachpit Press; 1 edition, 2012.
[6] 明日科技. Android 从入门到精通. 北京：清华大学出版社，2013.
[7] 李刚. 疯狂 Android 讲义. 北京：电子工业出版社，2013.
[8] 基于物联网操作系统 HelloX 的智慧家庭体系架构. http://blog.csdn.net/hellochina15/article/details/39610833.
[9] mbed OS. http://mbed.org/.

参考文献

[1] Juhani Lehtimaki. Smashing Android UI: Responsive User Interfaces and Design Patterns for Android Phones and Tablets. Wiley, 2012.
[2] 罗升阳. Android 系统源代码情景分析. 北京：电子工业出版社, 2012.
[3] Reto Meier. Professional Android 4 Application Development. Wrox 3 edition, 2012.
[4] Wei-Meng Lee. Beginning Android 4 Application Development. Wrox 1 edition 2012.
[5] Jason O.tander. Android UI Fundamentals: Develop & Design. Peachpit Press, 1 edition, 2012.
[6] 佘志龙. Android从入门到精通. 北京：清华大学出版社, 2012.
[7] 丰生强. Android软件安全. 北京：电子工业出版社, 2013.
[8] 基于一键即刷刷机新宠 Hellofx 出炉 支持多款机型. http://blog.csdn.net/bellochina13/article/details/8901833
[9] mbed OS. http://mbed.org.